STATISTICAL MODELS FOR OPTIMIZING MINERAL EXPLORATION

STATISTICAL MODELS FOR OPTIMIZING MINERAL EXPLORATION

T. K. Wignall and **J. De Geoffroy**
University of Guam
Mangilao, Guam
Consultant, Geostatistics
Eden, N.S.W., Australia

PLENUM PRESS • NEW YORK AND LONDON

Library of Congress Cataloging in Publication Data

Wignall, T. K.
  Statistical models for optimizing mineral exploration.

  Includes bibliographies and index.
  1. Prospecting—Mathematical models. I. De Geoffroy, J. II. Title.
TN270.W5 1987   622′.1   86-30681
ISBN-13: 978-1-4612-9038-4   e-ISBN-13: 978-1-4613-1861-3
DOI: 10.1007/ 978-1-4613-1861-3

© 1987 Plenum Press, New York
Softcover reprint of the hardcover 1st edition 1987
A Division of Plenum Publishing Corporation
233 Spring Street, New York, N.Y. 10013

PREFACE

        After the spectacular successes of the 1960's and 1970's, the
mineral exploration business is at a crossroads, facing uncertain times in
the decades ahead. This situation requires a re-thinking of the philosophy
guiding mineral exploration if it is to emulate its recent performance. The
main argument of a previous volume titled "Designing Optimal Strategies for
Mineral Exploration", published in 1985 by Plenum Publishing Corporation of
New York, is that a possible answer to the challenge facing mineral
explorationists lies in the philosophy of optimization. This new approach
should help exploration staff make the **best achievable** use of the
sophisticated and costly technology which is presently available for the
detection of ore deposits.

        The main emphasis of the present volume is placed on the
mathematical and computational aspects of the optimization of mineral
exploration. The seven chapters making up the main body of the book are
devoted to the description and application of various types of computerized
geomathematical models which underpin the optimization of the mineral
exploration sequence. The topics covered include:

(a) the optimal selection of ore deposit types and regions of search, as
    well as prospecting areas within the regions (Chapters 2, 3, 4, 6),
(b) the designing of airborne and ground field programs for the optimal
    coverage of prospecting areas (Chapters 2, 3, 4),
(c) delineation and evaluation of exploration targets within prospecting
    areas by means of optimized models (Chapter 5).

        Finally, all the computerized models used in the previous chap-
ters are assembled into an **Expert System** for the optimization of mineral
exploration, which is described in Chapter 7, and the computer listings
of all the programs are included in six appendices (2, 3, 4, 5, 6, 7). Many
of these programs are innovative, and should prove useful in various fields
other than mineral exploration. For example the improved generalized negative binomial fitting program (Appendix 2) and the POPMIX program (Appendix 7)
should find many applications in the biomedical field. The Linear Programming
computer program, using a dual simplex method (Appendix 6) should be valuable in business and economic fields. Lastly, the improved Normal, Student-t,
Chi-squared and F-distribution programs (Appendix 2) will provide a more
accurate tabling than presently available in most books on Statistics.

        The examples offered as applications of geomathematical models
cover six main types of base and precious metal deposits, as well as
deposits of other mineral resources such as uranium, petroleum, and aluminum
(bauxite). The geographic coverage of the present volume comprises two
regions of the North American continent, one region of the South American
continent, the Scandinavian & Mediterranean regions of Europe, three regions
of Australasia, and two regions of the eastern portion of the Asian
continent. The handling of the data collected from the 900 ore deposits
included in the data base required the use of a main frame Honeywell
computer, level 66/60, for data storage and retrieval as well as data
manipulation, editing and complex analyses, a Hewlett-Packard HP 9121

computer and HP 7580 plotter, as well as APPLE II, LISA & IBM personal
computers for simpler computations, table composition, word-processing and
printing.

This volume seeks to fill the gap existing between the very few
available books which deal with the application of Geomathematics to mineral
exploration on a world-wide basis. We would hope that the result of our
effort will be welcomed by academic teaching and research personnel as well
as by mineral exploration research personnel both in private and
governmental organizations. The dual purpose of the book is illustrated by
the logo featured on the book cover. The logo emphasizes the world-wide
scope and the nature of our methodology by combining an aircraft towing a
"geophysical bird" as a symbol of modern prospecting technology on the one
hand, and the normal curve, a corner-stone of modern Statistics,
representing the application of computerized Geomathematics, on the other
hand.

One of the main aims of the book is to facilitate the task of
Tertiary Education staff teaching graduating and post-graduate classes in
the fields of Earth Sciences and Operations Research. As a result, the
sequence of topics covered in the book features a progression from
relatively simple models of deterministic and heuristic types through
univariate statistical models to complex multivariate stochastic models.
Furthermore, we are submitting exercises of graded difficulty at the end of
several chapters, with answers to selected ones to be found in Appendix 10.

For the sake of mineral exploration research personnel, we have
tried to follow as closely as possible the logics of the exploration
sequence. The special requirements of these workers have been well catered
for by including numerous statistical tables providing probabilities of
economic worth, occurrence, and detection of six types of ore deposits in
four continents of the World, as well as optimal survey designs and other
informations which are essential for effective exploration planning. We made
a special effort to assist researchers by carefully screening a total of 360
journal and book titles pertinent to topics of the first six chapters of the
book, and by segregating them into sub-sections of special interest.

ACKNOWLEDGEMENTS

The building of the large data base required by this project could not have been successfully achieved without the generous assistance of many individuals as well as private and governmental organizations. First and foremost, we owe much gratitude to D. A. Barr, Vice-President, Du Pont of Canada Exploration, Vancouver, Canada, and his staff for providing much of the data on North American deposits required for this project.

Amongst the many people who assisted us in the arduous task of gathering data on Australian ore deposits, we should like to thank particularly the Chief Geologist, Geological Survey of Queensland, Brisbane, Australia; the Director, Geological Survey of Western Australia, Perth, Australia; D. W. Suppel, Principal Geologist, Department of Mineral Resources of New South Wales, Sydney, Australia; and finally, Dr. W. P. Laing, Senior Lecturer in Economic Geology, James Cook University, Townsville, Queensland, Australia.

Regarding Asian ore deposits, we are pleased to cite the staff of the Embassy of the Philippines, Canberra, Australia, for facilitating contacts with the Philippines Bureau of Mines, Manila, the Philippines. We are also grateful to the Head of the Metallic Division, Korea Institute of Energy & Resources, Seoul, Korea, and the staff of the Society of Mining Geology of Japan, Tokyo, Japan, for their help in securing data on two types of East Asian ore deposits. Data on European ore deposits were obtained from various sources made available to the joint author at the library of the Department of Mineral Resources of New South Wales, Sydney, Australia.

Much of the basic research required for this volume was carried out at the excellent Dixson Library, the University of New England, Armidale, New South Wales, Australia. Most of the additional research work was conducted at the library of the New South Wales Institute of Technology, the Physical Sciences Library of the University of New South Wales, both in Sydney, Australia, and the California Polytechnic State University Library, San Luis Obispo, California. We are grateful to the Head Librarians of these institutions and their staffs for assisting us in our difficult and time-consuming task, and for providing the full use of their excellent facilities.

We are greatly indebted to Dr. B. S. Thornton, Dean, Faculty of Mathematical & Computing Sciences, the New South Wales Institute of Technology, Sydney, Australia, for his unflagging support for this project. An undertaking of such magnitude requiring the storage, retrieval and complex processing of more than 10,000 data bits would not have been possible without Dr. Thornton granting us full access to the very modern computing facilities of his faculty. We also appreciate the assistance given us in the preparation and illustration of this book by the following personnel of the New South Wales Institute of Technology: Beth Cook, the Mathematics Departmental Secretary, Jeff Pickering, Visual Aids Officer, and Charles Evans, Senior Technical Officer.

We are very grateful to Dr. James C. Daly, Head, Department of Statistics, California Polytechnic State University, San Luis Obispo,

California, for facilitating access to the University's modern computers. We also wish to thank Pat Fleischauer, his Departmental Secretary. Finally our thanks are due to Cecelia Page, who assisted with the index and final proof-reading.

CONTENTS

CHAPTER 6

Multivariate Bayesian classification models: Application to the
optimal selection of prospecting areas and exploration targets

CHAPTER 7

The **Expert** computerized system for optimizing mineral exploration

APPENDICES

CHAPTER ONE

APPLICATION OF COMPUTERIZED GEOMATHEMATICAL MODELS TO THE OPTIMIZATION
OF EXPLORATION PROGRAMS

FOREWORD

We hope that the geomathematical models introduced in this chapter and
further detailed in the following chapters will serve as a foundation course
in applied Mathematical Geology. While the models are presented in a
sequence of increasing complexity for didactic purposes, we have tried as
much as possible to respect the logics of the mineral exploration sequence
which we wish to optimize.

1.1. GENERAL STATEMENT

1.1.1. Optimization of the mineral exploration sequence

1.1.1.1. The mineral exploration sequence

The main purpose of mineral exploration is to achieve success in the
discovery of ore deposits, subject to time, budget and skill constraints.
There are two sides to mineral exploration success: one is technical, the
other one is commercial. Technical success is achieved by the discovery of a
natural mineral concentration sufficiently attractive to justify the cost of
testing its commercial potential. Commercial success is attained when the
testing proves that the deposit meets all the criteria required for a
profitable operation in the prevailing geographic and economic environment.
The mineralized material which can be mined, processed and sold at a profit
is then termed "ore".

It is widely recognized that mineral exploration operates in an
environment of risk and uncertainty. While holding the prospect of larger

Table 1.1. Summary of mineral exploration sequence.

(a) Planning phase:

    (1) Selection of ore deposit type

    (2) Selection of region for ore search

    (3) Selection of prospecting areas within region

    (4) Selection of methodology of coverage for ore detection

    (5) Design of exploration programs

    (6) Exploration budget allocation between prospecting areas

(b) Field implementation phase:

    (1) Field data acquisition

    (2) Data processing & evaluation

    (3) Delineation of exploration targets

    (4) Selection of exploration targets for testing

    (5) Allocation of funds for the testing of exploration targets

(c) Evaluation phase:

    (1) Design of target evaluation programs

    (2) Data acquisition & processing

    (3) Selection of targets for development

    (4) Engineering & financial planning of development phase

---

financial returns than could be expected in most other business ventures, mineral exploration is affected by a high probability of failure. Risk may result from geological, technological, economic or political uncertainties; most mineral exploration decisions are based on information that is limited both in quality and quantity.

The hall-mark of good management is to strive to minimize the incidence of uncertainty and risk, because of their harmful effects on the success of exploration ventures. The sequential approach has been successfully applied for many decades for such a purpose, though in an intuitive rather than systematic and objective manner. Table 1.1 summarizes the standard mineral exploration sequence which is arbitrarily divided into three main phases and broken into fifteen steps for the readers' convenience. Each step is in turn structured along a logical time sequence designed to enmesh it closely with both the previous and the succeeding steps. This is shown in Figure 1.1, which further details the three-phase sequence shown in Table 1.1 into seven stages, each affected by three kinds of decision-making. An important feature of the procedure is the systematic reduction in acreage under increasingly detailed and costly investigations, as we proceed through the stages of the sequence (right-most column of Figure 1.1). This is accompanied by the systematic reduction of the number of exploration targets outlined at one stage to be subjected to more detailed and costly testing at the following stage.

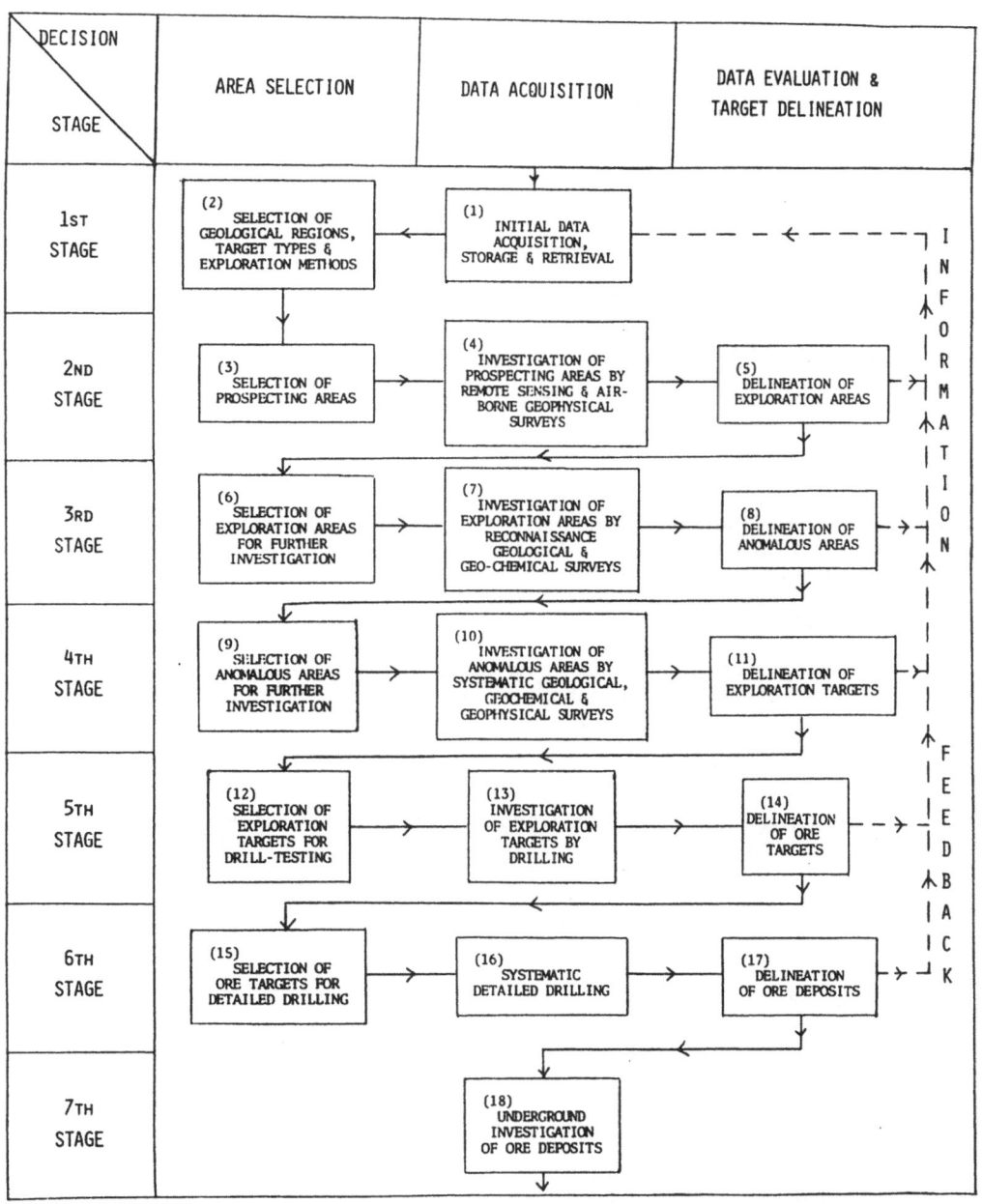

Figure 1.1. Flow diagram of sequential decision-making in mineral exploration.

1.1.1.2. Optimization of the mineral exploration sequence

At each stage of the sequence, decisions have to be made in conditions of uncertainty, requiring explorationists to choose a specific course of action among several options whose consequences are not fully known in advance. Explorationists must be satisfied to make a choice, which, if not the best choice in absolute terms, is the **best achievable** course of action within the context of corporate policies and under prevailing constraints. This is the definition of the Optimization Approach in a nutshell.

As indicated in Chapter 4 of the first volume, optimization is an objective procedure which codifies the selection of options in conditions of uncertainty through the use of quantified goals (criteria) and formal rules of action (strategies), in order to carry out pre-specified corporate policies in the most efficient manner. Through the use of dynamic programming methodology, the overall efficiency of the mineral exploration sequence is maximized by optimizing each stage while taking full advantage of the optimization of the preceding stages. In the past, the optimization of various stages of the sequence was carried out intuitively and in a qualitative manner, based on the subjective judgement of experts. In recent times, explorationists have been able increasingly to take advantage of the flexibility of modern computers and thus make full use of objective quantitative methods of optimization.

Since the coming of age in the early 1950's of the new science of Operations Research which is devoted to the optimization of complex systems, analytical methods of optimization based upon a mathematical framework have grown into use quite rapidly. The fundamentals of the methodology of mathematical optimization are briefly described in Chapter 4 of the first volume. Various types of methods may be applied, depending upon the nature of the systems to be optimized. Mathematical optimization is quite versatile, being equally applicable to deterministic situations which prevail in engineering, and to stochastic situations which are the rule in mineral exploration.

In the case of single stage systems, an indirect search method based on calculus or direct search method of a heuristic and sequential nature may be used when the number of variates and constraints are modest, while linear programming is effective when many variates and linear constraints are to be processed. When dealing with multi-stage, sequential systems which are commonly found in many business situations including mineral exploration, the dynamic programming model has proved invaluable whenever there are relatively few variables and constraints; each stage is optimized through an appropriate algorithmic approach.

1.1.2. Application of computerized models to the optimization of the mineral exploration sequence

The steady progress of computing sciences and the rapid development of computer capabilities of increasing sophistication through the 1960's and 1970's have greatly assisted the inroads of the mathematical methodology into systems optimization as a response to the very large computational requirements of the analytical approach, particularly in probabilistic situations. The main purposes of the present volume are to describe

(a) the mathematical framework of the models which are used to underpin the optimization of the various stages of the mineral exploration sequence, and

(b) the way the models can be computerized for a rapid and efficient application to various mineral exploration situations.

Regarding the explorations planning phase listed in Table 1.1, the optimization of the selection of ore deposit types and regions of search is based upon models which are described in Chapters 2 and 3 and computerized in Appendix 2 (MODFIT) and Appendix 3 (SEQPOOL). Chapters 5 and 6 introduce some complex models which are required to guide the optimal selection of prospecting areas within the previously chosen regions. These programs are computerized in Appendix 7 (CLASSIFICATION). The procedure of optimization of field program designs and the required models are described in Chapter 4 and the computer programs are presented in Appendix 4 (OPTGRID).

When dealing with the field implementation phase, Chapter 5 describes the complex models needed to optimize the delineation and evaluation of targets uncovered within the previously selected prospecting areas. These models are computerized by means of the FACTOR-TREND programs listed in Appendix 5. The models required to optimize the selection of exploration targets are described in Chapter 6, and computerized in Appendix 7 (POPMIX & CLASSIFICATION). Finally, in the evaluation phase, Chapter 5 introduces various optimized models which can be used to advantage for economic evaluation of tested targets and for development planning. The five main suites of computer programs listed above are broken into a total of sixteen sub-components which are assembled sequentially into an **Expert System** for the optimization of the mineral exploration sequence, as described in the concluding Chapter 7.

1.2. TYPES OF GEOMATHEMATICAL MODELS USED FOR OPTIMIZING MINERAL EXPLORATION

1.2.1. Types of models and their use in geomathematics

1.2.1.1. Rationale of modelling and nature of models

As pointed out by Chorley & Hagget (section 1, 12), the traditional approach in all Natural Sciences including Geo-Sciences is to decompose the apparent complexity of the real world into simplified statements termed "models", which are then organized in a logical and coherent manner into systems. Models are analogies of the real world used by their builders to re-formulate some features of interest into a more familiar and simplified form from which conclusions may be deduced; these conclusions may be re-applied to the real world in order to check the quality of the model before applying it.

The most fundamental characteristic of models is that their construction requires a "filtering" of "noise" in order to bring out the essential "signals", if we wish to use Information Theory terminology. The filtering necessarily involves a selective attitude to information. As a result, the models are often only subjective approximations representing the real world with varying degrees of probability of validity, legitimacy and adequacy. Models must be seen as a compromise, simple enough for ease of manipulation and comprehension by their users, yet complex enough to represent accurately the systems under study.

1.2.1.2. Types of models

The term "model" has been used in a great variety of contexts, which makes it difficult to define the various types of models without overlaps and ambiguities. Models can be broadly classified as "descriptive" or "normative", also as "deterministic" or "stochastic" including such sub-types as "heuristic" or "experimental design" models. On the one hand, descriptive models are concerned with a stylized description of reality, and can be termed "static", i. e. concentrating on structural features in conditions of equilibrium, or "dynamic", i. e. involving the time factor. Of the many kinds of static models, the most familiar example to geo-scientists is that of "maps", (section 1, 37). Maps are first thought of as line drawings making up a generalized picture of a territory, but they are also symbolic abstractions of it. Normative models, on the other hand, are mainly concerned with what might be expected to occur under certain stated conditions; they have a stronger predictive connotation than the descriptive models.

Another view of models is to classify them according to their contents rather than their form only. Chorley and Haggett distinguish between "iconic" models (from the Greek: ikon meaning image) and "mathematical" models. In the former kind, relevant properties of the real world are represented either by the same properties but with a change of scale (scale models) or by different properties easier to manipulate (simulation models). The mathematical models are concerned with symbolic assertions formulated in logical terms.

1.2.1.3. Mathematical models

According to Krumbein (section 1, 32), mathematical models are abstractions replacing real world objects or situations by expressions containing mathematical variables, parameters and constants. Mathematical models can be further classified into deterministic or stochastic models according to the certainty or probability which is associated with the predictions derived from their use.

Deterministic mathematical models are based upon exactly predictable relationships between independent and dependent variables from which unique consequences can be derived by logical mathematical argumentation from initially known conditions, so that the development of some system can be completely predicted in time and/or in space. All we need now are the initial conditions and a set of uniquely determined relationships. In many cases, the relationships can be rigorously demonstrated and mathematical proofs given. However, in other cases, the mathematical assertions at the core of the model are derived from definitions based upon experience or intuition, so that no formal proof may be offered: we are dealing with "heuristic" models.

Few deterministic statements can completely specify all the variables actually involved in many natural situations, so that discrepancies combine with random effects to produce "noises", which tend to obscure simple deterministic relationships. These random effects may become so important as to require the construction of other types of models, termed "statistical" or "stochastic" models. According to Krumbein (section 1, 32), stochastic models are expressions involving mathematical variables, parameters and constants with random components. The random variables are used to summarize complex situations or events whose components may be individually determined rigorously but whose aggregate effects are random, or those in which solely random factors operate.

An important type of stochastic models is that termed by Krumbein as "experimental design" models. This design is derived from a combination of past observations and logical deductions referred to as "control". Collected "study data" are statistically analyzed within the control structure to produce generalizations which overcome the inherent variability of the data, thus leading to a much clearer overall statement of the properties of the data. A commonly used experimental design model involves the fitting of data to regression models, through which a "data matrix" is analyzed to produce workable correlations which identify the direction and intensity of assumed causation.

The choice between deterministic and stochastic models is partly a function of ease of use. On the one hand, stochastic models are more flexible than deterministic models and often prove more realistic, particularly when a large number of factors have to be considered. However, many stochastic models entail considerable mathematical difficulties to the point of intractability; fortunately, these difficulties may be by-passed under certain conditions by use of computer simulations based on the "Monte Carlo" methods. Deterministic models, on the other hand, are easy to operate, but are rather narrowly constrained in their application because of the rigour of the assumptions built into them. Almost all deterministic models have their stochastic counterparts, and both types may be included to advantage in complex model systems.

1.2.1.4. Models and geomathematics

Krumbein, already cited above, and Agterberg, repeatedly cited throughout the book, have often and prominantly advocated the use of mathematical models in the geo-sciences during the past twenty years. Among the various types of models mentioned above, the mathematical models are of great interest to modern geo-scientists. While deterministic models are particularly suitable mainly in engineering and business situations, we will make only modest use of them (Chapter 2, section 2.2). However, the heuristic variety will be employed on several occasions, particularly when dealing with optimization problems (Chapters 4, 5 and 6).

Stochastic models will be the ones to be constantly used throughout the book, including relatively simple univariate ones (Chapter 2) and the more complex multivariate variety (Chapters 3, 4, 5 and 6). One kind of stochastic model, the experimental design type involving regressions, is used to advantage in multivariate situations in Chapters 5 and 6.

1.2.2. Models used for the optimization of mineral exploration

1.2.2.1. Application of three types of models to the optimization of mineral exploration

Many examples of three types of models listed above are used in various combinations and sequences in order to carry out the complex task of optimizing the mineral exploration business. For didactic purposes which is the goal pursued in Chapter 2, the most appropriate sequence is to start from the simplest and end with the most complex models on the standpoint of analytical methodology and level of mathematical knowledge required, as follows: (1) deterministic, (2) heuristic, (3) univariate stochastic, and (4) multivariate stochastic models.

However, if we have in mind practical computational considerations, the logical sequence in which we would list the models would be quite different from that mentioned above. Table 1.2 shows that univariate stochastic models come first, while heuristic ones come last. The reason for this reversal is as follows: subject to the results of the statistical testing carried out to reduce the computational work-load (Chapter 3), the

Table 1.2. Tabulation of types of geomathematical models used for optimizing exploration planning.

---

(1) UNIVARIATE STOCHASTIC MODELS
Frequency & Probability distribution models for economic, geometric and occurrence parameters of ore deposits

(2) STATISTICAL TESTING MODELS
MANOVA (multivariate analysis of variance); ANOVA (univariate) tests & multiple range tests (univariate) for the pooling of regions & ore deposit types into homogeneous groups based on economic & geometric parameters of the ore deposits

(3) DETERMINISTIC MODELS
General field detection models

(4) GEOMETRIC PROBABILITY MODELS
for the computation of detection probabilities of ore deposits of specific types by airborne and ground surveys and drilling programs in groups of regions outlined by test results

(5) MULTIVARIATE STOCHASTIC MODELS
Regression and classification models for the selection of prospecting areas and exploration targets for drill testing

(6) HEURISTIC OPTIMIZATION MODELS
Cost function models for the detection of ore deposits by airborne and ground programs

Optimization of detection of ore deposits by the Dynamic programming approach based on efficiency and payoff models

Optimization of the selection of prospecting areas and exploration targets based on the payoff model

Optimization of development planning through the Networking and Linear Programming approaches

---

univariate stochastic models (Chapter 2) and deterministic detection models (Chapter 2) are combined in the context of the theory of geometric probability in order to calculate the probabilities of detection of ore deposits by various types of surveys & programs in groups of regions (Chapter 4). Simple heuristic models such as that of cost functions (Chapter 2) are used in conjunction with geometric probability models to construct more complex heuristic models optimizing the detection of ore deposits (Chapter 4).

The flow diagram of Figure 1.2 illustrates how the planning of the optimized search for a specific type of ore deposit in (N) regions of the World can be organized in a rational manner for computational purposes. The storage file containing the economic and geometric data for each region is accessed, and either lognormal, normal or circular normal models are fitted to the frequency distributions which are derived for each parameter. The fitted distributions are used to infer probability distributions for general planning purposes, and the expected values and their confidence intervals are used for the optimization of search programs (Chapter 2).

The commonality procedure, SEQPOOL, given in Appendix 3, is carried out on the whole set of regions based upon the geometric parameters and the economic parameters. This leads to a pooling of (N) regions into a smaller number (n) of statistically homogeneous group of regions, thus resulting in a substantial saving in computational effort (Chapter 3). The detection

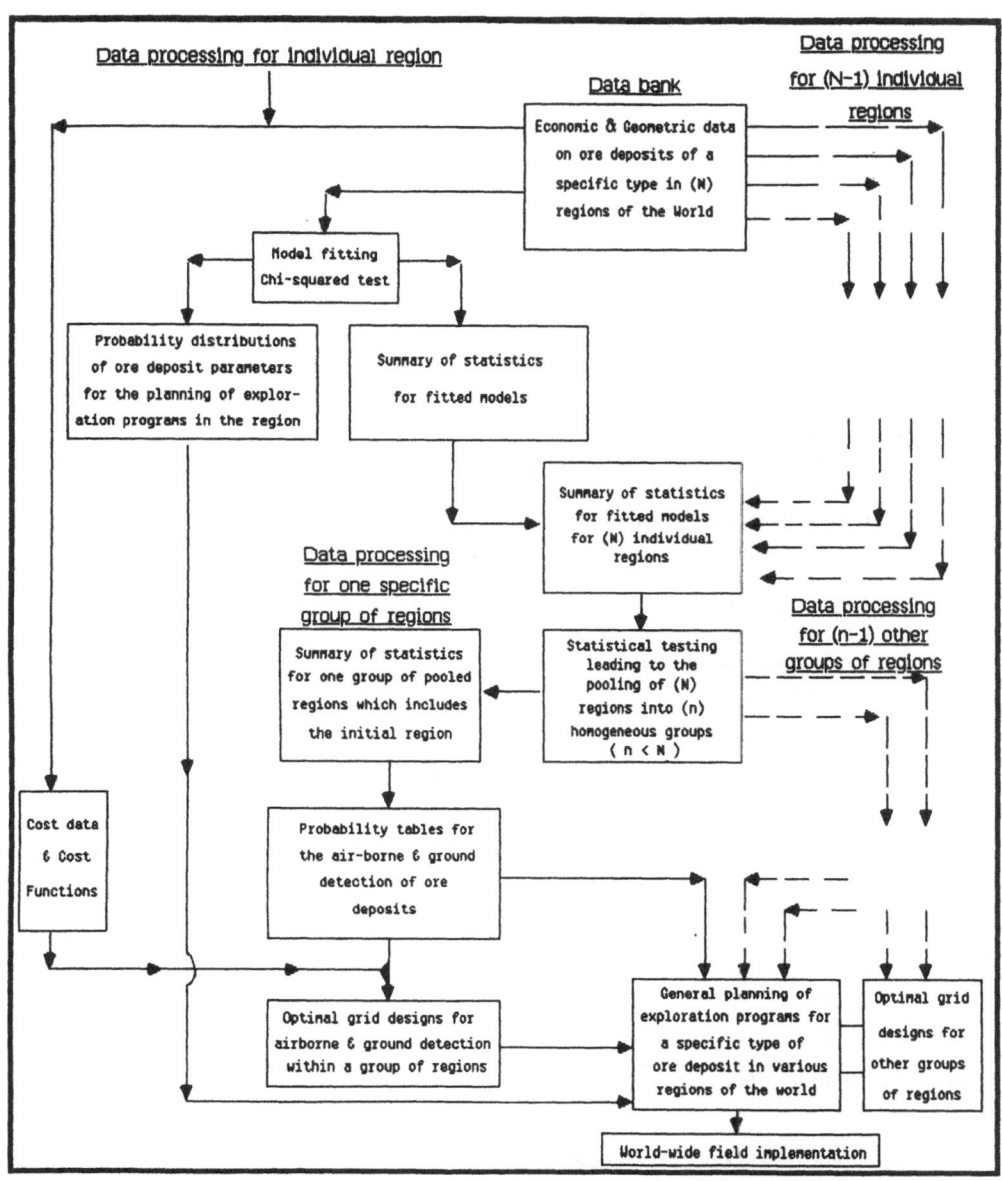

Figure 1.2. Flow diagram of optimized world-wide exploration planning for a specific type of ore deposit.

Table 1.3. Types of geomathematical models used for the optimization of the mineral exploration sequence.

PHASE & STAGE OF EXPLORATION SEQUENCE	GEOMATHEMATICAL MODELS
(a) Planning phase	
(a1) Selection of ore deposit type	economic & geometric models: lognormal, normal, circ. normal
(a2) Selection of region	occurrence model: Poisson, exponential, neg. binomial test procedures: MANOVA, ANOVA Duncan's multiple range
(a3) Selection of prospecting areas within region	multivariate regression model Bayesian classification model Pay-off model
(a4) Design of exploration programs	geometric probability models heuristic model: cost functions dynamic programming
(a5) Budget allocation between prospecting areas	linear programming
(b) Field implementation phase	
(b1) Initiation or termination of field program	Bayesian decision tree model dynamic programming
(b2) Delineation of exploration targets	Factor analysis, trend factor & residual trend model
(b3) Selection of exploration targets based on control locations	multivariate regression model Bayesian classification model Pay-off model
(b4) Selection of exploration targets without control	pre-classification analysis model Bayesian classification model
(b5) Allocation of funds for testing of selected targets	linear programming
(c) Evaluation phase	
(c1) Design of sampling plans for ore target evaluation	statistical error model heuristic models: cost functions, precision efficiency function
(c2) Ore estimation	trend factor analysis model
(c3) Engineering & financial planning	correlation & regression models network model

probability tables are then calculated for each pool of regions, using the program suite, OPTGRID, which is based upon geometric probability laws, for both airborne and ground surveys, and for drilling programs (Chapter 4). The detection probabilities, together with the cost functions for each field program, are then used to optimize the grid design by means of the Efficiency model using the dynamic programming approach (Chapter 4). However, Figure 1.2 covers only a portion of the optimization of the whole exploration sequence, which we will consider in the following section.

1.2.2.2. Types of models used for the optimization of the overall exploration sequence

The sequential structuring of mineral exploration activities was previously described in Figure 1.1 and Table 1.1 by means of three main phases which are broken into 15 stages. If we are to minimize the large risk involved in mineral exploration, the whole sequence has to be optimized by optimizing each stage through a recurrence of various models which belong to the three types mentioned above. The models used in the optimization of each of the 15 stages of the exploration sequence are listed in Table 1.3.

The optimization of the selection of ore deposit types and regions of search, the first task of an exploration planner, is based upon univariate stochastic models of the economic, geometric and occurrence characteristics of deposits (Chapter 2) supplemented by statistical testing requiring MANOVA and ANOVA models, in order to pool statistically similar regions (Chapter 3). The optimal selection of prospecting areas to be covered within the selected groups of regions requires the use of complex multivariate stochastic models (Chapter 6) and linear programming algorithms which ensure an optimal allocation of budget between selected areas (Chapter 5). Finally, the designing of optimal search programs, the chief topic of the previous volume, relies on a complex combination of univariate stochastic models (geometric parameters of ore deposits), geometric probability models, and heuristic models within the context of the dynamic programming approach (Chapter 4).

When we turn to the optimization of the field implementation, the first task is to decide on initiation or termination of data-gathering field programs; this is done by means of graphical models (decision-tree models) within the context of the dynamic programming approach. The next task is the optimal selection of targets delineated through the statistical analysis of survey results. This crucial procedure makes use of complex multivariate stochastic models combined with heuristic (pay-off) models (Chapters 5 and 6). The optimal allocation of effort between targets selected for testing is done through linear programming algorithms.

The optimization of the sampling plans required for the evaluation of the economic potential of tested targets is based upon univariate stochastic models (error models) combined with heuristic models. Based on the results of optimized sampling, the ore estimation (grade and tonnage) is carried out by means of multivariate stochastic models (Chapter 5). Finally, the engineering and financial planning ensuring an optimal development of ore deposits prior to production rests on the use of multivariate stochastic models (regression) and Operations Research models (networking) (Chapter 5).

1.3. CONSTRUCTION OF A DATA BASE FOR THE OPTIMIZED SEARCH FOR SIX TYPES OF ORE DEPOSITS IN FIVE CONTINENTS

1.3.1. General considerations

The initial stage of the project comprises several steps including:
(1) selection of ore deposit types and regions to be covered,
(2) acquisition of data,
(3) preparation of data-files for storage and retrieval for statistical analyses at each stage of the processing.

There are many difficulties involved in the construction of a reliable data base, with locating the data, availability of data for private use and quality control being three of the many obstacles to overcome. We were fortunate enough to be able to secure nearly all the data required for the North American deposits included in our coverage from two main sources, one private and the other governmental, which are fully acknowledged in the previous volume. However, the documentation of European, Australasian and Asian ore deposits proved much more difficult, particularly as far as the geometric data are concerned. Almost the whole data base is listed in Appendices 1 and 8, which are printed out directly from the storage files of the main frame computer. Appendix 1 provides a list of the names of 900 ore deposits included in the data base, grouped by regions under the headings of six main genetic types covered by the study. Appendix 8 provides most of the data describing economic and geometric characteristics of the European and Australian deposits. Only a portion of the North American data are included because these deposits have been fully covered in the previous volume.

Another important consideration is concerned with practical statistical and computational requirements. The former involve the statistical quality of samples and availability of appropriate statistical techniques required for the processing. The latter concern the availability of computer memory space, and the capabilities of the various types of computers available. Two types of computers were at our disposal to store, summarize and process a total exceeding 10,000 data bits at the New South Wales Institute of Technology, Sydney, Australia. A main frame Honeywell computer level 66/60 provided the storage of the data base, and was used for most of the complex multivariate analyses and calculation of detection probabilities. A Hewlett-Packard HP9816 and HP7580 plotter were available for the simpler calculations and graphical methods in the preliminary processing stage, and in the construction of detection optimization tables. Finally, three types of Word Processors, including an APPLE II, an IBM PC, and an APPLE LISA were put to work for the composition and printing of the more complex tables and diagrams.

1.3.2. Geologic and geographic coverage of the data base for the present study

The study covered a total of 900 economic and sub-economic ore deposits, whose names are listed in Appendix 1. These deposits belong to six main genetic types. Five of them, namely: porphyry-Cu-Mo, contact metasomatic, Ni-Cu ultramafic, volcanogenic massive sulfides and Mississippi Valley-type Pb-Zn, are important sources of base metals, while the sixth, vein gold, is a main source of precious metals. These six types of deposits were purposely selected because they are the most commonly sought targets on a world-wide basis.

Table 1.4 itemizes by continent and region the size of samples used as statistical controls for each genetic type. The largest groups include the porphyry-Cu-Mo and volcanogenic sulfide deposits, amounting to 60% of the data base. The contact metasomatic, Mississipi Valley Pb-Zn and Archean vein gold deposits make up 34%, and the Ni-Cu ultramafic, the remaining portion of the data base. Table 1.5 displays a geographic summary of the locations of the 900 deposits regardless of geological types. About 60% of the deposits are located in four regions of North America, reflecting the greater availability and quality of the data. Two regions of Europe and three regions of Australasia evenly share about 33% of the deposits included in the data base. East Asia and South America account for the small remainder, which underscores the great difficulty in securing adequate data for these regions.

Table 1.4. Geologic summary of the data base.

Genetic type of ore deposit	Continent	Region	Number of deposits
Porphyry-Cu-Mo	North America	Cordillera Belt	67
	South America	Cordillera Belt	23
	Asia	Central Asia (USSR)	25
	Australasia	S. Pacific Island Arc	31
		Tasman Geosyncline	12
Sub-totals	4	5	158
Contact metasomatic	North America	Cordillera Belt	116
	Asia	East Asia	10
Sub-totals	2	2	126
Ni-Cu ultramafic	North America	N. American Shield	25
	Europe	Scandinavian Shield	18
	Australasia	W.Australian Shield	25
Sub-totals	3	3	68

Genetic type of ore deposit	Continent	Region	Number of deposits
Volcanogenic sulfides	North America	Cordillera Belt	29
		N.American Shield	102
		Appalachian Belt	39
	Europe	Scandinavian Shield	28
		Scandin. Caledonides	19
		Iberian Peninsula	25
		East Mediterranean	27
	Asia	Japan	31
	Australasia	N. Australian Shield	14
		Tasman Geosyncline	45
Sub-totals	4	10	359
Mississippi Valley-type Pb-Zn	North America	Arctic Paleo. Platform	12
		Missouri - Tri-State	22
		Upper Mississippi	61
Sub-totals	1	3	95
Archean Vein-gold	North America	N. American Shield	52
	Australasia	W. Australian Shield	42
Sub-totals	2	2	94
6 types of deposits	Total number of deposits		900

13

Finally, Table 1.6 provides a more detailed account of the geologic and geographic structure of the data base. The control samples for the porphyry-Cu-Mo deposits and associated pyritic halos are drawn from five regions of the World. The largest grouping, both on the geographic and geological standpoints, consists of volcanogenic sulfide deposits which represent eleven regions for the acid volcanic pile sub-type, two regions for the ophiolitic sub-type, and two regions for the sedimentary pile (exhalative) sub-type. The next largest group is made up of contact metasomatic deposits which belong to four sub-types, sampled in two regions. Finally, the least represented types of deposits are the Archean vein-gold deposits drawn from two regions, and the Mississipi Valley Pb-Zn deposits representing three sub-regions of the North American Paleozoic Platform. The information displayed in Table 1.5 is generalized, and displayed on a world map shown in Figure 1.3 in two portions, one covering North and South America, and the second covering the rest of the World.

1.3.3. Types of deposit descriptors used in the data base

1.3.3.1. Introduction

The qualitative concept of "ore deposit" has to be translated into a quantitative one through the use of numerical descriptors of the various characteristics of the deposits. Two groups of characteristics of ore deposits may be considered, including:

(a) economic and geometric properties which are intrinsic to the deposits themselves and

(b) those describing the spatial distribution of ore deposits considered as discrete entities.

In this section we will concentrate on the descriptors of the first group of characteristics of ore deposits, while the second group will be dealt with in section 2.6 of Chapter 2.

As is well-known in mineral industry circles, there are many complex interactions between the geologic, economic and geometric factors involved in the quantitative description of ore deposits for economic evaluation purposes. We have attempted to depict these relationships in a simplified manner in the flow diagram of Figure 1.4. The initial concept is that of "mineral deposits", which is defined as a natural concentration of "economic minerals", i.e. of possible use in the present state of technology. The geometry of a mineral deposit as expressed by dimensions and attitude is governed by geological boundaries which, in turn, are controlled by a combination of stratigraphic, structural or petrographic and geochemical factors. These factors also control the concentration of economic elements within the deposit, as expressed by the average grade in percentage form.

The second basic concept, that of "ore deposit" is derived from the first one. An ore deposit is defined as the portion of a mineral deposit which can be mined at a profit in the prevailing geographic, economic, technological and political environment. The combination of all these extraneous factors determines a threshold or "cut-off" grade below which the mineralized material ceases to be "ore". Prevailing metal prices which reflect the economic environment are used to convert both the cut-off grade and the average grade of the "ore body" from percentage grade to unit value grade ($/tonne).. Finally, the geometry of the "ore body", as expressed by size, shape and attitude, is defined within that of the mineral deposit on the basis of the cut-off grade.

Table 1.5. Geographic summary of the data base.

North American Continent		
Western Cordillera	212	
Canadian Shield	179	
Appalachian Belt	39	
Paleozoic Platform	95	
Sub-total	525	deposits
South American Continent		
Western Cordillera	23	deposits
Asian Continent		
Siberia	17	
Japan & Korea	41	
Sub-total	58	deposits
European Continent		
Scandinavian Caledonides	19	
Scandinavian Shield	46	
Mediterranean Paleozoic & Mesozoic Fold Belts	60	
Sub-total	125	deposits
Australasia		
Tasman Geosyncline	57	
Western Australian Shield	81	
South Pacific Island Arc	31	
Sub-total	169	deposits
Grand total	900	deposits

1.3.3.2. Economic descriptors of ore deposits

The three basic descriptors used to describe the economic characteristics of ore deposits in this study include:

(a) the aggregate unit value, combining the various elements of economic interests expressed in US$ per tonne,

Table 1.6. Combined geologic and geographic summaries
of the data base.

Porphyry-Cu-Mo

(*) N. American Cordillera	Canada	40
	U. S. A.	27
Sub-total		67

South American Cordillera	Mexico	1
	Peru	5
	Ecuador	1
	Chili	11
	Argentina	5
Sub-total		23

S.E. Europe	Caucasus	8
Central Asia	W. Siberia	15
	E. Siberia	2
Sub-total		25

Australasia	Papua-N.Gui. & outlying islds	9
	Philippines	20
	Borneo	1
	Fiji	1
Sub-total		31

Tasman Geosyncline	Queensland	6
	New South Wales	6
Sub-total		12

| Total | | 158 |

Ni-Cu Ultramafic

(*) N. American Shield	Quebec	8
	Ontario	11
	Manitoba	6
Sub-total		25

Scandinavian Shield	Finland	10
	Norway	8
Sub-total		18

| Australasia | W.Austral.Shield | 25 |
| Total | | 68 |

Mississippi Valley-type Pb-Zn

North America	Arctic Paleozoic Platform (*)	12
	Missouri - Tri-State	22
	Upper Mississippi Valley	61
Total		95

Vein - Gold

(*) N. American Shield	Quebec	16
	Ontario	33
	N. W. T.	3
Sub-total		52

| Australasia | Western Australian Shield | 42 |
| Total | | 94 |

Contact - metasomatic

	sub-types:	Cu-Fe	Pb-Zn-Ag	Cu-Mo-Au	W-Mo
(*) North American Cordillera	Canada	13	5	3	6
	U.S.A.	10	36	11	31
	Mexico			1	
Sub-totals		23	41	15	37
East Asia	Japan				2
	South Korea				8
Sub-total					10
Total					126

(*) N.B. * indicates deposit types & regions already covered in previous volume
titled "*Designing optimal strategies for Minerals Exploration* "

(b) the tonnage of material considered to be ore, i. e. whose grade is above the cut-off, and

(c) aggregate gross dollar value. All economic data are expressed in constant dollars based on metal prices prevailing in a given year to ensure uniform evaluations (1982 in this study).

Volcanogenic sulfides		
Sub-type: Acid volcanic piles		

North America			Europe		
North American Shield (*)	Quebec	52	Scandinavian Shield	Finland	12
	Ontario	15		Sweden	16
	Manitoba	31		Sub-total	28
	N. W. T.	4			
	Sub-total	102			
North American Cordillera (*)	Canada	12	Scandinavian Caledonides		19
	U. S. A.	9			
	Sub-total	21			
Appalachian	Canada	33	Iberian Peninsula	Portugal	7
	U. S. A.	3		Spain	18
	Sub-total	36		Sub-total	25
Australasia					
Tasman Geosyncline	Queensland	10	Eastern Mediterranean	Yugoslavia	7
	New South Wales	12		Greece	3
	Tasmania	15		Turkey	4
	Sub-total	37		Sub-total	14
Western Australian Shield	W. Australia	14	East Asia	Japan	31
Total		210	Total		117

Total	327 deposits

Sub-type: Ophiolitic		

North America			Europe		
Appalachian	Quebec	2	Eastern Mediterranean	Cyprus	12
	Newfoundland	1		Turkey	1
	Sub-total	3		Sub-total	13

Total	16 deposits

Sub-type: Sedimentary piles (exhalative)		

North America	Cordillera	8	Australasia	Tasman Geosyn.	8

Total	16 deposits

Grand total	900 deposits

(*) N.B. * indicates deposit types & regions already covered in previous volume
titled *Designing optimal strategies for Minerals Exploration*

The grade of single commodity ore is expressed as the amount of element per tonne, which is usually expressed as a % ratio and translated into the $ value per tonne by weighting the grade by the price of the metal per unit of weight. For base metals, the weighting factor is 1000 times the price per kilogram of the metal. The expression for the grade in dollars per tonne is

$$grade = (ratio \%) \times 1000 \times (dollar\ price\ per\ kilogram).$$

For precious metals, such as gold and silver, the grade is usually given in troy ounces per short ton or grams per tonne. For polymetallic ores involving differently priced elements, the $ grade/tonne is a very convenient way of obtaining a common expression, although it is subject to many variations with time due to international economic conditions and inflation. A way of resolving this difficulty in the case of commonly-associated metals such as: Cu-Mo, Pb-Zn, Cu-Ni, Au-Ag, Au-Pt, is to use ratios.

Tonnage is another economic parameter commonly used to describe the

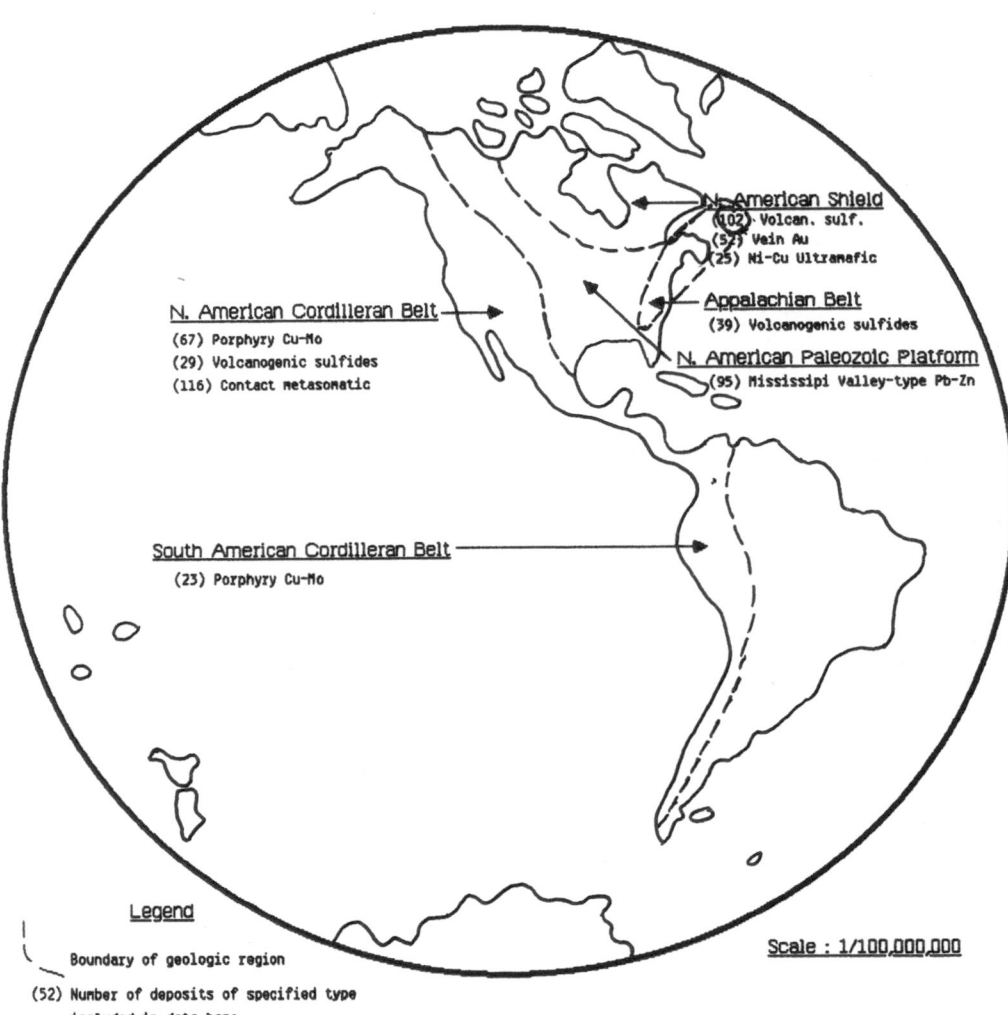

Figure 1.3 (a) World map showing the geographic and geologic
coverage of the data base: North and South
American Continents.

economic features of ore deposits. Tonnage is affected by the influence of some economic factors as well as purely geometric ones. In the case of mono-metallic ores such as gold ores, the tonnage parameter may be expressed as tonnage of ore above the cut-off grade, or as tonnage of metal derived by multiplying the average % grade by the tonnage of material above the cut-off grade. For most polymetallic ore deposits, the first definition of tonnage prevails, since tonnages of different metals are not additive.

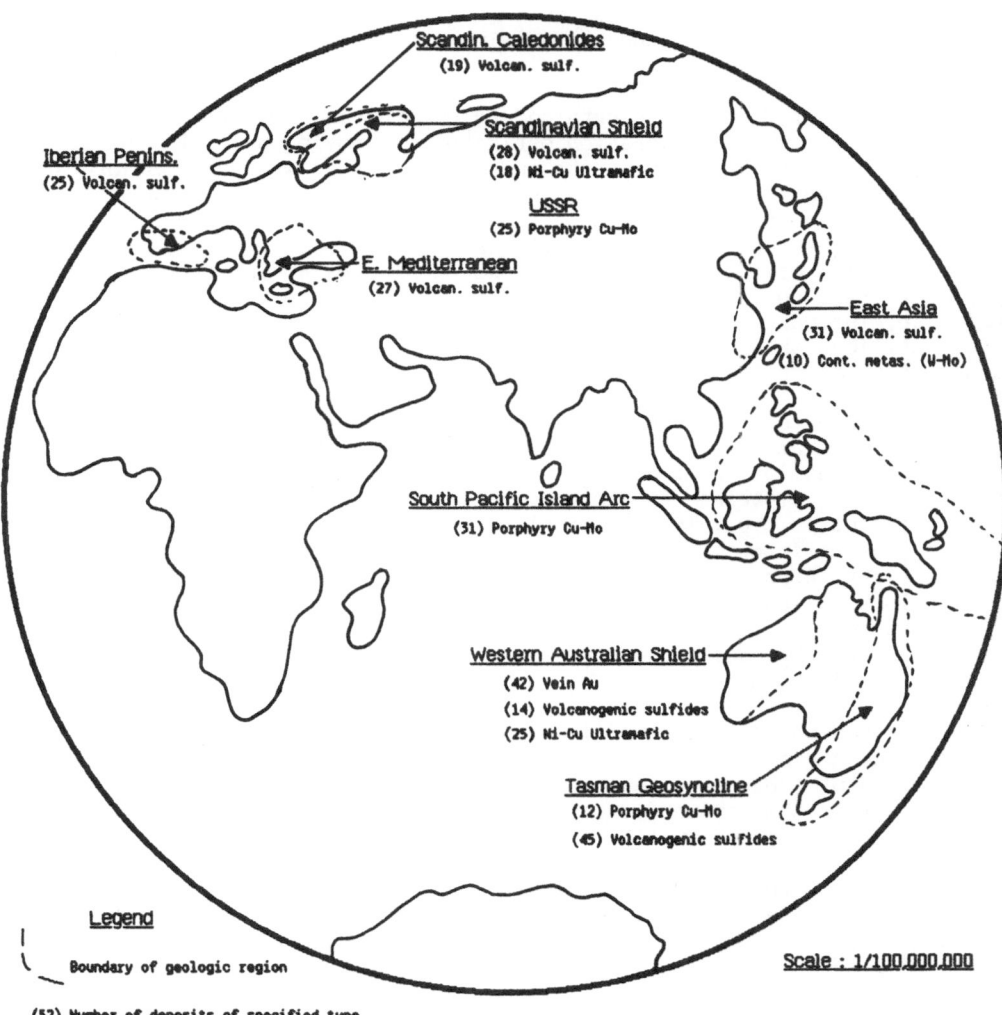

Figure 1.3 (b) World map showing the geographic and geologic coverage of the data base: Europe, Asia and Australasia.

The tonnage computation is relatively simple, involving the calculation of the volume of material above the cut-off grade, multiplied by a conversion factor, known as the tonnage factor, which is the average specific gravity of the material (Section 1, 35).

In the case of ore deposits being mined, the concept of ore reserves is a very important one on which the life of the operations and its economic fortunes depend. To be included in reserves, ore bearing material must not only exist, but it must also be economically extractable. The ore reserves of any mine consist only of the mineralized material that can be turned into concentrate by the plant already installed at the time of the calculation; anything else is waste material.

Let us now consider the gross value parameter. The term "gross value of a deposit" is not in any way a measure of its true economic value, which is a measure of the profit that could be expected from mining it when it was discovered: in many cases the gross value stated could be produced only at a loss.

The expected profit is obtained by subtracting from the gross value estimates the cost of exploration, estimates of proving up costs and of mining, processing, transportation and marketing, as well as cost of capital and tax charges. Furthermore, for the sake of valid financial comparisons and projections, the expected profit has to be considered in a time context using the concept of present value for all yearly profits generated during the whole life of the mine, as explained in Chapter 4. The gross value parameter is a derived parameter obtained by multiplying the tonnage estimate, including reserves, by the average aggregate $ grade, subject to the qualifications outlined above. Both components are affected by economic conditions, as reflected by metal prices and cut-off grade.

In order to accept the addition of tonnages of ores of different metals in mineral deposits, the metal tonnages must first be converted into the common denominator of dollars. The gross metal value is a convenient yardstick for comparison of various types of deposits or various deposits of the same genetic type. The net metal value, however, is a more meaningful figure, taking into account the smelting charges, and losses due to dilution and recovery. A uniform procedure of evaluation based upon the 1982 metal prices in U. S. dollars was used to transform the tonnage and grade figures into gross dollar value for each orebody. The dollar worth of the deposit was calculated by adding the value of known reserves to that of the total recorded production.

For the purpose of economic planning, it is best to consider first the gross value parameter as a criterion for the selection of ore deposit types and regions of occurrence. However, since gross value is a derived parameter resulting from the product of the other two, a more refined selection will result if the planners qualify their intitial choice by considering the tonnage parameter as a second criterion and the grade as a third one. The measurements of the three parameters were calculated for all 900 deposits included in the data base, except for the U. S. S. R. porphyry-Cu-Mo deposits for which only the gross value parameter was available.

1.3.3.3. Geometric descriptors of ore deposits

The determination of geometric characteristics of ore deposits such as size, shape and attitude is affected by the complex interaction of geologic, economic and technological factors, as reflected in the calculation of the cut-off grade (see Figure 1.4).

The descriptors of the geometry of ore deposits may be grouped into two categories including: (1) dimensional parameters, and (2) attitudinal parameters. The basic dimensional parameters are length, breadth and thickness; they are combined in various ways to define shape ratios (breadth/length), or areas of principal sections, or volumes. The attitudinal parameters are angular in nature and include: strike orientation, dip angle and angle of plunge; these parameters were shown on Figure 2.1 of the first volume. The strike angle is measured clockwise from true North within the range 0° to 180°. The true dip angle, (a_o), is

measured within the range 0° to 90° below the horizontal, within the vertical plane which is perpendicular to the strike direction. The apparent dip angle, (a), measured in any other vertical plane is related to the true dip angle by the

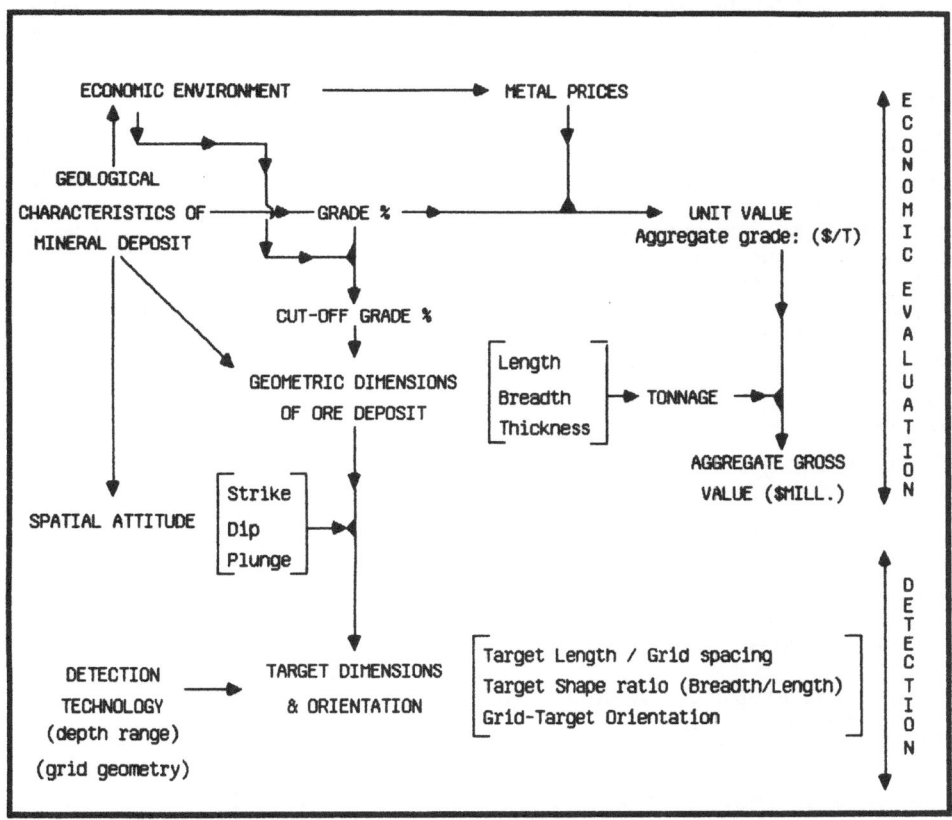

Figure 1.4. Relationship between econmic parameters of ore deposits for evaluation purposes and geometric parameters for detection purposes.

following relationship:

$$Tan(a) = Tan(a_o)*sin(d),$$

where (d) is the angle measured between the vertical plane section considered and the strike line of the target. Finally the angle of plunge measures the inclination of the ore deposit within the vertical plane containing the strike line, such as the dip does in the plane perpendicular to the strike.

However, in this study we are interested in the geometric parameters of ore deposits essentially for the purpose of calculating the probabilities of detection by various types of control grids. Therefore, we are not talking necessarily about the ore deposits themselves, but, instead, about the targets they offer to detection. The geometry of the latter has to be considered in terms of the horizontal section of ore deposits which lies within the maximum detection range of the various types of detectors used in the field, whether airborne or ground geophysical sensors, or drilling machines.

Under these circumstances, the geometric parameters used for detection purposes include the following two types of parameters: (a) dimensional parameters comprising the length and breadth of the horizontal section of that portion of the ore deposits which lies within the detection range (100 meters). These two are combined into the shape ratio,
R = breadth/length; and (b) attitudinal parameters including the unorientated true dip of deposits, and the strike direction usually measured in degrees from true North in the right semi-circle. These target parameters are combined with search grid parameters, namely grid spacing and orientation, in a manner indicated in Figure 1.4 to provide all elements required to calculate probabilities of detection and to optimize search grid designs after the introduction of cost factors. All five parameters were available for four types of deposits included in the data base, but strike direction data were inadequate for porphyry-Cu-Mo and Mississipi Valley Pb-Zn types as well as the volcanogenic sulfide deposits of the Kuroko Belt of Japan and the Eastern Mediterranean, which were then assumed to be randomly orientated.

1.3.4. Conclusion

Following the selection of six types of ore deposits as targets for a world-wide search, we proceeded with the construction of a data base including 900 control deposits occurring in a total of fourteen geologic-geographic regions in five continents. Eight parameters, including three economic and five geometric ones, were selected as descriptors of the control deposits and measured whenever available. The resulting 10,000 data bits were stored in the memory of the main-frame computer for further statistical processing.

The next step was the selection and formulation of suitable models of the three following kinds:
(a) deterministic, (b) heuristic, and (c) stochastic.
The models were combined to organize the optimized search for six types of ore deposits in pools of regions, which were statistically grouped in order to reduce the processing work-load. The theoretical background and application to our data base of some of the simpler models are given in Chapter 2, while Chapters 4, 5, and 6 cover some of the more complex models. The statistical procedure resulting in the labour-saving pooling of regions based upon the economic or geometric attributes is described in Chapter 3.

BIBLIOGRAPHY FOR CHAPTER 1

Section 1: Geo-model building; general topics

1. ABRY, C.G, 1975, Geostatistical model for predicting oil: Tatum Basin, New Mexico; The Am. Assoc. Petrol. Geol. Bull., V.59, No.11.

2. ACKOFF, R.L., GUPTA, S.K. and MINAS, J.S., 1962, Scientific Method: optimizing research decisions, New York.

3. AGTERBERG, F.P, 1967, Mathematical models in ore evaluation; J. Can. Oper. Res Soc., V.5, pp. 144-158.

4. AGTERBERG, F.P, & DIVI, S.R, 1978, A statistical model for the distribution of Copper, Lead and Zinc in the Canadian Appalachian Region; Econ. Geol. V.73, pp.230-245.

5. AGTERBERG, F.P. & DIVI, S.R, 1979, Geochemical crustal abundance models; preprint of paper presented at the Fall meeting, A.I.M.E., Tucson, Arizona, October.

6. AGTERBERG, F.P. & KELLY, A.M, 1971, Geomathematical methods for use in prospecting; Can. Min. J., V.92, No.5, pp.61-72.

7. AGTERBERG, F.P. & ROBINSON, S.C, 1972, Mathematical problems of Geology; Proc. 38th Intern. Stat. Inst., ISI Bull. 44, Book 1, pp.567-595.

8. APOSTEL, L., 1961, Towards the formal study of models in the non-formal sciences; In Freudenthal, H., (Ed.), The Concept and the Role of the Model in Mathematics and Natural and Social Sciences, Dordrecht, Holland, pp.1-37.

9. BRAITHWAITE, R.B., 1962, Models in the empirical sciences; In Nagel, E., Suppes, P. and Tarski, A., (Eds.), Logic, Methodology and Philosophy of Science, Stanford, pp.224-231.

10. CAMP, G.D., 1961, Models as approximations; In Banbury, J. and Maitland, J., (Eds.), Proceedings of the Second International Conference on Operational Research, Aix-en-Provence, pp.20-25.

11. CHORLEY, R.J. and HAGGETT, P., 1965, Trend-surface mapping in geographical research; Transactions of the Institute of British Geographers, No.37, pp.47-67.

12. CHORLEY, R.J. & HAGGETT, P. (edit.), 1967, Models in Geography; Methuen, London.

13. CHUNG, C.F., 1978, Computer program for the logistic model to estimate the probability of occurrence of discrete events; Geol. Surv. Canada, Paper 78-11.

14. CHUNG, C.F. & AGTERBERG, F.P, 1980, Regression models for estimating mineral resources from geological map data; Jour. Math. Geol., V.12, pp.458-473.

15. COLE, J.P. & KING, C.A.M, 1968, Quantitative Geography, techniques and theories in Geography; (Chpt 2); J. Wiley, New York.

16. CRAMER, H, 1964, Model building with the aid of stochastic processes; Technometrics, Vol.6, pp.133-159.

17. DAHLBERG, E.C, 1969, Use of model for relating geochemical prospecting data to geological attributes of a region, South Mountain, Pennsylvania; Quarterly of the Colo. Sch. of Mines, V.64, No.1, pp.195-216.

18. DE GEOFFROY, J.G. & WIGNALL, T.K, 1973, Statistical models for porphyry-Cu-Mo deposits of the Cordilleran Belt of North and South America; Can. Inst. M. & M., Bull., May 1973, pp.84-90.

19. DIVI, S.R, 1980, Deposit modeling and resource estimation of stratiform massive sulphide deposits in Canada; Comput. & GeoSc., V.6, No.2, pp.163-174.

20. FAVINI, G & ASSAD, R, 1979, An optimized decision model for area selection in massive sulphide exploration; Can. Inst. M. & M. Bull., April, pp.118-126.

21. GETIS, A. & BOOTS, B, 1978, Models of spatial processes; An approach to the study of point, line and area patterns, (Chpt. 2 & 3); Cambridge Univ. Press, U.K.

22. GRAYBILL, F.A, 1961, An introduction to linear statistical models; McGraw-Hill, New York.

23. GRIFFITHS, J.C. & DREW, L.J, 1964, Simulation of exploration programs for natural resources by models; Quart. Colo. Sch. of Mines, V.59, pp.187-207.

24. GRIFFITHS, W.C, 1967, Scientific method in the analysis of sediments; (Chpt. 13-21); McGraw-Hill, New York.

25. HARRIS, D.P. & BROCK, T.N, 1973, A conceptual Bayesian geostatistical model for metal endowment; Proc. A.P.C.O.M. Symposium No. 11, Tucson, pp.B113-184.

26. HAZEN, S.W. Jr., & MEYER, W.L, 1966, Using probability models as a basis for making decisions during mineral deposit exploration; U.S. Bureau of Mines Report of Investig. No.6778.

27. HESSE, M., 1953-54, Models in physics; British Journal of the Philosophy of Science, V.4, pp.198-214.

28. HUTTON, E.H., 1953-54, The role of models in physics; British Journal of the Philosophy of Science, V.4, pp.284-301.

29. KENDAL, M.G., 1968, An introduction to model building and its problems; Hafner, New York.

30. KOCH, G.S., SCHUENMEYER, J.H. and LINK, R.F., A mathematical model to guide the discovery of ore bodies in a Coeur d'Alene lead-silver mine; U.S. Bur. of Mines, Inf. Circ., No.7989.

31. KRUMBEIN, W.C. and GRAYBILL, F.A., 1965, An introduction to statistical models in Geology; (Chpts. 2, 5, 9, & 12 to 15); McGraw-Hill, New York.

32. KUIPERS, A., 1961, Model and insight; in Freudenthal, H., (Ed.), The Concept and the Role of the Model in Mathematics and Natural and Social Sciences, Dordrecht, Holland, pp.125-132.

33. LEWONTIN, R.C., 1963, Models, mathematics and metaphors; Synthese, V.15, pp.222-244.

34. MEADOWS, P., 1957, Models, system and science; American Sociological Review, V.22, pp.3-9.

35. PETERS, W.C., 1978, Exploration and mining geology, (Chpt.18, section 5);J. Wiley, New York.

36. ROBINSON, A.H. and BARTZ-PETCHENIC, M., 1976, The nature of maps; essay toward understanding maps and mapping; Univ. of Chicago Press, Chicago, Ill., U.S.A.

37. ROSENBLUETH, A. and WIENER, N., 1945, The role of models in Science; Philosophy of Sciences, V.12, pp.316-321.

38. SINDING-LARSEN, R. and VOKES, F., 1978, The use of deposit modeling in the assessment of potential resources as exemplified by Caledonian strata and sulphide deposits; Jour. Math. Geology, V.10, No.5, pp.565-579.

39. SUPPES, P., 1961, A comparison of the meaning and uses of models in mathematical and empirical sciences; In Freudenthal, H., (Ed.), The Concept and the Role of the Model in Mathematics and Natural and Social Sciences, Dordrecht, Holland, pp.163-177.

40. SUPPES, P., 1962, Models of data; In Nagel, E., Suppes, P. and Tarski, A. (Eds.), Logic, Methodology and Philosophy of Science, Stanford, pp.252-261.

41. THEOBALD, D.W., 1964, Models and method; Philosophy, V.39, pp.260-267.

42. UHLER, R.S. and BRADLEY, P., 1970, Stochastic models for determining the economic prospects of petroleum exploration over large regions; Journ. Am. Stat. Assoc., V.65, No.330, pp.623-630.

43. WHITTEN, E.H.T , 1966, Representative models in the economic evaluation of rock units illustrated with the Donegal Granite and Witwatersrand Conglomerates; Trans. Inst. M.& M., V.75, pp.B181-B198.

Section 2: Construction of data base; general topics

1. BAILEY, R.W. and CHILDERS, M.O., 1977, Applied mineral exploration with special reference to Uranium; Westview Press, Boulder, Colorado.

2. BEUKES, N.J., 1973, Precambrian iron-formations of Southern Africa; Econ. Geol., V.68, pp.960-1004.

3. BRONNER, G. and CHAUVEL, J.J., 1979, Precambrian banded iron-formations of the Ijil Group (Kediat Ijil, Reguibat Shield, Mauritania); Econ. Geol., V.74, pp.77-94.

4. CHURCH, W.R., 1972, Ophiolite: its definition, origin as oceanic crust, mode of emplacement in orogenic belts, with special reference to the Appalachians; Publications of the Earth Physics Branch, Dept. Energy, Mines and Resources, Canada, V.42, No.3, pp.71-86.

5. DE VILLIERS, J., 1959, The mineral resources of the Union of South Africa; Geol. Surv., South Africa, p.622.

6. DORR, J.V.N., II, 1973, Iron-formations in South America, Econ. Geol., V.68, pp.1005-1022.

7. EMMONS, W.H., 1937, Gold deposits of the World; New York and London, McGraw-Hill, 562p.

8. GOLE, M.J. and KLEIN, C., 1981, Banded iron formations through much of the Precambrian Time; J. of Geol., V.89, pp.169-181.

9. KILBURN, L.C., WILSON, H.D.B., GRAHAM, A.R., OGURA, Y., COATS, C.J.A. and SCOATES, P.F.J., 1969, Nickel sulphide ores related to ultra-basic intrusions in Canada; Symposium on Magmatic Ore Deposits, edit. H.D.B. Wilson, Econ. Geol. Publ. Co., pp.276-293.

10. KUZWART, M. and BOHMER, M., 1978, Prospecting and exploration of mineral deposits; Elsevier Publ. Amsterdam, Holland.

11. LAURENT, R., 1977, "Ophiolites from the Northern Appalachians of Quebec", North American Ophiolites, R.G. Coleman and W.P. Irwin, eds., Oregon Dept. Geology and Mineral Industries, Bulletin 95, pp.24-40.

12. LAZNICKA, P., 1972, The University of Manitoba file of World's non-ferrous metal deposits, MANIFILE; Dept. of Earth Sc., University of Manitoba, Winnipeg, Canada.

13. LIGHTFOOT, B., 1934, The larger gold mines of Southern Rhodesia; S. Rhod. Geol. Surv. Bull., No.26.

14. LINDGREN, W., 1933, Mineral deposits (4th ed.); McGraw-Hill Book Co., Inc., New York.

15. PARK, C.F., and MACDIARMID, R.A., 1975, Ore deposits, (3rd Edition); W.H. Freeman and Co., San Francisco, Calif. U.S.A.

16. RACLEY, R.I., 1976, Origin of Western States-type Uranium mineralizations in "Handbook of Stratabound and Stratiform Ore Deposits", K.H. Wolfe, Edit., V.7, Elsevier Publ. Co., New York, U.S.A.

17. RIDGE, J.D., 1972, Annotated bibliographies of mineral deposits in the Western Hemisphere; Geological Society of America Memoir 131, Boulder, Colo., U.S.A.

18. SANGSTER, D.F., 1972, Precambrian volcanogenic massive sulphide deposits: a review; Geol. Surv., Canada, Pap.72-22.

19. SCHAFFER, J.W., 1975, Bauxitic raw materials, pp.443-462, in industrial minerals and rocks, S.J. Lefont, edit.; A.I.M.E., New York.

20. WILSON, H.D.B. and LAZNICKA, P., 1972, Copper belts, lead belts and copper-lead lines of the World; 24th International Geological Congress, Montreal, Canada.

21. WOLFE, K.H., (Edit.), 1976, Handbook of stratabound and stratiform ores, Elsevier, Amsterdam, Holland.

22. WYLIE, P.J., (Edit.), 1967, A Review of the Geology of the ultra-mafic and related rocks; J. Wiley, New York.

Section 3: Construction of data base: Asia

1. ALEXANDROV, E.A., 1973, The Precambrian banded iron-formations of the Soviet Union; Econ. Geol., V.68, pp.1035-1062.

2. LAMBERT, I.B. and SATO, T., 1974, The Kuroko and associated ore deposits of Japan: a review of their features and metallogenies; Econ. Geol., V.69, pp.1215-1236.

3. LAZNICKA, P., 1976, Porphyry-copper and molybdenum deposits of the U.S.S.R. and their plate tectonic settings; Trans. Inst. Min. Metall., V.85, pp.B14-B32.

4. MATSUKUMA, T. and HORIKOSHI, E., 1970, Kuroko deposits in Japan, a review, pp.153-180; in "Volcanism and ore genesis", T. Tatsumi (ed.); Univ. of Tokyo Press, Tokyo.

5. RIDGE, J.D., 1976, Annotated bibliographies of mineral deposits in Africa, Asia (exclusive of the USSR), and Australia; Pergamon Press, Oxford, U.K.

6. SATO, T., 1974, Distribution and setting of the Kuroko deposits, in "Geology of Kuroko Deposits" (Ishihara, S., Edit.), Soc. Mining Geol., Japan, pp.1-10.

7. SMIRNOV, V.I., 1971, Essays on metallogeny (translated by E.A. Alexandrov); Queens College Press, New York.

8. UNITED NATIONS, 1960, Copper, lead and zinc ore resources of Asia and the Far East; Min. Res. Dev. Series, No.14.

Section 4: Construction of data base: Australasia

1. ALEXANDER, J. and HATTERSLEY, R., 1980, Australian mining, minerals and oil; David Ell Press, Sydney, Australia.

2. ALMOGELA, D.H., 1974, Phillipine porphyry-coppers: now 34 known deposits, 8 mines; World Mining, December, pp.29-33.

3. ARNOLD, G.O. and GRIFFITH, T.J., 1978, Intrusions and porphyry-copper prospects of the Star Mountain, Papua New Guinea; Econ. Geol., V.73, pp.785-795.

4. AUSTRALASIAN INSTITUTE OF MINING AND METALLURGY, 1965, Geology of Australian ore deposits, 8th Commonwealth Mining and Metallurgy Congress, Melbourne, Australia.

5. AUSTRALASIAN INSTITUTE OF MINING AND METALLURGY, 1975, Economic Geology of Australia and Papua New Guinea; Melbourne, Australia.

6. BRAITHWAITE, R.L., 1974, The geology and origin of the Rosebery ore deposits, Tasmania; Econ. Geol. V.69, pp.1086-1101.

7. CAMPANIA, B., HUGHES, F.H., BURNS, W.G., WHITCHER, I.G. and MUCENIEKAS, E., 1964, Discovery of the Hamersley iron deposits; Aust. Inst. Min. Met., Proceedings, V.210, pp.1-30.

8. COLLEY, H., 1976, Classification and exploration guide for Kuroko-type deposits based on occurrences in Fiji; Trans. Inst. M. & M., V.85, pp.B190-199.

9. DAVIS, L.W., 1977, The geophysical volcanogenic target in Eastern Australia; Bull. Austr. Soc. Expl. Geophys., V.8, No.3, Sept., pp.50-59.

10. ESPIE, F.F., 1971, The Bougainville copper project; Austr. I.M. & M. Proc. No.238, pp.1-10.

11. GEOLOGICAL SURVEY OF NEW SOUTH WALES, 1974, The mineral deposits of New South Wales, Centenary Volume, Sydney, Australia.

12. HESP, W.R. and RIGBY, D., 1975, Aspects of tin metallogenesis in the Tasman Geosyncline, Eastern Australia, as reflected by cluster and factor analysis; J. Geochem. Explor., V.4, pp.331-347.

13. HORTON, D.J., 1978, Porphyry-type copper-molybdenum mineralization belts in Eastern Queensland; Econ. Geol., V.73, pp.904-921.

14. HUTCHINSON, R.W., 1975, Geological environments of massive sulphide deposits in Tasmania (abstr.); Can. Inst. M.& M. Bull., V.68, p.49.

15. KINKEL, A.R., Jr., SANTOS-YINIGO, L.M. and SAMANIEGO, S., 1956, Copper deposits of the Philippines; Philip. Bureau of Mines, Spec. Proj. Series, No.16, Manila, Philippines.

16. LACY, W.C., 1976, Porphyry-copper deposits; Austral. Mineral. Foundation Inc.,

17. MacLEOD, W.N., De La HUNTY, L.E., JONES, W.R. and HALLIGAN, R., 1962, A preliminary report on the Hammersley Iron Province, Northwest Division; W.A. Geological Survey Ann. Report, Perth, Australia.

18. MASON, D.R. and McDONALD, J.A., 1978, Intrusive rocks and porphyry-copper occurrences of the Papua New Guinea-Solomon Island region; a Reconnaissance Study; Econ. Geol., V.73, pp.857-877.

19. McANDREW, J. and MADIGAN, R.T., (Edits.), 1965, Geology of Australian ore deposits, Vol.1, Eighth Commonwealth M.& M. Congress, Austr. I.M.& M., Melbourne, Australia.

20. REGISTER OF AUSTRALIAN MINING, 1982 Edition; Lodestone Publications, Sydney, Australia.

21. ROBERTS, P.J., 1977, Mineral exploration in Australia, 1965 to 1973; Aust. Min. Ind. Qtly., V.29, No.2, pp.48-59.

22. ROSS, J.R.and TRAVIS, G.A., 1981, Nickel sulphide deposits of Western Australia in global perspective; Econ. Geol. V.76, pp.1291-1329.

23. SAEGART, W.E. and LEWIS, D.E., 1976, Characteristics of Philippine porphyry-copper deposits and summary of current production and reserves; Soc. Mining Engineers of A.I.M.E., A.I.M.E. General Meeting, Las Vegas, Preprint No. 76-1-79, pp.47.

24. STANTON, R.L., 1978, Mineralization in island arcs with particular reference to the Southwest Pacific Region; Proc. Australasian Inst. Min. & Metall., V.265, Dec., pp.9-19.

25. TAYLOR, G.A.M., (Edit.), 1974, Porphyry-copper deposits of the southwestern Pacific; Geol. Soc. Australia - CSIRO Spec. Publ., Sydney, Australia.

26. TITLEY, S.R., 1975, Geological characteristics and environment of some porphyry copper occurrences in the southwestern Pacific; Econ. Geol., V.70, pp.490-514.

27. UNIVERSITY OF WESTERN AUSTRALIA (Extension), 1982, Publication No.7, Regional geology and nickel deposits of the Norseman - Wiluna Belt, Western Australia.

28. WILLIAMS, D.A.C. and HALLBERG, J.A., 1972, Archean layered intrusions of the Eastern Goldfields region, W.A.; Contr. Min. Petrol.

29. WOODALL, R., 1979, Gold - Australia and the World; Publ. No.3, Geology Dept. Extension Services, Univ. of W.A., Perth, Australia, pp.1-34.

Section 5: Construction of data base: Europe

1. CHECKLAND, S.G., 1967, The mines of Tharsis, Rio Tinto, Spain; G. Allen and Unwin Ltd., London, U.K.

2. HUTCHINSON, R.W. and SEARLE, D.R., 1971, "Stratibound pyrite deposits in Cyprus and relations to other sulfide ores". Soc. of Mining and Geology, Japan, Special Issue 3, pp.198-205.

3. JENKS, W.F., 1975, Origins of some massive pyritic ore deposits of Western Europe; Econ. Geol. V.70, pp.488-498.

4. RAMOVIC, M., 1968, Principles of Metallogeny; Geographical Institute, Univ. of Serajevo, Yugoslavia.

5. ROUTHIER, P., 1976, A new approach to metallogenic provinces: the example of Europe; Econ. Geol. V.71, pp.803-811.

6. STRAUSS, G.K. and MADEL, J., 1974, Geology of massive sulphide deposits in the Spanish-Portugese pyrite belt; Geol. Rundsch, V.63, pp.191-211.

Section 6: Construction of data base: North America

1. ANNIS, R.C., CRANSTONE, D.A. and VALLEE, M., 1978, A survey of known mineral deposits in Canada that are not being mined; Mineral Bulletin MR 181, Department of Energy, Mines and Resources, Ottawa, Canada.

2. BAYLEY, R.W. and JAMES, H.L., 1973, Precambrian iron-formations of the United States; Econ. Geol., V.68, pp.934-959.

3. BROBST, D.A. and PRATT, W.P., Edits., United States mineral resources; U.S. Geol. Survey, Prof Paper No. 820.

4. CANADIAN MINES HANDBOOK, 1978-1979, The Northern Miner Press, Toronto, Canada.

5. CRANSTONE, D.A. and WHILLANS, R.T., 1979, Canadian reserves of copper, nickel, lead, zinc, molybdenum, silver and gold as of January 1, 1978; Mineral Bulletin MR 185, Department of Energy, Mines and Resources, Ottawa, Canada.

6. DOUGLAS, R.J.W., (Editor), 1970, Geology and economic minerals of Canada; Geol. Survey Canada, Econ. Geol. Rept. No.1, Ottawa.

7. DREW, M.W., 1977, U.S. uranium deposits - a geostatistical study; Resources Policy, V.3, No.1, pp. 60-70.

8. EMMONS, W.H. and LANEY, F.B., 1926, Geology and ore deposits of the Ducktown mining district, Tenn.; U.S. Geol. Survey, Prof. Paper 139.

9. FOWLER, G.M. and LYDEN, J.P., 1932, The ore deposits of the Tri-State district; Am Inst. Min. Met. Engr. Trans., V. 102, pp. 206-251.

10. GERDERMAN, P.E. and MYERS, H.E., 1973, Relationship of carbonate facies pattern to ore distribution and to ore genesis in the S. W. Missouri Lead District; Econ. Geol., V.67, pp. 429-438.

11. GILBERT, C. (Edit.), 1957, Structural geology of Canadian ore deposits (Congress volume), Montreal, Canadian Inst. Mining Metall.

12. HEYL, A.V. Jr. and AGNEW, A.F., 1959, The geology of the Upper Mississippi Valley Zinc-lead district; U.S.G.S. Prof. paper No. 309.

13. HEYL, A.V. Jr. and BOZION, C.N., 1971, Some little-known types of massive sulfide deposits in the Appalachian Region, U.S.A.; Soc. Mining Geol. Japan, Special Issue No. 3, pp. 42-59.

14. HOLLISTER, V.F., POTTER, R.R. and BARKER, A.L., 1974, Porphyry-type deposits of the Appalachian Orogen; Econ. Geol., V.69, pp. 618-630.

15. S.A. and FOLINSBEE, R.E., 1969, The Pine Point Lead-zinc deposits, N.W.T., Canada; Introduction and Paleoecology of the Presqu'ile Reef; Econ. Geol., V.64, pp.711-717.

16. MACKENZIE, B.W., 1968, Nickel-Canada and the World; Mineral Rept. No. 16, Mineral Resources Div., Dept. Energy, Mines and Resources, Ottawa, p. 176.

17. McALLISTER, A.L., 1960, Massive Sulphide deposits in New Brunswick; Can. Min. & Met. Bull., February, pp. 88-98.

18. McKNIGHT, E.T. and FISHER, R.P., 1970, Geology and ore deposits of the Picher Field, Oklahoma and Kansas; U.S. Geol. Survey Prof. Paper no. 588, p. 165.

19. NALDRETT, A.J. and GASPARRINI, E.L., 1971, Archean nickel sulphide deposits in Canada: their classification, geological setting and genesis with some suggestions as to exploration; Geol. Soc. Aust., Spec. Pub. No.3, pp. 201-226.

20. RIDGE, J.D. (Edit.), 1968, Ore deposits of the United States, 1933-1967, Amer. Int. of Mining, Metallurg. and Pet. Eng., New York.

21. ROSS, F.G., 1935, Copper deposits of the Eastern United States: Copper resources of the World; 16th Int. Geol. Cong., V. 1, pp. 158.

22. SYNDER, F.G. and GERDEMANN, P.E., 1968, Geology of the Southeast Missouri Lead district, in "Ore deposits of the United States, 1933-1967", V. 1, Am Inst. Mining, Metall. Petroleum Engineers, pp. 326-328, New York.

23. SUTHERLAND BROWN, A., CATHRO, R.J., PANTELEYEV, A. and NEY, C.S., 1971, Metallogeny of the Canadian Cordillera, Can. Int. M. & M. Bull., V.64, No. 709, pp. 37-61.

Section 7: Construction of data base: South America

1. CLARK, A.H. and ZENTILLI, M., 1972, The evolution of a metallogenic province at a consuming plate margin: The Andes between 26 deg. and 29 deg. South; Can. Min. Met. Bull., V.65, No. 37 (abstr.)

2. HOLLISTER, V.F., 1973, Characteristics of porphyry-copper deposits of South America; Min. Eng., V.25, August, pp. 51-56.

3. LOWELL, J.D., 1974, Three new porphyry-copper mines for Chile; Min. Eng., V.26, November, pp.22-28.

4. PETERSEN, U., 1970, Metallogenic provinces in South America; Geol. Rundschau, V.59, No.3, pp. 834-879.

5. SILLITOE, R.HL, HALLS, C. and GRANT, J.N., 1975, porhyry-tin deposits in Bolivia; Econ. Geol., V.70, pp. 913-27.

6. SILLITOE, R.H., L976, A reconnaissance of the Mexican porphyry-copper Belt; Trans. Inst. Min. & Metall., V.85, pp.8170-8189.

7. ZANTOP, H., 198L, Argentina's porphyry-copper potential; Min. Eng., V.33, February, pp. 137-142.

CHAPTER TWO

DETERMINISTIC, HEURISTIC AND UNIVARIATE STOCHASTIC MODELS
USED FOR OPTIMIZING MINERAL EXPLORATION

2.1. FOREWORD

The optimization of mineral exploration is carried out through the use
of quantitative models which aim at representing the various stages of the
mineral exploration sequence as faithfully as possible, but in a simplified
manner in order to be tractable. The models belong to four broad categories
which are listed here in order of increasing complexity, for didactic
purposes, as follows:

(1) deterministic,
(2) heuristic,
(3) univariate stochastic,
(4) multivariate stochastic models.

It should be noted that this listing does not coincide with that in
which the models are applied in the exploration sequence. Models of one or
several kinds may be used repeatedly within each stage or within the whole
exploration procedure, often in an order quite different from that followed
in this chapter. The models are first briefly described in a theoretical
manner, and numerous examples of their application derived from our data-
base are presented throughout the following chapters. Finally, the models
are computerized, with all pertinent listings included in the appendices for
easy application to other projects by interested readers.

2.2. DETERMINISTIC MODELS

2.2.1. Introduction

Deterministic models are mathematical models which are predicated on
uniquely determined relationships between variables. Many of them can be
mathematically demonstrated in a rigorous manner. Because of their lack of
flexibility, these models are more useful in engineering situations than in
mineral exploration planning. We are considering below an example of a very
simple deterministic model which covers field detection situations.

2.2.2. Detection models

Detection models are uniquely determined relationships between
dimensional and angular variables describing actual spatial relationships
between target and detector which can be rigorously demonstrated in
mathematical terms. It should be noted that all the variates involved in the
models in this section are considered as deterministic in nature, although,
in practice, most of those used in the calculation of probabilities of

detection in Chapter 4 are actually stochastic in nature. We consider two detection situations, including:

(a) vertical detection, which covers airborne and ground geophysical surveys and many drilling programs,
(b) angled detection by drilling. The latter is restricted to a range of drilling angles between 35 and 90 degrees for surface drilling, while underground drilling which is not covered in this study may be carried out at any angle.

The parameters needed to calculate detection probabilities include the length of the horizontal section of the target (L_t meters), the target shape ratio ($R_t = B_t/L_t$) where B_t is the breadth of the horizontal section of the target, and the true dip is (a_o) degrees. Figure 2.1 illustrates the case of vertical detection by drilling or geophysical surveys, showing how the target breadth, B_t, is calculated from the horizontal section of the ore body, B_h, within the range of detection, D, taking into account the true dip and the thickness, H in meters, of the overburden. The expression is as follows:

$$B_t = B_o/\text{Sin}(a_o) + (D - H)/\text{Tan}(a_o),$$

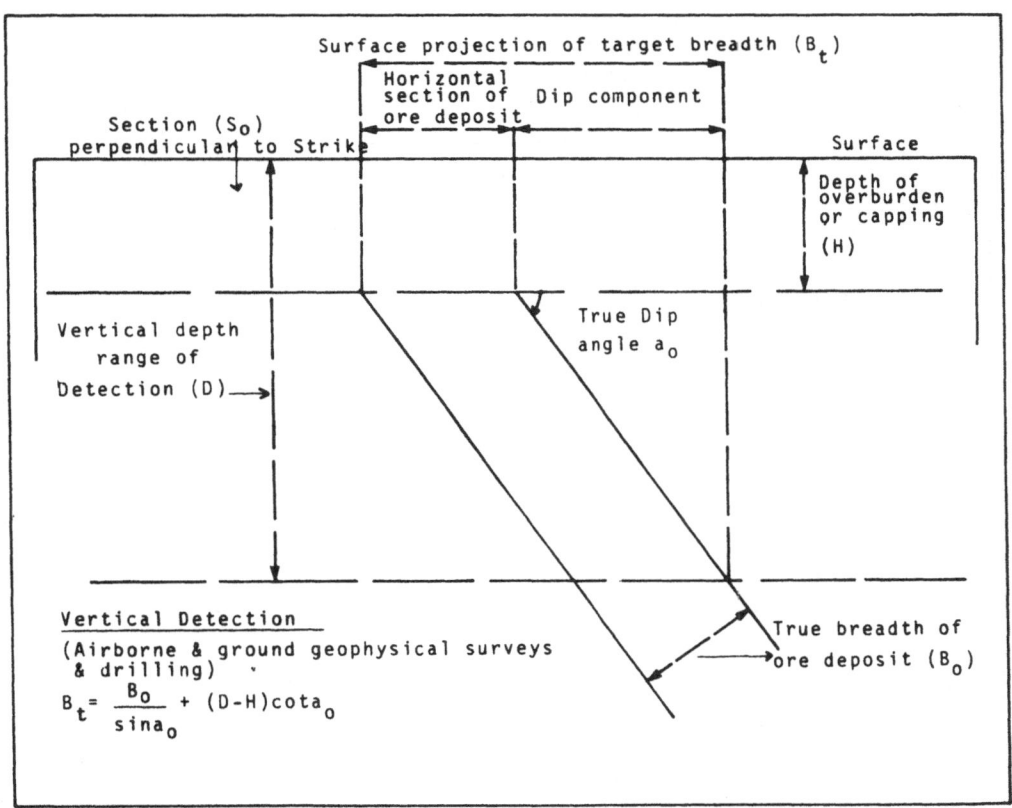

Figure 2.1. Example of deterministic model: vertical drilling model for the detection of ore deposits.

where B_o is referred to as the "true width" in drilling and reserve calculation terminologies. Whenever the deposits are subvertical with dip and plunge in the 85° to 90° range, the three parameters L_t, B_t, and R_t are approximately equal to the dimensions of the horizontal sections of the deposits L_h, B_h, and R_h. In the case of angled detection at an angle of (b) degrees to the horizontal, (Figure 2.2), the target width is augmented by $(D - H)/Tan(b)$, so that

$$B_t = B_h/\sin(a_o) + (D - H)/Tan(a_o) + (D - H)/Tan(b).$$

This results in a much larger horizontal target for detection purposes, B_t, which substantially increases the probability of detection, since the shape ratio R_t is much larger.

2.3. HEURISTIC MODELS

2.3.1. Introduction

Heuristic models are mathematical models which are based upon uniquely determined relationships between variables much like the deterministic

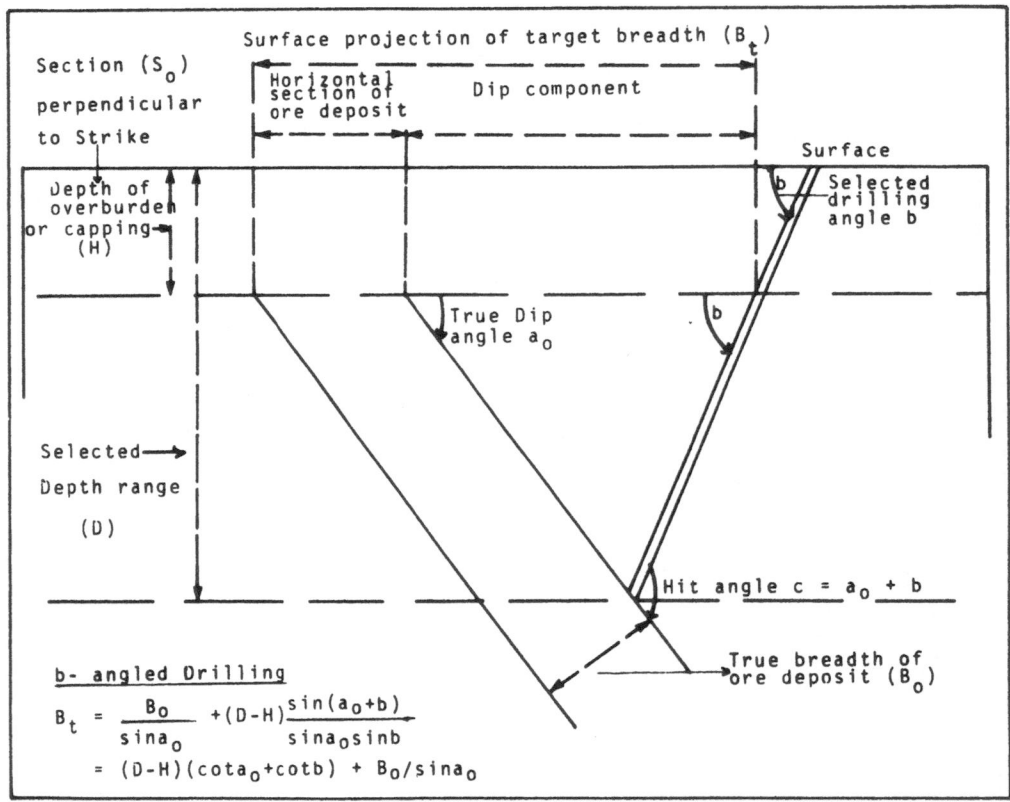

Figure 2.2. Example of deterministic model: angled drilling model for the detection of ore deposits.

models mentioned above, with the important difference that the relationships are derived from experience and intuition, and so cannot be proved mathematically. The example of cost functions for field surveys given below is a typical case of a simplified description of a very complex activity by means of a relationship between two factors, which is based entirely on experience and intuition.

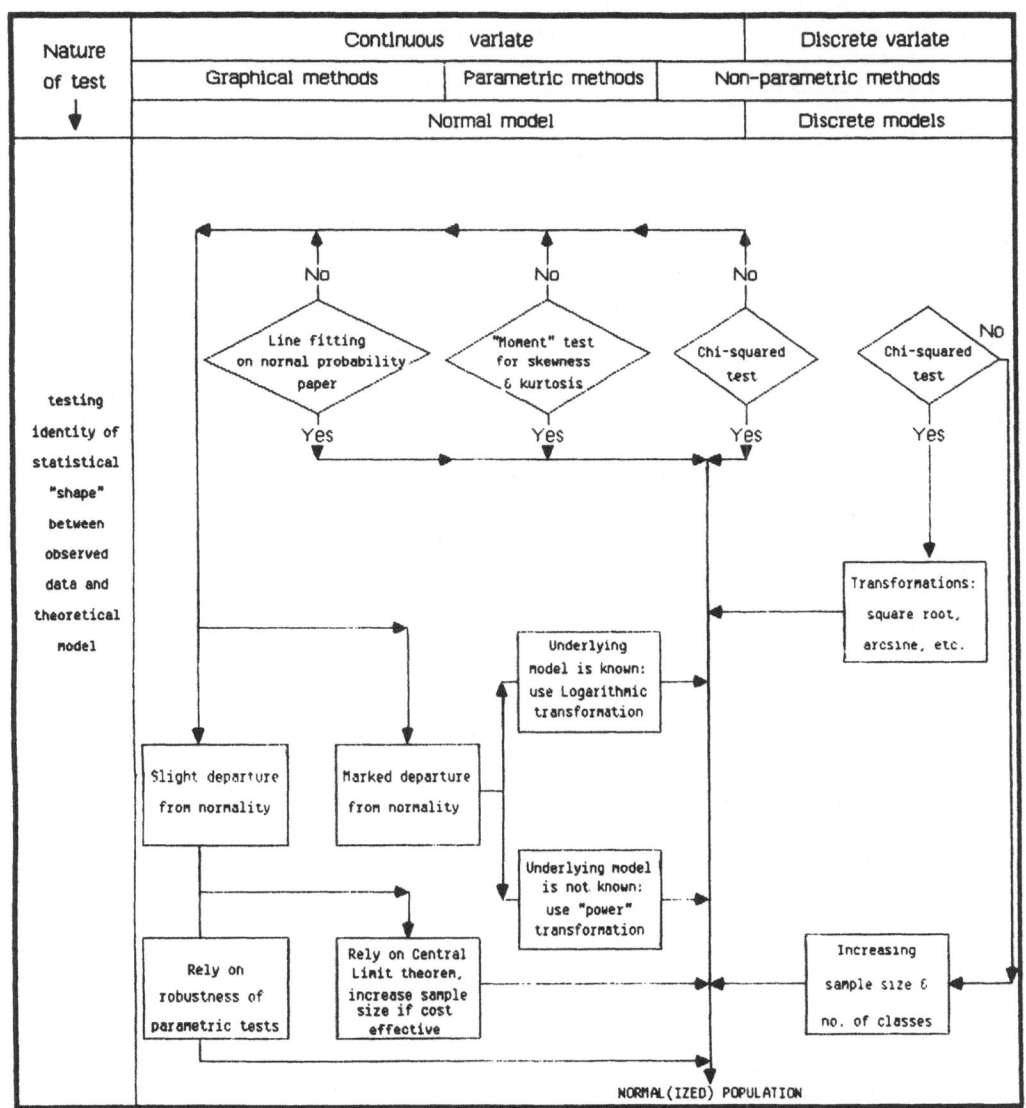

Figure 2.3. Example of heuristic model: exploration expenditure versus probability of detection of an ore deposit.

2.3.2. Cost models

2.3.2.1. General background

Mineral exploration being a business proposition, it is obvious that cost considerations must be paramount in the mind of explorationists, if they are to make optimal decisions at each stage of the exploration sequence. The construction of cost models representing in a simplified manner various exploration activities is of critical importance.

Cost function models for field exploration planning purposes are made up of two kinds of components including (a) distance related and (b) sampling related components. The models are based on the principles of unit costing and ratio costing (section 1, 8). They combine the two components mentioned above in a manner which is specific for each type of activity or stage of the exploration sequence. As a result, three main types of cost models are considered in this section, including

(1) overall exploration programs,
(2) airborne geophysical surveys,
(3) discrete ground sampling programs including drilling.

All three types are listed in the accompanying Table 2.1; however, only the last two are used in our study for the optimization of airborne and ground surveys and drilling programs (Chapter 4).

2.3.2.2. Cost functions for overall exploration programs

In regional exploration, distance is a significant component of the cost function at every stage, from initial reconnaissance to target drill-testing. The overall cost function is therefore expressed as a linear function of the distance from the nearest transport facility (railhead, air landing strip, road, river, port facility, etc.) to the center of the prospecting area. This type of model has been used by the authors for several regional programs in the eastern Canadian Shield. The cost function, C, is defined by the exponential model listed in the bottom row of Table 2.1. In a project located in the central portion of the Canadian Superior Province, we have used

c = 0.200 $million for a 100 sq km cell, and
k = 0.086,

while the exponent, α, is determined empirically in the range 1 to 2.

A very useful heuristic model describing the relationship between overall exploration cost, (C), and the probability of success, (P_s), can be simply structured as follows, using the exponential distribution function

$$P_s = 1 - \exp(-kC),$$

and is illustrated by Figure 2.3 which is based on the example of exploration programs in the Canadian Shield. The coefficient, (k), is determined empirically, depending on the mean number of successes per cell, and the cost is that of saturation coverage of a 36 square mile (85 sq. km) cell.

The model clearly illustrates the diminishing returns in terms of marginal exploration success when coverage cost increases to high levels.

Table 2.1. Expressions of cost functions for various types of exploration programs.

STAGE OF EXPLORATION SEQUENCE	METHODS & TECHNIQUES OF COVERAGE	PARAMETERS OF COST FUNCTIONS	TYPE & GEOMETRY OF SURVEY CONTROL	COST FUNCTION MODEL FOR AREA (A)	COST FUNCTION MODEL FOR ONE km SQUARE ($L = l = 1$)
REGIONAL COVERAGE	AIRBORNE GEOPHYSICAL SURVEYS	Area covered = A sq.km $A = length(L) \times width(l)$ Perimeter $P = 2(L+l)$ Control grid spacing(s) $=S$, (parallel lines, square grid) or S.I. (rectangular grid) C_0 = unit cost ($/line km)	Parallel lines or strips (s)	$C = C_0 (A/S + P/2)$	$C_1 = C_0 (1/S + 2)$
			Square Grid (SxS)	$C = C_0 (2A/S + P/2)$	$C_1 = 2C_0 (1/S + 1)$
			Rectangular Grid (SxT)	$C = C_0 [A(1/S+1/T) + P/2]$	$C_1 = C_0 (1/S + 1/T + 2)$
	GROUND RECONNAISSANCE: Geological, Geochemical, Geophysical	C_o = overhead cost C_d = cost of travelling to station C_s = cost of observation at the station n = number of stations	No systematic grid	$C = C_o + \int nC_d + nkC_s$	(not considered)
SELECTIVE & LOCAL COVERAGE	SYSTEMATIC GROUND SURVEYS	Area covered = $A = L \times l$ Perimeter $P = 2(L+l)$ Grid spacing(s): S,T C_d = "distance" unit cost C_s = cost of observation at the station $C_s = 20 C_d$	Square Grid (SxS)	$C = C_s\,(A/S + \tfrac{1}{2}[L/S+2L+1])$	$C = C_{1s}\,(1/S + 1)(1/S + 2l)$
			Rectangular Grid (SxT)	$C = C_s\,(A/T + \tfrac{1}{2}[L/S+2L+1])$	$C = C_{1s}\,(1/T + 1)(1/S + 2l)$
	SYSTEMATIC		Square Grid (SxS)	$C = C_s\,(A/S+\tfrac{1}{2})[L/S+0.02L+1]$	$C = C_{1s}\,(1/S+1)(1/S+1.02)$
	DRILLING	C_d = drilling cost per hole C_s = "distance" unit cost $C_s = 50\,C_d$	Rectangular Grid (SxT)	$C = C_s\,(A/T+\tfrac{1}{2})[L/S+0.02L+1]$	$C = C_{1s}\,(1/T+1)(1/S+1.02)$
OVERALL EXPLORATION PROGRAM	OVERALL COVERAGE	D = distance from centre of exploration area to nearest transportation facilities C_o = overall cost of unit area in easily accessible region	(Not considered)	$C = C_0 \times A \times (1+kD^{\hat{a}})$ $0 < k \le 1,\ 1 \le \hat{a} \le 2$	(not considered)

37

2.3.2.3. Cost functions for airborne geophysical surveys

The unit cost of airborne geophysical surveys depends upon the type of aircraft used and on the survey technology. Combined magnetic, electromagnetic and radiometric surveys are most widely used. Geophysical airborne surveys are run along flight lines on a continuous sampling basis, so that the sampling and distance related components are confounded.

There are three kinds of grid designs used for the control of airborne geophysical surveys:

(a) parallel flight-lines with spacing s meters apart,
(b) orthogonal grids including square grids with with spacings s x s meters,
(c) rectangular grids with spacings s x t, with t > s.

The cost models for three types of grids are shown in the upper portion of Table 2.1, first for the coverage of a region in terms of the area (A) in sq. kms and its perimeter, (P), and the flight line spacing (s), and (t) when it applies; and secondly for the coverage of a unit area of one kilometer square.

2.3.2.4. Cost functions for ground surveys

When we deal with ground surveys (central portion of Table 2.1), the structure of the cost function is quite different to the continuous line airborne surveys. There are now three components to consider in defining the cost, which include

(1) sampling-related,
(2) distance-related (grid),
(3) distance-related (travel) components.

All three terms of the cost function are weighted by different unit costs. Ground reconnaissance surveys, whether geological or geochemical, are carried out without the benefit of a permanently established control grid, therefore the number of sampling stations and distance travelled between stations have to be considered as in the model displayed in the table.

Systematic ground surveys require the laying out of a permanent control grid for accurate identification of sampling locations. Table 2.1 lists the cost models required for the coverage of a rectangular region of dimensions L and ℓ by means of a square grid with spacings (s) and (s) or by a rectangular grid with spacings (s) and (t). The reader should note that the expressions are considerably simplified by the introduction of ratio costing involving unit costs for distance related component, (C_d), and the sampling related component, (C_s). The relationship is expressed empirically as $C_d = 20 * C_s$, based upon exploration experience in the Canadian Shield.

2.3.2.5. Cost functions for drilling programs

The structure of the cost function model chosen for drilling programs differs markedly from that for ground surveys, because of the much greater importance of the sampling-related component, (C_s), in relation to the distance-related one, (C_d). The former reflects the spacing between drill-holes, and is weighted by a unit cost which depends on the vertical depth and angle of inclination of drill holes as well as the type of drilling technology. The distance-related component reflects the moving cost between set-ups of the drilling rig, and the cost of establishing a control grid for the drilling program as well as surveying in collar locations. Based upon past experience in Canadian Shield exploration, we are using the following ratio-costing expression for the unit sample drilling cost, (C_s):

$$C_s = 50 * C_d.$$

2.3.3 Optimizing models for mineral exploration

2.3.3.1. Introduction

In the mineral exploration business, most situations confronting management require the choice of the best course of action based on information which is generally inadequate both in quality and quantity. In order to tackle the problem of optimal choice under uncertainty, two main concepts have to be established:

(1) goals, and
(2) criteria.

Goals are a qualitative and subjective reflection of corporate policies. Goals are designed to guide the acquisition of rewards under the constraint of limited resources. Rewards can be defined in monetary terms such as gross value or profit. Other kinds of reward are efficiency, effectiveness and precision, which are eventually converted into monetary units. The quantifying of qualitative goals into criteria is a necessary step in the optimization process. The structure of models representing the criteria are of a varied nature. The most useful models are those designed as a linear contrast between a weighted reward function and a weighted cost function. In this study, we have used extensively the Efficiency model and the Pay-off model (Chapter 4).

2.3.3.2. Efficiency model

The efficiency model is convenient and flexible and quite suitable for use in business, engineering and mineral exploration. It is defined by

$$\text{Efficiency function}(x, y, z) = k_1 \text{Reward function}(x, y, z) - k_2 \text{Cost function}(x, y, z),$$

where (x, y, z) are variates in units such as \$, m, etc, and k_1, k_2 are scaling factors. The reward function may be expressing gross value, probability of detection or sampling precision, etc. The detection efficiency model is a key-stone of our methodology for the optimization of search grids for specific types of ore deposits. In the application of the Efficiency model to ore detection optimization, the grid spacing, s, is the independent variate, so that the efficiency function is written

$$F(s) = k_1 P(s) - k_2 C(s),$$

The scaling factors, k_1 and k_2, are determined in such a way as to keep the probability of detection function, (P), and the cost function, (C), within compatible domains (see Chapter 4). Figure 2.4 shows how the detection efficiency model is constructed.

The detection reward function is used throughout this study to optimize the world-wide search for six types of ore deposits. The criterion sets the probability of detection expressed in terms of grid spacings, or orientation with respect to the target, or drilling angle, against the cost function expressed in terms of grid spacing or angular intervals.

2.3.3.3. Pay-off model

The pay off model is the standard one employed in business managements. It is also very useful in some aspects of mineral exploration planning. The reward function is the EMV, which stands for Expected Monetary Value. The EMV is defined as follows: the economic value of success,(gross value or gross profit), weighted by its probability, (P_s) is contrasted with the expected loss value weighted by the probability of failure, P_f which is

$$P_f = 1 - P_s = \exp(-kC).$$

In the context of mineral exploration, the pay-off model is not widely used when we seek to optimize individual stages of the exploration sequence. It is not rewarding to match the expected gross value of exploration success against the cost of failure of an airborne or ground geophysical survey or even a drilling program of average size, due to the imbalance in order of magnitude of the two terms of comparison. However the pay-off model may be used to advantage to optimize the selection of areas for coverage as shown in the diagram of Figure 2.5, or in the selection of exploration targets for

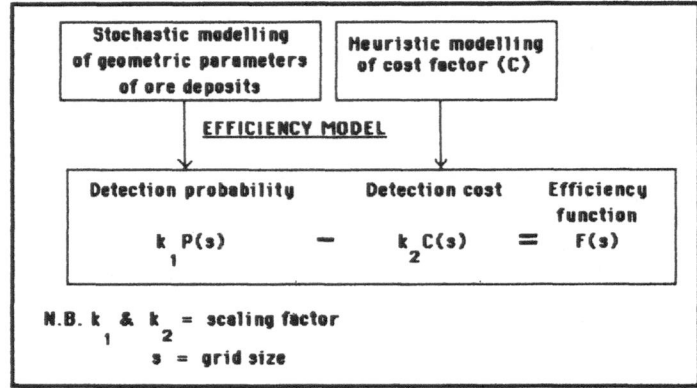

Figure 2.4. Heuristic model for the optimization of efficiency of detection of ore deposits.

further testing, as shown in Figure 2.6. The multivariate approach shown in Figures 2.5 and 2.6 is fully described in Chapter 6.

2.4. UNIVARIATE STOCHASTIC MODELS

2.4.1. Introduction

2.4.1.1. Normal, Student-t, χ^2, and F-distributions

The probability values, termed p-values, for these distributions are given in many standard texts in Statistics, but most tables are designed for

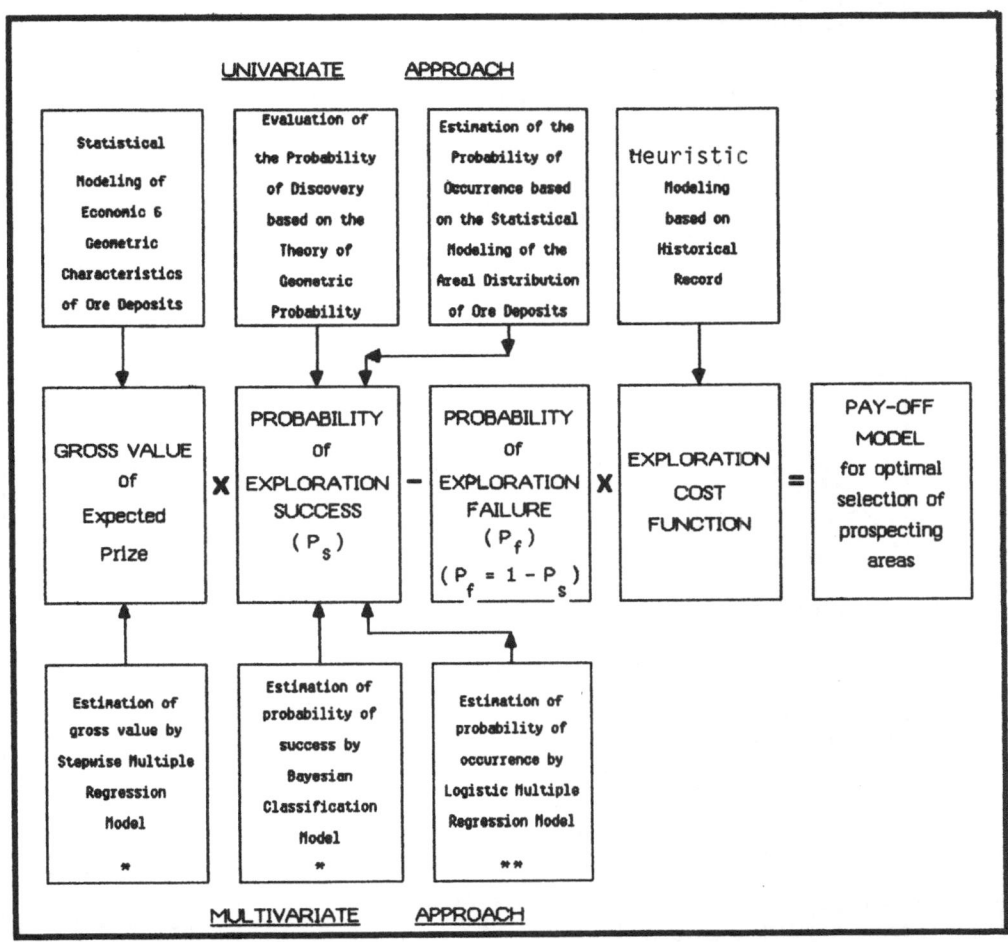

N.B. * See chapters 5 & 6
    ** See references 2, 3, 4 and 22, chapter 5

Figure 2.5. Pay-off model for the optimal selection of prospecting areas.

small sample sizes usually up to 30. We decided, therefore, to produce our own accurate tables of p-values, using the precise integral of the probability density functions wherever possible and the integral of the Taylor series expansion of these functions otherwise. Due to format restrictions, details of the integration of the density functions can not be entered into here, as a full account has been given in a comprehensive table package and book. The resulting programs are given in Section 2 of Appendix 2 and tables are given in Appendix 9. More complete tables based upon these programs are the subject of a new book and computer package.

2.4.1.2. The stochastic modelling of ore deposit parameters

The goal of mineral exploration is the discovery of ore deposits. Ore deposits of a specific genetic type, or rather the variates describing them, can be considered as a multivariate population in the statistical sense. These populations may be composed of sub-sets of deposits occurring in various groups of regions or deposits belonging to various genetic sub-types which occur within a specific region.

A population of deposits of a specific type comprises two portions: one portion of known finite size is composed of deposits of that type which

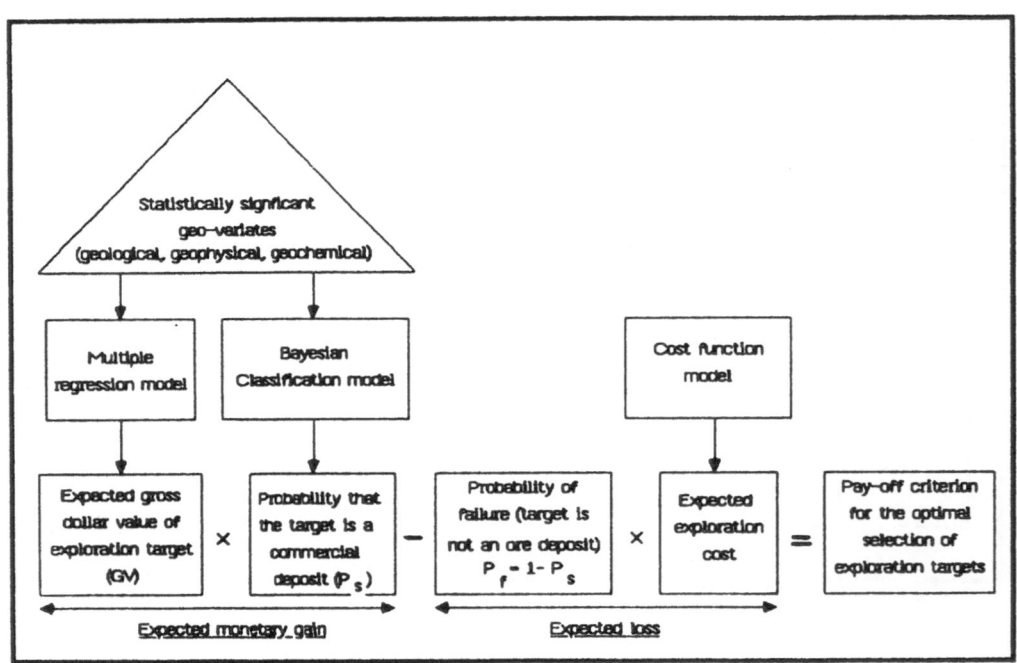

Figure 2.6. Pay-off model for the optimal selection of exploration targets for drill testing.

have already been discovered. The known portion of the population, or, generally, a representative sample of it, is used as a control for the study of the unknown portion. The sampling control set should be given the utmost attention since the quality of the inferences and predictions concerning the unknown portion largely depends on that of the control sample. Much has been written on the requisites for valid statistical sampling (Section 2, 15 and 23). Once the sampling has been carried out, however imperfectly because of the constraints of diverse kinds affecting the control data, the next step is to describe the control data by means of frequency distributions which are summarized by various statistical parameters.

2.4.2. Stochastic description of ore deposits by means of frequency distributions and summarizing statistical parameters

The first task is to construct frequency distributions of the observed data. In order to do this, we first consider the range of the sample data. For each geometric or economic variate described an ore deposit. The range is expressed as the difference between the maximum and minimum values:

range = largest magnitude - smallest magnitude.

The class widths are then found by dividing the range into 5 to 15 equal intervals, depending on the sample size. A frequency distribution is established for each ore deposit variate, and is displayed in tabular form, which is supplemented by a frequency histogram or polygon or cumulative frequency graph.

The frequency distributions may then be summarized by means of statistics describing the three following concepts:

(a) central location,
(b) variability about the central location, and
(c) "shape".

The statistics related to the central location concept include the mean, median and mode. The arithmetic mean is particularly appropriate for normal distributions, and the geometric mean is more suitable when dealing with asymmetric distribution, such as lognormal types. The median is also useful, when dealing ordered of data. It is easily defined from the frequency polygon, as it occurs at the point which divides the area under the polygon into two equal parts, and for this reason is also referred to as the 50th percentile of the distribution. Similarly, the 3rd quartile, Q_3, is the 75th %ile, and occurs with 75% of the area to its left. Finally the mode is another important measure of central location, which is defined by the peak(s) of the frequency polygon. As a result, the mode may may not be unique (polymodal distributions), which often infer mixed populations. Graphical techniques are often used in splitting such populations into two or more component populations(Chapter 6, section 2, ref. 3, 5 and 10); but a far more satisfactory methodology using Fisher k-statistics is given in Chapter 6, and computerized in Appendix 7, (POPMIX program).

The concept of dispersion or spread of the data about the mean is introduced to describe the variability. The main measures are the variance or its square root, the standard deviation. Sometimes it is more convenient to express the dispersion by a dimensionless measure, called the coefficient of dispersion which is defined by

Coefficient of dispersion = standard deviation/mean.

Finally, the concept of "shape" of the distribution is introduced to fully describe the nature of the frequency distribution. The mean, mode, and median coincide in symmetrical distributions, such as the normal distribution. Many distributions are asymmetrical with negative or positive skew. The coefficient of skewness is conveniently defined by:

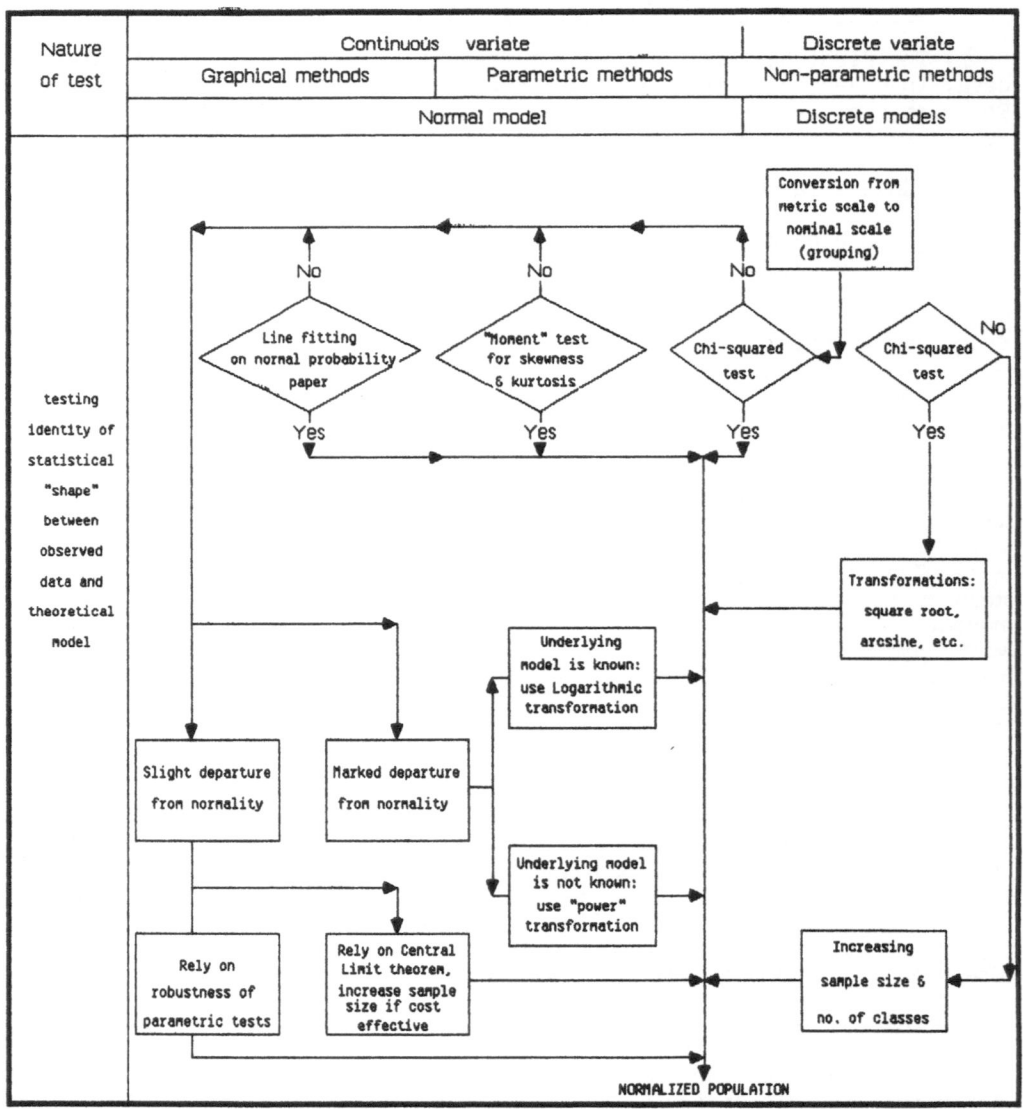

Figure 2.7. Flow-diagram of the "goodness of fit" testing of models and "normalization" of univariate populations.

skewness = (mean - mode)/standard deviation.

The reader will immediately note that the skewness = 0 for symmetrical
distributions, such as the normal. Mathematical transformations may often be
devised for normalizing skewed distributions. For example, in the case of
lognormal distributions, a logarithmic transformation results in a normal
distribution for the logarithms of the data. The next step is to assign each
observed frequency distribution to some category of well-known theoretical
distributions which fits it most closely. This is often referred to as
"model fitting".

2.4.3. Foundation of univariate model fitting: χ^2-test, transformations

Fitting various theoretical statistical models to the frequency
distributions of observed data is a most important step in statistical
processing. One of the many advantages derived from model fitting is the
detection of skewness, a sign of deviation from a normal distribution, since
the assumption for the validity of most parametric tests and confidence
interval predictions is that the distribution is normal.

The test to determine whether deviations from the hypothesized
distribution are significant is the χ^2-test. The χ^2 expression is defined
as

$$\chi^2 = \Sigma_{i=1}^{k} (o_i - e_i)^2/e_i,$$

where

k is the number of classes or combined classes for those with an expected
class size of < 2,
o_i is the observed number and e_i the expected number in $class_i$, i = 1, 2,
..., k.
We further assume the hypothesis of normality for each term in the
summation, so that χ^2 is χ^2- distributed with k - 1 - j degrees of freedom,
where j = the number of parameters used to compute the expected class sizes,
e_i, (2 for normal and lognormal distributions, 1 for Exponential and
Poisson).

In the present study, the specified significance level is set at 0.05.
Almost all the distributions were found to fit the lognormal distribution,
which implies that the log-transformed data are normal. Since this is the
basic assumption of most parametric test-statistics and confidence interval
predictions, the logtransform of the data was performed before every such
test and prediction.

Some geochemical data have been found to be too skewed even after a
log-tranform, and several authors including Boot & Cox (3, Chapter 3), have
recommended a more complex transform referred to as the "power" transform
which is expressed as

$$z = (x^\lambda - 1)/\lambda,$$

where λ is chosen to minimize the deviation from normal. However, the
minimization is tedious and requires an extensive use of dynamic
programming, so its application is rather restricted. In general, in this

study, the validity of the logtransform has been highly satisfactory as proved by the results of the χ^2 goodness-of-fit test. Figure 2.7 summarizes the normalization transforms of data, and the χ^2 goodness-of-fit test.

2.4.4. Types of models used for the fitting of observed distributions of ore deposit parameters

As seen in Chapter 1, section 2, several types of parameters are used to characterize ore deposits. They can be grouped into three categories including (a) economic, (b) geometric, and (c) occurrence parameters. The first two describe the intrinsic characteristics of ore deposits; the geometric parameters are, in turn, subdivided into dimensional and attitudinal parameters. The third category describes the spatial distribution of ore deposits. The first two are of a continuous nature while the third is obviously discrete (we cannot considered fractions of ore deposits).

As mentioned earlier, the observed distributions of most economic and geometric parameters of ore deposits are satisfactorily fitted by lognormal models. The attitudinal parameters, however, are best fitted by normal models for unoriented angular data such as dip, while oriented data such as strike directions are best fitted by circular or semi-circular normal models. Among the discrete variates describing ore deposit occurrence, the density of occurrence is the most commonly used measure. Various types of discrete models are used: they include Poisson and the related negative exponential models, and the negative binomial models.

The relationships between these various models are complex, and an attempt to summarize this difficult matter has been made by means of a flow diagram shown in Figure 2.8. The diagram emphasizes quite well the central role played by the normal distribution in modern Statistics. These relationships belong to three main types including

(a) transforms,
(b) asymptotic convergence,
(c) functional.

The lognormal transform, already mentioned several times above, is most commonly used. However, arcsine transformations can be used to normalize proportion distributions of the binomial type, and the square root transformation may be used to advantage to normalize Poisson-type data. These two types of discrete random variable distributions are said to converge asymptotically to the normal distribution when a statistic or sample size becomes very large. Finally, there is a direct relationship between the Poisson occurrence model with expected success rate, μ, and the exponential model for the areal extent between deposits, which has a mean equal to $1/\mu$.

2.4.5. Statistical applications of model fitting

2.4.5.1. General output of statistical processing

The main aim of statistical modelling is to define the actual population distribution based on a sample of observed data. Since ore deposit parameters are random variables, in order to estimate statistically the parameters of expected prizes, we need to rely on the parameters derived from the observed data of ore deposits of the same genetic type as the prizes. One of the main results derived from the goodness-of-fit test is the ability to obtain an accurate and a reliable estimate of the mean and variance to ensure accurate predictions, and furthermore, confidence intervals for the mean, as well as probability distribution models for ore deposit parameters. A very important additional purpose is to ensure a normalization of the data when carrying out such tests as ANOVA, and the Multiple Range Tests, which are dealt with in Chapter 3.

2.4.5.2. Construction of confidence intervals for expected values of parameters of ore deposits

The concept of confidence intervals for the mean is crucial for planning and predictive purposes because the point estimate does not provide enough information. The distribution of the means of samples when (σ), the population standard deviation is not known, but estimated from the sample as (s), is the Student t-distribution with n - 1 degrees of freedom, so that a 95% confidence interval for μ, the population mean, is given by

$$95\% \text{ confidence interval} = m \pm t_{0.025, \ n-1}(s^2/n)^{0.5},$$

where m is the mean of the sample, n is the sample size and $s/\sqrt{n}$ is the standard error (standard deviation) of the means of samples, size n.

If the sample size, n, is large (n > 50), then the central limit theorem is often applied: the distribution of sample means is asymptotically Normal when the population standard deviation is known, and the 95% confidence interval for μ, the population mean, is given by

$$95\% \text{ confidence interval} = m \pm 1.96* (s^2/n)^{0.5}, \text{ with the same}$$

terminology as above. However, if the standard deviation of the sample is used, the student-t value with n-1 degrees of freedom should be used (see Appendix 9).

2.4.5.3. Ordering of populations based on expected values

Gibbons and Olkin (9, section 2) point out the problem of choosing the best from among several alternatives in a statistical context. In any decision making, we must select one or more items or courses of action from k others, so a sampling survey is conducted until sufficient information has been collated to form a basis for the choice. In other cases, the populations may need to be ordered, in which case a statistical ranking procedure can be implemented, such as ranking according to the arithmetic means or geometric means in the case of lognormal populations, and secondly according to variability statistics such as the dispersion coefficient. In the former case, the ranking is done in order of decreasing magnitude of the mean; in the latter case, in order of increasing magnitude of the coefficient; obviously, variability is an "adverse" factor which should be minimized.

2.4.5.4. Construction of probability tables for ore deposit parameters

In probability distribution theory, the probability distribution function, F, evaluated at a point x, is the integral from $-\infty$ to x, which can be represented as the area under the density curve to the left of the ordinate x. For example the 75th %ile occurs at the point such that 0 .75 of the area is to its left.

The Table 2.12, presented below in section 2.5, describes how a probability distribution for the gross value of porphyry Cu-Mo deposits of the South Pacific Island Arc, a lognormal distribution, is derived from the fitted model via the concept of expected relative cumulative frequencies. From the table, it can be seen that for example that the probability that the gross value exceeds $400 million is 0.798.

A similar method is used in Section 6 of this chapter to calculate the probability of occurrence of ore deposits. The discrete data is modelled by the Poisson or generalized negative binomial distributions. The formulation of the latter requires a much fuller treatment to be given in Section 2.6. The reader will be given a full account of the use of the maximum likelihood criterion combined with Taylor's theorem to estimate the distribution parameters in order to satisfactorily fit actual field data.

2.4.5.5. Conclusion

Summarizing statistical estimates, confidence intervals, and probility tables for the parameters of ore deposits of specific genetic types for various regions are indispensable to guide the optimal planning of mineral exploration. Modelling and derived statistics are essential as a preliminary step for the selection of ore deposit types & regions of search. As a further spin-off from modelling it will be possible, as explained in Chapter 3, to use techniques of statistical commonality testing in order to pool regions which are statistically similar on economic parameters of ore deposits in order to facilitate the selection. The modelling and derivation of geometric and attitude parameter estimates and confidence intervals are also essential for the calculation of probabilites of detection and for the optimal design of search grids.

2.5. MODELLING OF ECONOMIC & GEOMETRIC PARAMETERS OF ORE DEPOSITS

2.5.1. Foreword

In the following sections, we will describe the application of lognormal fitting, testing the goodness-of-fit and derivation of probability distributions from observed statistical distributions. Three kinds of economic parameters (grade, tonnage, and gross value) and two geometric parameters for three genetic types of deposits in five regions of the World are analyzed by processing the data which is contained in Appendix 8. We will also describe the fitting or lack thereof of attitudinal geometric parameters of three types of deposits in three regions of the World by means of normal and circular normal models.

2.5.2. Lognormal model

2.5.2.1. General properties

The characteristics of the lognormal distribution are well described in Krumbein (section 2, 15), Johnson & Kotz (section 2, 11), and mostly in

Aitchkinson & Brown (section 2, 4). An important property of the lognormal distribution is that the variance is proportional to the mean.

The variate x is lognormal with geometric mean Exp(m), if log(x) is normal with mean m, and standard deviation, s. The 95% confidence interval

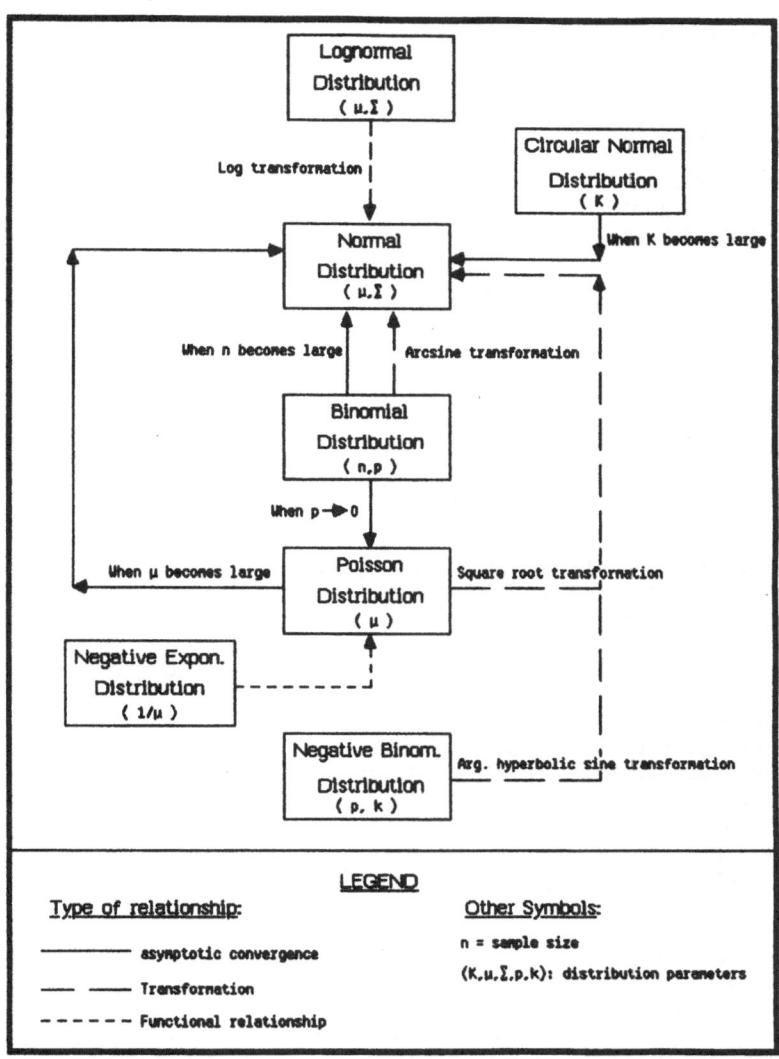

Figure 2.8. Diagram showing the relationship between statistical distributions used for modelling of economic and geometric parameters and areal distribution of ore deposits

for $\mu_{\log(x)}$ is given by

   $m \pm 1.96s/\sqrt{n}$,

for large n, when the Central Limit Theorem applies. For smaller samples, the value 1.96 is replaced by $t_{0.025, n-1}$.

Hence the 95% confidence interval for x is given by

   $Exp(m \pm 1.96s/\sqrt{n})$

so that the lower 95% confidence limit for x is given by

   $Exp(m)/Exp(1.96s/\sqrt{n})$

and the upper 95% confidence limit by

   $Exp(m)*Exp(1.96s/\sqrt{n})$

that is the Lognormal distribution for a continuous variate is almost analogous to the geometric distribution of a discrete variate, which is the result of random arrivals and random departures.

2.5.2.2. Application of lognormal model fitting to distributions of economic parameters of ore deposits

Allais (1, section 3) states that the gross values of individual deposits are lognormally distributed, that is, if x is the log-transformed variate then its distribution function, F, is defined by

$$F(z) = (2*\pi)^{-0.5} \int_{-\infty}^{z} e^{-y^2/2} dy, \text{ where } y = (x - m_x)/s.$$

This applies also to tonnages as shown by fitting a straight line on logarithmic graph paper for the tonnages of gold deposits of South Africa: all points fell within a 95% confidence belt determined by Krige.

Boldy (section 2, 6) and Divi (section 2, 8) also confirmed the satisfactory fitting of lognormal models to the distribution of gross values and tonnages of volcanogenic sulfide deposits of the Canadian Shield. Allais gives a definition of the truncated lognormal distribution which was used by De Geoffroy & Wu (section 3, 19) to estimate the number of deposits in the Canadian Shield with gross values between $20 million and $2000 million. Further lognormal models were successfully fitted by the authors (section 3, 20, 21) to derive the distribution of the gross value of porphyry-Cu-Mo deposits of the North and South American Cordilleran Belt, and of the Upper Mississipi Valley Pb-Zn deposits.

Figure 2.9 shows the fitting of lognormal models to the observed distributions of economic and geometric parameters of the porphyry-Cu-Mo deposits of the Cordilleran Belt of North and South America. Figure 2.10. shows the results of the fitting of the logarithmic transforms of gross value data of deposits in the S. W. Wisconsin region of the U. S. A., and in the N. W. Quebec - N. E. Ontario area of the Canadian Shield by the authors, (section 3, 19 & 20). Sarma (section 2, 19) established that the economic parameters of the Cu deposits at Ingladahl, India, were lognormally distributed, and applied the model to the Cu prospects of adjacent areas.

Uhler & Bradley (22, section 2) examined the data on the oil reservoirs occurring in the Province of Alberta, Canada, to test the size distribution for lognormality. Because they claimed the graphical test and χ^2- test were unreliable (due to its inability to test the tail values satisfactorily), they decided to use the method of comparison based on the first 4 moments of the distribution of the logarithmic transforms of the data:

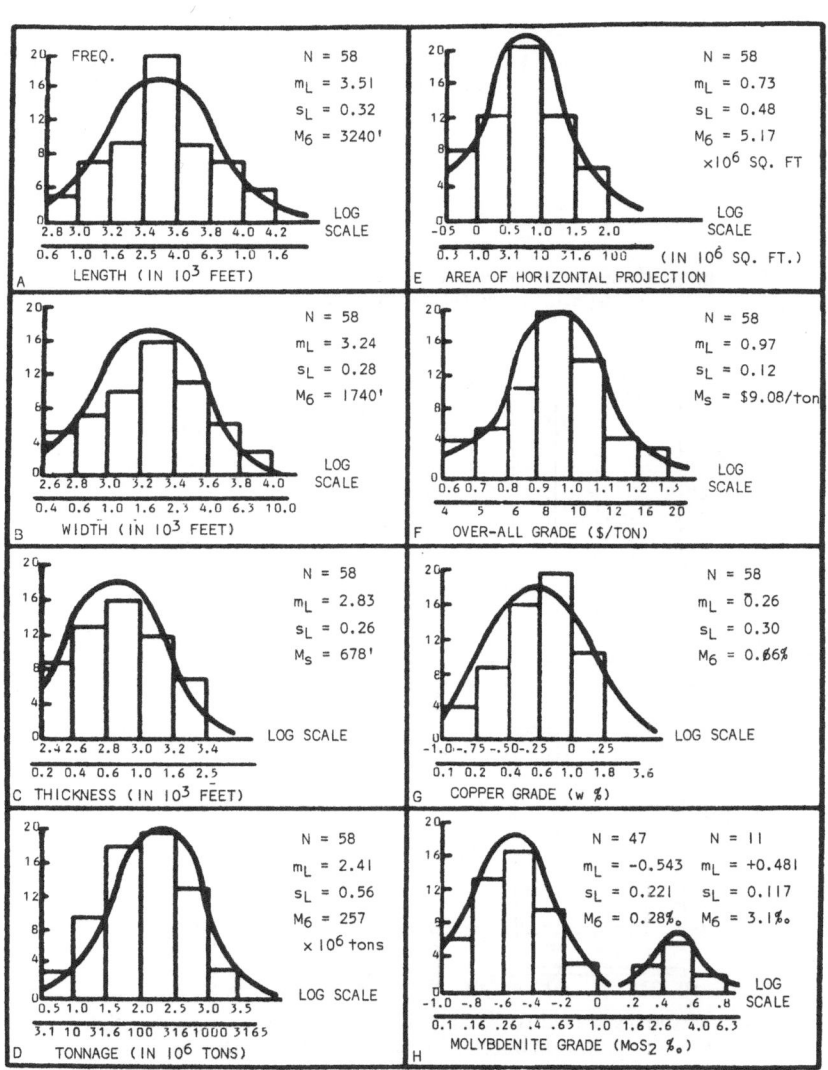

Figure 2.9. Lognormal models fitted to the observed distributions of economic and geometric parameters of porphyry-Cu-Mo deposits of the North and South American Cordillera.

$$m_k = [\Sigma_{i=1}^{n=315}(x_i - \overline{x})^k]/315,$$

for k = 1, 2, 3, 4; where n = 315 is the sample size.

The results were:

Statistic	Observed value	Expected value Lognormal	Upper (Lower) 5% value
$M.D/\sqrt{m_2}$	0.8201	0.7979	0.8185
$m_3/m_2^{3/2}$	0.3989	0.0000	0.2300
m_4/m_2^2	2.6908	3.0000	(2.5900)

where M.D is the mean deviation.
It should be noted that both the graphical and the moment tests confirmed the hypothesis of lognormality, but the third test rejected it.

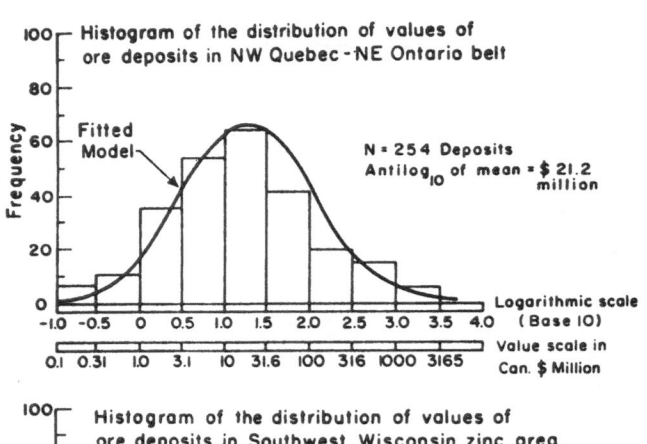

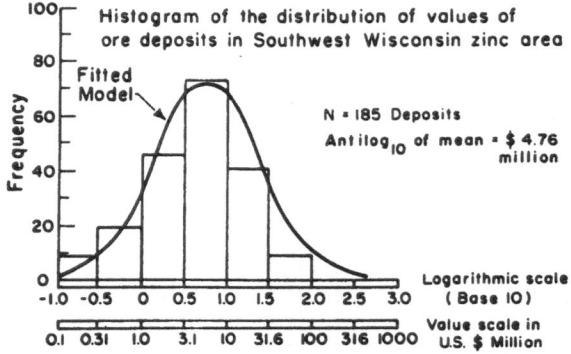

Figure 2.10. Lognormal models fitted to the gross value distribution of ore deposits of all types in the N. W. Quebec - N. E. Ontario section of the Canadian Shield (top), and of Pb-Zn deposits of the Upper Mississipi Valley district of the North American Paleozoic Platform (bottom).

Table 2.2. Example of successful lognormal model fitting to
the observed distribution of the grade parameter
(gram/tonne) of vein-gold deposits of the Western
Australian Shield.

Grouping Interval log 10	Grouping Interval antilog	Observed frequency	Expected frequency lognormal	Probability grade exceeds lower class limit	Statistic: antilog
0.5 - 0.3	3.0 - 4.2	9	7.04	1.00	
0.3 - 1.0	4.2 - 8.4	27	23.36	0.86	25%lle - 6.0
1.0 - 1.1	8.4 - 12.0	5	11.58	0.42	95%Lcl - 6.6
1.1 - 1.3	12.0 - 16.8	8	6.69	0.19	G. mean - 7.5
1.3 - 1.5	16.8 - 33.0	3	3.33	0.07	95%Ucl - 9.0
		52	52.00		75%lle - 10.5

Results of chi-squared significance test at the δ = 0.05 significance level:

Chi-squared = 4.73 < Chi-sq. (0.05, 3 d.f.) = 7.81, which implies the
distribution is accepted as lognormal at the 95% confidence level

<u>Statistics of observed data</u> :

Arithmetic mean	- 8.7
Mode	- 7.8
Standard deviation	- 5.4

2.5.2.3. Fitting of lognormal models to the distributions of economic and geometric parameters of our data base

The results of lognormal fitting to the economic parameters, grade,
tonnage and gross value of the 900 deposits included in our data-base were
mainly good, some excellent, and few showed significant deviations from the
model. Obviously, because of format constraints, we could not possibly show
all the results in this book. We are including here three tables displaying
the lognormal fitting of economic parameters and three tables covering the
fitting of geometric parameters.

Examples given in Tables 2.2 & 2.3 show the distribution of grades of
vein gold deposits of the Western Australian Shield and the tonnage
distribution of the Ni-Cu ultramafic deposits of the Scandinavian Shield.
Table 2.4. displays the lognormal distribution of the gross values of
volcanogenic sulfide deposits of the Iberian Peninsula Pyrite Belt. The
χ^2 value for testing the significance of the difference of these
distributions from lognormal was not significant at the 0.05 level in any of
the three cases.

Table 2.3. Example of successful lognormal model fitting to
the observed distribution of the tonnage parameter
(millions of tonnes) of Ni-Cu deposits of the
Scandinavian Shield.

Grouping Interval log 10	Grouping Interval antilog	Observed frequency	Expected frequency lognormal	Probability tonnage exceeds lower class limit	Statistic: antilog	
-1.3 - -0.7	0.1 - 0.2	3	3.16	1.00		
-0.7 - -0.1	0.2 - 0.8	6	4.29	0.96	25%lle	= 0.2
-0.1 - 0.5	0.8 - 3.2	4	3.54	0.86	95%Lcl	= 0.5
0.5 - 1.1	3.2 - 12.5	2	2.55	0.71	G. mean	= 1.5
1.1 - 1.6	12.5 - 43.0	3	4.46	0.50	95%Ucl	= 3.0
		18	18.00		75%lle	= 4.5

Results of chi-squared significance test at the ∂ = 0.05 significance level:

Chi-squared = 1.34 < Chi-sq. (0.05, 3 d.f.) =7.81, which implies the

distribution is accepted as lognormal at the 95% confidence level

Statistics of observed data :

Arithmetic mean	= 6.2
Mode	= 0.5
Standard deviation	= 11.4

The results of lognormal fitting to the distributions of dimensional
geometric parameters are shown in Tables 2.5, 2.6 and 2.7. Table 2.5 shows
the lognormal fit of the lengths of the horizontal section of vein gold
deposits of the North American Shield, again with a low χ^2- test value.
Table 2.6 shows the successful lognormal fit of the breadths of the
horizontal section of vocanogenic sulfide deposits of the Kuroko Belt of
Japan. Table 2.7 displays the fitting to the shape-ratios of the
volcanogenic sulfide deposits of the North American Shield.

2.5.3. Normal and Circular Normal models

2.5.3.1. General Properties

The general properties of the normal model, the linchpin of modern
Statistics, are so well known and described in so many treatises that it
would be superfluous to mention the topic again here. We have, however given

Table 2.4. Example of successful lognormal model fitting to the observed distribution of the gross value parameter (US $millions) of volcanogenic massive sulfide deposits of the Iberian Peninsula Pyrite Belt.

Grouping Interval log 10	Grouping Interval antilog	Observed frequency	Expected frequency lognormal	Probability value exceeds lower class limit	Statistic: antilog
1.90 - 2.80	80 - 630	4	4.88	1.00	25%ile = 900
2.80 - 3.10	630 - 1260	5	6.31	0.80	95%Lcl = 932
3.10 - 3.40	1260 - 2500	10	6.90	0.54	G. mean = 1393
3.40 - 4.00	2500 - 10,000	6	6.90	0.26	95%Ucl = 2082
		25	25.00		75%ile = 2500

Results of chi-squared significance test at the $\partial = 0.05$ significance level:

Chi-squared = 1.94 < Chi-sq. (0.05, 2 d.f.) = 5.99, which implies the distribution is accepted as lognormal at the 95% confidence level

Statistics of observed data :

Arithmetic mean	= 2500
Mode	= 2500
Standard deviation	= 2700

a computer program in Appendix 2, NORMAL program 2.3, which gives the normal distribution function probability values for any z value, and as already mentioned our standard normal distribution table in Appendix 9. The circular normal model is more complex, and not so well known, so additional comments are fully justified.

The circular normal model domain is 0 to 2π, and the mean and variance calculations are based on vector methods, with the summations carried out over the horizontal projections, cos(x), and vertical projections, sin(x). The sample mean, $\bar{x}$, is given by

$$\bar{x} = \text{Arctan}\{[\textstyle\sum \sin(x)]/[\textstyle\sum \cos(x)]\},$$

and the other parameter, α, to which the variance is related is expressed by

$$\alpha^2 = \{[\textstyle\sum \cos(x)]^2 + [\textstyle\sum \sin(x)]^2\}/n^2.$$

Table 2.5. Example of successful lognormal model fitting to the observed distribution of the length parameter (in meters) of the horizontal section of vein-gold deposits of the North American Shield.

Grouping Interval log 10	Grouping Interval antilog	Observed frequency	Expected frequency lognormal	Probability length exceeds lower class limit	Statistic: antilog
2.00 - 2.30	100 - 200	7	5.33	1.00	
2.30 - 2.45	200 - 285	4	6.89	0.91	25%lle - 340
2.45 - 2.60	285 - 400	11	10.02	0.79	95%Lcl - 470
2.60 - 2.75	400 - 570	14	11.67	0.62	G. mean - 570
2.75 - 2.90	570 - 800	9	10.64	0.42	95%Ucl - 670
2.90 - 3.05	800 - 1150	4	7.58	0.24	75%lle - 900
3.05 - 3.50	1150 - 2500	10	6.87	0.12	
		59	59.00		

Results of chi-squared significance test at the ∂ = 0.05 significance level:

Chi-squared = 5.66 < Chi-sq. (0.05, 5 d.f.) = 11.07, which implies the distribution is accepted as lognormal at the 95% confidence level

Statistics of observed data :

Arithmetic mean - 690

Mode   - 485

Standard deviation - 450

The circular normal density function, f, is expressed as

$$f(x) = [2\pi\ I_o(k)]^{-1} \exp[k\cos(x)],$$

where I_o is a Bessel function of the first kind of purely imaginary argument. From Agterberg (1, section 2), the maximum likelihood estimator for k may be found from Gumbel's variance tables, given in terms of the parameter, α. The integral can be approximated for x < 1, by substituting

$$\cos(x) \approx 1 - x^2/2,$$

where x is measured in radians. However, if x < 1, the normal distribution is a good approximation for the circular normal.

Table 2.6. Example of successful lognormal model fitting to the observed distribution of the breadth parameter (in meters) of the horizontal section of volcanogenic massive sulfide deposits of the Kuroko Belt of Japan.

Grouping Interval log 10	Grouping Interval antilog	Observed frequency	Expected frequency lognormal	Probability width exceeds lower class limit	Statistic: antilog
1.20 - 1.80	15 - 65	4	3.80	1.00	25%ile = 85
1.80 - 2.10	65 - 125	12	10.23	0.88	95%Lcl = 110
2.10 - 2.40	125 - 250	10	11.39	0.55	G. mean = 140
2.40 - 2.70	250 - 500	5	5.58	0.18	95%Ucl = 170
					75%ile = 240

Results of chi-squared significance test at the $\delta = 0.05$ significance level:

Chi-squared = 0.55 < Chi-sq. (0.05, 2 d.f.) = 5.99, which implies the distribution is accepted as lognormal at the 95% confidence level

Statistics of observed data :

Arithmetic mean	- 170
Mode	- 95
Standard deviation	- 110

2.5.3.2. Examples of normal fitting for unorientated dips

A normal model was fitted to the unorientated dip data of a sample of 21 vein gold deposits of the Western Australian Shield, and was found to be an excellent fit, with $\chi^2 = 0.99$, as shown in Table 2.8. However, when a normal model was fitted to the unorientated dip data of a sample of 15 contact metasomatic deposits of the North American Cordillera, the null hypothesis was rejected, with

$$\chi^2 = 12.4 > \chi^2_{2, \; 0.05} = 5.99;$$

detailed results are given in Table 2.9.

Table 2.7. Example of successful lognormal model fitting to
the observed distribution of the shape ratio
parameter of the horizontal section of volcanogenic
massive sulfide deposits of the North American Shield.

Grouping Interval log 10	Grouping Interval antilog	Observed frequency	Expected frequency lognormal	Probability ratio exceeds lower class limit	Statistic: antilog
-3.0 - -2.1	0.00 - 0.01	6	4.45	1.00	
-2.1 - -1.8	0.01 - 0.02	6	9.57	0.96	25%lle = 0.02
-1.8 - -1.5	0.02 - 0.03	18	15.18	0.86	95%Lcl = 0.05
-1.5 - -1.2	0.03 - 0.06	23	19.92	0.71	G. mean = 0.06
-1.2 - -0.9	0.06 - 0.12	15	20.02	0.50	95%Ucl = 0.08
-0.9 - -0.6	0.12 - 0.25	20	15.30	0.30	75%lle = 0.12
-0.6 - -0.0	0.25 - 1.00	11	14.54	0.15	
		99	99.00		

Results of chi-squared significance test at the ∂ = 0.05 significance level:

Chi-squared = 8.11 < Chi-sq. (0.05, 5 d.f.) = 11.07, which implies the
distribution is accepted as lognormal at the 95% confidence level

Statistics of observed data :

Arithmetic mean = 0.14

Mode   = 0.06

Standard deviation = 0.20

2.5.3.3. Examples of circular normal model fitting to strike directions

The observed strike directions of volcanogenic massive sulfide
deposits of the North American Shield had an arithmetic mean very close to
the circular mean, with a standard deviation = $\pi/4$ (45 degrees); and again
the ensuing fit was satisfactory (Table 2.10). The observed strike
directions of the contact metasomatic deposits of the North American
Cordillera had a wider dispersion, with a standard deviation of 57 degrees;
the null hypothesis was rejected. Details are given in Table 2.11.

2.5.4. Derivation of probability distributions for economic parameters

2.5.4.1. Introduction

The availability of accurate probability tables for the three economic
parameters of ore deposits of various genetic types in various regions of
the World is of great importance for the sound economic planning of mineral
exploration programs. For the purpose of the optimal selection of ore

Table 2.8. Example of successful normal model fitting to
the observed distribution of the unorientated
dip angle parameter of vein gold deposits of the
Western Australian Shield.

Grouping Interval degrees	Observed frequency	Expected frequency normal	Probability dip exceeds lower class limit	Statistic: degrees	
25 - 35	3	2.42	1.00	25%ile	= 40
35 - 45	4	3.15	0.88	95%Lcl	= 49
45 - 55	3	4.37	0.73	Mean	= 57
55 - 65	5	4.53	0.52	95%Ucl	= 65
65 - 85	6	6.53	0.31	75%ile	= 78
	21	21.00			

Results of chi-squared significance test at the $\delta = 0.05$ significance level:

Chi-squared = 0.99 < Chi-sq. (0.05, 3 d.f.) = 7.81, which implies the
distribution is accepted as normal at the 95% confidence level

Statistics of observed data :

Arithmetic mean	- 57 degrees
Mode	- 80 degrees
Standard deviation	- 18 degrees

deposit types as exploration targets and regions of occurrence, the gross
value parameter is of cardinal importance, immediately followed by tonnage
and by aggregate dollar grade. Here again, due to format constraints, we can
display only a partial coverage of our world-wide data base by means of a
set of four tables. The lognormal model was successfully fitted to the
economic parameters of each region-ore deposit type of our data base, and an
example of the derivation of a probability distribution for the gross values
of porphyry-Cu-Mo deposits of the South Pacific Island Arc is also given in
Table 2.12. The expected lognormal frequency of each class was calculated
and divided by the sample size, 31, to give the expected relative frequency.

The probability distribution, F, is defined in this study by

$$F(x) = \text{Probability } \{\text{gross value} > x\},$$

where the gross value is the variate on which the probability is defined.
For example from the table, the probability that the gross value exceeds 400
is given by

$$F(400) = \text{Probability } \{x > 400\} = 1 - 0.202 = 0.798.$$

Table 2.9. Example of unsuccessful normal model fitting to the observed distribution of the unorientated dip angle parameter of contact metasomatic deposits (Cu-Au-Mo type) of the Cordilleran Belt of North America.

Grouping Interval degrees	Observed frequency	Expected frequency normal	Probability dip exceeds lower class limit	Statistic: degrees		
0 - 5	11	4.92	1.00	25%ile	=	0
5 - 50	0	6.76	0.67	95%Lcl	=	3
50 - 65	2	1.51	0.23	Mean	=	22
65 - 90	2	1.80	0.13	95%Ucl	=	41
	15	15.00		75%ile	=	65

Results of chi-squared significance test at the ∂ = 0.05 significance level:

Chi-squared = 12.4 > Chi-sq. (0.05, 2 d.f.) = 5.99, which implies the distribution is not accepted as normal at the 95% confidence level

Statistics of observed data :

Arithmetic mean	= 22 degrees
Mode	= 0 degrees
Standard deviation	= 37 degrees

2.5.4.2. Application to specific types of deposits included in the data-base

The results of applying the theory given in 2.5.4.1. to each porphyry-Cu-Mo region are given in Tables 2.12 & 2.13 for each of the economic parameters: grade, tonnage and gross value.

Only the gross value data were available for the U. S. S. R., but the three parameters are covered in full for the North American Cordillera, South American Cordillera and South Pacific Island Arc regions. The probability tables for the Western Australian Shield and Scandinavian Shield Ni-Cu ultramafic deposits are given in Table 2.14. The economic parameters for seven regions of occurrence of volcanogenic sulfide deposits are given in Table 2.15, so that they can quickly be compared on their three economic parameters. The deposits of the Iberian Peninsula Pyrite Belt obviously stand out in comparison to the remainder. Table 2.16 gives the results for the vein-gold deposits of the Western Australian Shield.

Table 2.10. Example of successful circular normal model fitting
to the observed distribution of the strike direction
parameter of volcanogenic massive sulfide deposits
of the North American Shield.

Grouping Interval degrees	Observed frequency	Expected frequency normal	Probability strike exceeds lower class limit	Statistic: degrees	
0 – 15	10	5.70	1.00		
15 – 45	12	13.73	0.94	25%ile	– 62
45 – 60	8	10.27	0.79	95%Lcl	– 84
60 – 90	26	24.41	0.68	Circ. mean	– 93
90 – 120	15	21.09	0.41	95%Ucl	= 102
120 – 180	21	16.81	0.17	75%ile	= 130
	92	92.00			

Results of chi-squared significance test at the δ = 0.05 significance level:
Chi-squared = 6.60 < Chi-sq. (0.05, 3 d.f.) = 7.81, which implies the
distribution is accepted as circular normal at the 95% confidence level

Statistics of observed data :

Arithmetic mean	– 92 degrees
Mode	– 97 degrees
Standard deviation	– 45 degrees

2.5.5. Optimal selection of ore deposit types and regions of occurrence

2.5.5.1. Introduction

One of the principal tasks of the exploration planner is to select the
most appropriate type of deposit to be considered as the exploration target
in the context of corporate goals and policies. The second one is to choose
regions of search which will maximize the probabilities of success for the
search. A rather simple selection approach is to rank the deposit types and
regions, based on the decreasing magnitude of the means of the three
economic parameters mentioned above which were derived from the model
fitting exercise of section 2.4.

However, since we are dealing with stochastic quantities, a ranking
based solely upon the mean parameters cannot be expected to yield always a
reliable ranking to guide the selection. The variability of the parameters
should be considered as an additional criterion for the ranking. This is
done here by using the coefficient of dispersion (standard deviation/mean).
Obviously, we will be choosing regions and deposit types which show the

Table 2.11. Example of unsuccessful circular normal model fitting
to the observed distribution of the strike direction
parameter of contact metasomatic (Cu-Au-Mo type) deposits
of the Cordilleran Belt of North America.

Grouping Interval degrees	Observed frequency	Expected frequency normal	Probability strike exceeds lower class limit	Statistic: degrees	
0 - 20	5	2.48	1.00	25%ile	= 0
20 - 80	1	5.51	0.83	95%Lcl	= 46
80 - 110	3	2.94	0.44	Circ. mean	= 82
110 - 150	6	4.07	0.27	95%Ucl	= 105
	15	15.00		75%ile	= 115

Results of chi-squared significance test at the ∂ = 0.05 significance level:

Chi-squared = 6.21 > Chi-sq. (0.05, 2 d.f.) = 5.99, which implies the

distribution is not accepted as circular normal at the 95% confidence level

Statistics of observed data :

Arithmetic mean = 75 degrees

Mode = 110 degrees

Standard deviation = 57 degrees

smallest coefficient of dispersion, which signifies the greatest reliability
of the mean. Therefore, the ranking based on the dispersion coefficient is
carried out by listing the elements to be ranked in order of increasing
magnitude of the coefficient of variation.

Ranking for the purpose of assisting the selection of ore deposit
types and regions of occurrence may be considered in two aspects, including:
(a) ranking regions of occurrence throughout the World within each type of
ore deposit, and
(b) ranking the combination of ore deposit type-region on a continent by
continent basis.

2.5.5.2. World-wide ranking of regions of occurrence within specified genetic types of ore deposits

We first consider the ranking of four regions of occurrence of
porphyry-Cu-Mo deposits based on the three economic parameters (Table 2.17).
Regarding the gross value parameter, the ranking based on the mean is
similar to that obtained for the other two economic parameters, despite
introducing a new region (Central Asia, U. S. S. R.). However, ranking based
on the dispersion coefficients markedly differs from that based on the

Table 2.12. Example of the derivation of a probability distribution from fitting a lognormal model to the gross value parameter (US $millions) of porphyry-Cu-Mo deposits of the South Pacific Island Arc.

Grouping Interval log 10	Grouping Interval antilog	Observed frequency	Expected frequency lognormal	Expected relative frequency lognormal	Cumulative relative frequency: Probability Distribution
2.00 - 2.60	100 - 400	6	6.26	0.202	1.000
2.60 - 2.90	400 - 800	8	6.03	0.194	0.798
2.90 - 3.20	800 - 1600	8	6.97	0.224	0.604
3.20 - 3.50	1600 - 3200	2	5.87	0.190	0.380
3.50 - 4.20	3200 - 15260	7	5.87	0.190	0.190
					0.000
		31	31.00	1.000	

Results of chi-squared significance test at the δ = 0.05 significance level:
Chi-squared = 3.58 < Chi-sq. (0.05, 3 d.f.) = 7.81, which implies the
distribution is accepted as lognormal at the 95% confidence level

means: the least variable deposits concerning gross value are those of the North American Cordillera and the most variable are those of the U. S. S. R. Tonnage and grade-wise, the ranking lists are exactly the same for the two parameters both on the mean and on the dispersion coefficient statistics, with South America ranked first and the Tasman Geosyncline last.

Table 2.18 displays the ranking of three regions of occurrence of Ni-Cu ultramafic deposits based on the three economic parameters. The Western Australian deposits come first and the Scandinavian Shield last in the ranking lists for all three parameters on both statistics, with the exception of a reversion in the grade ranking on the dispersion coefficient.

A total of eight regions of occurrence of volcanogenic sulfide deposits based on three economic parameters on the mean statistic show some marked differences, with the exception of the common first position of the North American Cordillera exhalative deposits (see Table 2.19).

A similar picture emerges when considering the dispersion coefficient statistic criterion of selection. Furthermore, there are marked differences in the ranking based on the mean and dispersion coefficient for each of the three economic parameters. For example, while the volcanogenic sulfide exhalative deposits of the North American Cordillera rank first on the means of the three parameters, they rank only second on the dispersion coefficient for tonnage and third for gross value.

Table 2.13. Probability distribution table for the economic
parameters (grade, tonnage and gross value)
of porphyry-Cu-Mo deposits of the Cordilleran
Belt of North and South America.

Region : N. & S. American Cordillera			South American Cordillera		
Sample size : 58			23		
Parameter	Antilog: grade $ per tonne	Probability grade exceeds class value	Antilog: grade $ per tonne	Probability grade exceeds class value
Grade:	5	1.00	5	1.00
aggregate	7	0.92	10	0.89
US $ value	9	0.82	14	0.64
per tonne	11	0.69	16	0.53
	14	0.40	22	0.16
	17	0.14		
Parameter	Antilog: tonnage in millions of tonnes	Probability tonnage exceeds class value	Antilog: tonnage in millions of tonnes	Probability tonnage exceeds class value
Tonnage:	5	1.00	40	1.00
	20	0.93	80	0.94
millions of	40	0.83	160	0.82
tonnes	80	0.66	315	0.62
	160	0.46	630	0.38
	320	0.27	1260	0.19
Parameter	Antilog: gross value US $ millions	Probability gross value exceeds class value	Antilog: gross value US $ millions	Probability gross value exceeds class value
Gross value:	290	1.00	1000	1.00
	1150	0.85	2000	0.87
1982 US	2300	0.67	4000	0.70
$ millions	4540	0.48	8000	0.47
	9000	0.22	16000	0.25
	18000	0.07		

Table 2.13. (Continued) Probability distribution table for the
economic parameters (grade, tonnage and gross value)
of porphyry-Cu-Mo deposits of the South Pacific
Island Arc and the U. S. S. R.

Region : Sample size :	South Pacific Island Arc 31		U. S. S. R. 24	
Parameter	Antilog: grade $ per tonne	Probability grade exceeds class value	Antilog: grade $ per tonne	Probability grade exceeds class value
Grade: aggregate US $ value per tonne	4 8 11 16	1.00 0.82 0.52 0.20	not	available
Parameter	Antilog: tonnage in millions of tonnes	Probability tonnage exceeds class value	Antilog: tonnage in millions of tonnes	Probability tonnage exceeds class value
Tonnage: millions of tonnes	10 40 80 160 320	1.00 0.78 0.56 0.33 0.15	not	available
Parameter	Antilog: gross value US $ millions	Probability gross value exceeds class value	Antilog: gross value US $ millions	Probability gross value exceeds class value
Gross value: 1982 US $ millions	100 400 800 1600 3200	1.00 0.80 0.59 0.38 0.19	100 200 400 800 3150	1.00 0.84 0.66 0.44 0.24

Table 2.14. Probability distribution table for the economic
parameters (grade, tonnage and gross value)
of Ni-Cu ultramafic deposits of the Scandinavian
Shield and the Western Australian Shield.

Region : Western Australian Shield Sample size : 21			Scandinavian Shield 18		
Parameter	Antilog: grade $ per tonne	Probability grade exceeds class value	Antilog: grade $ per tonne	Probability grade exceeds class value
Grade: aggregate US $ value per tonne	10 80 115 160	1.00 0.79 0.46 0.24	10 40 50 60	1.00 0.63 0.47 0.15
Parameter	Antilog: tonnage in millions of tonnes	Probability tonnage exceeds class value	Antilog: tonnage in millions of tonnes	Probability tonnage exceeds class value
Tonnage: millions of tonnes	0.2 0.8 1.6 3.2 6.3 12.6	1.00 0.82 0.69 0.52 0.34 0.20	0.1 0.2 0.8 3.2 12.6	1.00 0.82 0.59 0.39 0.17
Parameter	Antilog: gross value US $ millions	Probability gross value exceeds class value	Antilog: gross value US $ millions	Probability gross value exceeds class value
Gross value: 1982 US $ millions	90 160 320 630 1260	1.00 0.71 0.53 0.35 0.19	2 16 63 250	1.00 0.78 0.47 0.18

Table 2.15. Probability distribution table for the economic
parameters (grade, tonnage and gross value)
of volcanogenic massive sulfide deposits of the
of the Tasman Geosyncline and the Western Australian
Shield.

| Region   :   Tasman Geosyncline | | | Western Australian Shield | |
| Sample size : 37 | | | 14 | |
Parameter	Antilog: grade $ per tonne	Probability grade exceeds class value	Antilog: grade $ per tonne	Probability grade exceeds class value
Grade: aggregate US $ value per tonne	20	1.00	40	1.00
	28	0.86	56	0.65
	40	0.77	112	0.24
	56	0.66	158	0.00
	80	0.53		
	112	0.40		
	160	0.28		
	224	0.18		
Parameter	Antilog: tonnage in millions of tonnes	Probability tonnage exceeds class value	Antilog: tonnage in millions of tonnes	Probability tonnage exceeds class value
Tonnage: millions of tonnes	0.2	1.00	0.2	1.00
	0.5	0.87	0.5	0.75
	1.0	0.74	2.0	0.37
	2.0	0.58	24.0	0.00
	4.0	0.41		
	8.0	0.25		
	16.0	0.13		
Parameter	Antilog: gross value US $ millions	Probability gross value exceeds class value	Antilog: gross value US $ millions	Probability gross value exceeds class value
Gross value: 1982 US $ millions	10	1.00	10	1.00
	40	0.88	40	0.68
	80	0.76	160	0.38
	160	0.61	630	0.24
	315	0.43		
	630	0.26		
	1260	0.14		

Table 2.15. (Continued) Probability distribution table for the
economic parameters (grade, tonnage and gross value)
of volcanogenic massive sulfide deposits
of the Scandinavian Shield and the Scandinavian
Caledonides.

Region : Sample size :	Scandinavian Shield 32		Scandinavian Caledonides 19	
Parameter	Antilog: grade $ per tonne	Probability grade exceeds class value	Antilog: grade $ per tonne	Probability grade exceeds class value
Grade: aggregate US $ value per tonne	10 20 35 70 200	1.00 0.96 0.67 0.26 0.00	10 20 30 50 80	1.00 0.80 0.61 0.33 0.09
Parameter	Antilog: tonnage in millions of tonnes	Probability tonnage exceeds class value	Antilog: tonnage in millions of tonnes	Probability tonnage exceeds class value
Tonnage: millions of tonnes	1.0 2.0 4.0 8.0 16.0	1.00 0.87 0.67 0.41 0.18	0.5 2.5 4.5 10.5 22.5	1.00 0.62 0.39 0.13 0.00
Parameter	Antilog: gross value US $ millions	Probability gross value exceeds class value	Antilog: gross value US $ millions	Probability gross value exceeds class value
Gross value: 1982 US $ millions	20 80 160 320 1260	1.00 0.83 0.66 0.45 0.20	10 60 160 360 910	1.00 0.72 0.45 0.18 0.00

Table 2.15. (Continued) Probability distribution table for the economic parameters (grade, tonnage and gross value) of volcanogenic massive sulfide deposits of the Eastern Mediterranean and the Kuroko Belt of Japan.

Region :	Eastern Mediterranean		Kuroko Belt of Japan	
Sample size :	27		14	
Parameter	Antilog: grade $ per tonne	Probability grade exceeds class value	Antilog: grade $ per tonne	Probability grade exceeds class value
Grade: aggregate US $ value per tonne	15 30 40 70 140	1.00 0.74 0.53 0.31 0.00	40 56 80 112 224	1.00 0.85 0.66 0.43 0.23
Parameter	Antilog: tonnage in millions of tonnes	Probability tonnage exceeds class value	Antilog: tonnage in millions of tonnes	Probability tonnage exceeds class value
Tonnage: millions of tonnes	1.0 4.0 7.0 13.0 40.0	1.00 0.80 0.57 0.32 0.00	0.5 1.4 5.6 8.0 31.5	1.00 0.73 0.48 0.23 0.00
Parameter	Antilog: gross value US $ millions	Probability gross value exceeds class value	Antilog: gross value US $ millions	Probability gross value exceeds class value
Gross value: 1982 US $ millions	40 160 320 630 1260	1.00 0.77 0.53 0.28 0.16	40 315 630 1260 5800	1.00 0.67 0.44 0.24 0.00

Table 2.15. (Continued) Probability distribution table for the economic parameters (grade, tonnage and gross value) of volcanogenic massive sulfide deposits of the Iberian Peninsula Pyrite Belt.

Region :	Iberian Peninsula Pyrite Belt	
Sample size :	25	
Parameter	Antilog: grade $ per tonne	Probability grade exceeds class value
Grade: aggregate US $ value per tonne	20	1.00
	28	0.95
	40	0.73
	50	0.36
	106	0.00
Parameter	Antilog: tonnage in millions of tonnes	Probability tonnage exceeds class value
Tonnage: millions of tonnes	2.0	1.00
	4.0	0.96
	16.0	0.73
	64.0	0.20
	110.0	0.00
Parameter	Antilog: gross value US $ millions	Probability gross value exceeds class value
Gross value: 1982 US $ millions	80	1.00
	630	0.80
	1260	0.54
	2520	0.26
	10,000	0.00

Table 2.16. Probability distribution table for the economic
parameters (grade, tonnage and gross value)
of vein-gold deposits of the Western
Australian Shield.

Region :	Western Australian Shield	
Sample size :	41	
Parameter	Antilog: grade $ per tonne	Probability grade exceeds class value
Grade: aggregate US $ value per tonne	40 110 160 225 310	1.00 0.81 0.57 0.30 0.11
Parameter	Antilog: tonnage in millions of tonnes	Probability tonnage exceeds class value
Tonnage: millions of tonnes	0.1 0.4 0.8 1.6 3.2	1.00 0.77 0.57 0.34 0.16
Parameter	Antilog: gross value US $ millions	Probability gross value exceeds class value
Gross value: 1982 US $ millions	10 20 40 80 160 320	1.00 0.91 0.77 0.58 0.36 0.18

N.B. gold price US$14/gram

Table 2.17. Ranking of five regions in three continents based on three economic parameters of porphyry-Cu-Mo deposits.

Type of deposit: Porphyry-Cu-Mo					
Parameter	Region	Geometric mean	Rank	Dispersion cofficient	Rank
Aggregate grade in US $ / tonne	South American Cordillera	16	1	0.12	1
	North American Cordillera	13	2	0.26	3
	South Pacific Island Arc	12	3	0.16	2
	Tasman Geosyncline	11	4	0.29	4
Tonnage in millions of tonnes	South American Cordillera	450	1	0.18	1
	North American Cordillera	257	2	0.27	3
	South Pacific Island Arc	96	3	0.26	2
	Tasman Geosyncline	39	4	0.28	4
Gross value in Us $ millions	South American Cordillera	7290	1	0.12	2
	North American Cordillera	3490	2	0.06	1
	South Pacific Island Arc	1092	3	0.17	3
	U. S. S. R. (Central Asia)	1015	4	0.23	5
	Tasman Geosyncline	409	5	0.22	4

2.5.5.3. Continent-wide ranking of ore deposit types and regions of occurrence

A continent-wide ranking of a combination of ore deposit types and regions is more practical for the purpose of exploration planning than the previous exercise. Most mineral industry concerns operate largely on a continent-wide basis, while only relatively fewer large international corporations spread their activities world-wide. As an example, we are ranking below three types of ore deposits occurring in five regions throughout Europe (Table 2.20), and in eight regions throughout Australasia and East Asia (Table 2.21). The ranking of ore deposit types and regions for North America was covered in the previous volume referred to in the preface.

Table 2.20 shows that the Iberian Peninsula volcanogenic sulfide deposits rank first for all three parameters based on the mean statistic, while the Ni-Cu deposits of the Scandinavian Shield share the last position with the Scandinavian Caledonides volcanogenic sulfide deposits. The Scandinavian Shield volcanogenic sulfide deposits occupy the median position. Similar top and bottom ranking positions are obtained for the

Table 2.18. Ranking of three regions in three continents based on three economic parameters of Ni-Cu ultramafic deposits.

Type of deposit: Ni-Cu Ultramafic					
Parameter	Region	Geometric mean	Rank	Dispersion cofficient	Rank
Aggregate grade in US $ / tonne	Western Australian Shield	100	1	0.11	1
	North American Shield	92	2	0.15	3
	Scandinavian Shield	47	3	0.13	2
Tonnage in millions of tonnes	Western Australian Shield	3.9	1	0.95	1
	North American Shield	2.1	2	1.89	2
	Scandinavian Shield	1.3	3	7.73	3
Gross value in US $ millions	Western Australian Shield	395	1	0.20	1
	North American Shield	198	2	0.26	2
	Scandinavian Shield	63	3	0.40	3

Iberian Peninsula volcanogenic sulfides and Ni-Cu deposits of the Scandinavian Shield, when we consider the ranking based upon the dispersion coefficient, for all three economic parameters. However, we can see that the intermediate positions in the ranking lists based on the two statistics differ quite markedly.

The next table, 2.21, shows that, based on both statistics, the porphyry Cu-Mo deposits of the South Pacific Island Arc stand prominently first in the gross value and tonnage sequences, followed by those of the Tasman Geosyncline, while the East Asian W-Mo contact metasomatic deposits stand last. The sequences of deposit-regions of intermediate rankings are also very similar if we consider the mean statistic; however, if we take the coefficient of dispersion as the yardstick, there are substantial discrepancies between the ranking based on gross values and that based on tonnages. Finally, we find that the sequence based on the mean dollar grade differs considerably from that based on the other two parameters: porphyry-type deposits, star performers of the previous rankings, come last, contrasting with the East Asia W-Mo deposits which take first rank based on the dispersion coefficient, instead of last as mentioned above.

Table 2.19. Ranking of ten regions in four continents based on three economic parameters of volcanogenic sulfide deposits.

Type of deposit: Volcanogenic sulfides					
Parameter	Region	Geometric mean	Rank	Dispersion cofficient	Rank
Aggregate grade in US $ / tonne	North American Cordillera (exhalative)	109	1	0.07	1
	Kuroko Belt of Japan	101	2	0.22	7
	Tasman Geosyncline	86	3	0.19	5
	Western Australian Shield	72	4	0.21	6
	Iberian Peninsula Pyrite Belt	50	5	0.10	2
	Scandinavian Shield	43	6	0.18	4
	Eastern Mediterranean	42	7	0.13	3
	Scandinavian Caledonides	36	8	0.21	6
Tonnage in millions of tonnes	North American Cordillera (exhalative)	29.5	1	0.43	2
	Iberian Peninsula Pyrite Belt	28.2	2	0.25	1
	Eastern Mediterranean	8.1	3	0.47	3
	Scandinavian Shield	6.0	4	0.64	4
	Kuroko Belt of Japan	5.2	5	0.98	6
	Scandinavian Caledonides	3.4	6	0.90	5
	Tasman Geosyncline	2.8	7	1.25	7
	Western Australian Shield	1.3	8	7.67	3
Gross value in Us $ millions	North American Cordillera (exhalative)	3207	1	0.18	3
	Iberian Peninsula Pyrite Belt	1393	2	0.14	1
	Kuroko Belt of Japan	531	3	0.32	7
	Eastern Mediterranean	328	4	0.15	2
	Scandinavian Shield	271	5	0.22	4
	Tasman Geosyncline	239	6	0.23	5
	Scandinavian Caledonides	120	7	0.26	6
	Western Australian Shield	92	8	0.53	8

Table 2.20. Ranking of three types of ore deposits occurring
in five regions of Europe based upon three
economic parameters (gross value, tonnage and grade).

Parameter	Region	Ore Deposit type	Geometric mean	Rank	Dispersion cofficient	Rank
Gross value in US $millions	Iberian Peninsula Pyrite Belt	Volc. sulf.	1393	1	0.14	1
	U. S. S. R. (Caucasus)	Porphyry Cu-Mo	1015	2	0.22	4
	Eastern Mediterranean	Volc. sulf.	328	3	0.15	2
	Scandinavian Shield	Volc. sulf.	272	4	0.21	3
	Scandinavian Caledonides	Volc. sulf.	120	5	0.26	5
	Scandinavian Shield	Ni-Cu Ultramaf.	57	6	0.40	6
Tonnage in million tonnes	Iberian Peninsula Pyrite Belt	Volc. sulf.	28.0	1	0.25	1
	Eastern Mediterranean	Volc. sulf.	8.1	2	0.47	2
	Scandinavian Shield	Volc. sulf.	6.0	3	0.64	3
	Scandinavian Caledonides	Volc. sulf.	3.4	4	0.90	4
	Scandinavian Shield	Ni-Cu Ultramaf.	1.2	5	7.73	5
Grade in US $ per tonne	Iberian Peninsula Pyrite Belt	Volc. sulf.	50	1	0.10	1
	Scandinavian Shield	Ni-Cu Ultra.	47	2	0.13	2
	Scandinavian Shield	Volc. sulf.	43	3	0.18	4
	Eastern Mediterranean	Volc. sulf.	42	4	0.14	3
	Scandinavian Caledonides	Volc. sulf.	36	5	0.21	5

Table 2.21. Ranking of five types of ore deposits occurring
in six regions of Australasia and East Asia
based upon three economic parameters (gross value,
tonnage and grade).

Parameter	Region	Ore Deposit type	Geometric mean	Rank	Dispersion cofficient	Rank
Gross value in US $millions	South Pacific Island Arc	Porphyry-Cu-Mo	1092	1	0.17	1
	Kuroko Belt of Japan	Volc. sulf.	531	2	0.32	6
	Tasman Geosyncline	Porphyry-Cu-Mo	409	3	0.22	4
	Western Australian Shield	Ni-Cu Ultramaf.	359	4	0.20	2
	Tasman Geosyncline	Volc. sulf.	339	5	0.23	5
	Western Australian Shield	Vein Au	131	6	0.21	3
	Western Australian Shield	Volc. sulf.	92	7	0.53	8
	East Asia Region	Cont.met.(W-Mo)	86	8	0.35	7
Tonnage in million tonnes	South Pacific Island Arc	Porphyry-Cu-Mo	96.0	1	0.47	1
	Tasman Geosyncline	Porphyry-Cu-Mo	38.0	2	0.11	2
	Kuroko Belt of Japan	Volc. sulf.	5.3	3	0.22	4
	Western Australian Shield	Ni-Cu Ultramaf.	3.4	4	0.10	3
	Tasman Geosyncline	Volc. sulf.	2.8	5	0.19	5
	Western Australian Shield	Volc. sulf.	1.2	6	0.21	6
	Western Australian Shield	Vein Au	1.1	7	0.16	8
	East Asia Region	Cont.met.(W-Mo)	0.9	8	0.29	7
Grade in US $ per tonne	Western Australian Shield	Vein Au	156	1	0.47	8
	Western Australian Shield	Ni-Cu Ultra.	106	2	0.11	2
	Kuroko Belt of Japan	Volc. sulf.	101	3	0.22	6
	East Asia Region	Cont.met.(W-Mo)	95	4	0.10	1
	Tasman Geosyncline	Volc. sulf.	86	5	0.19	4
	Western Australian Shield	Volc. sulf.	72	6	0.21	5
	South Pacific Island Arc	Porphyry-Cu-Mo	12	7	0.16	3
	Tasman Geosyncline	Porphyry-Cu-Mo	11	8	0.29	7

2.6. MODELLING OF OCCURRENCE PARAMETERS OF ORE DEPOSITS

2.6.1. Introduction

Modelling occurrence parameters or spatial distributions is usually
formulated using the quadrat method in which the area being studied is
divided up into contiguous square or rectangular cells. The choice of
dimensions of the cells depends on the size of the survey area. The number
of occurrences in each cell is then recorded and the result compared with a
theoretical discrete distribution such as the Poisson or negative binomial

distribution. An important property of the Poisson distribution is that the mean equals the variance. As a result, only one parameter which is referred to as the expected rate of occurrence per cell is required to specifiy the Poisson distribution.

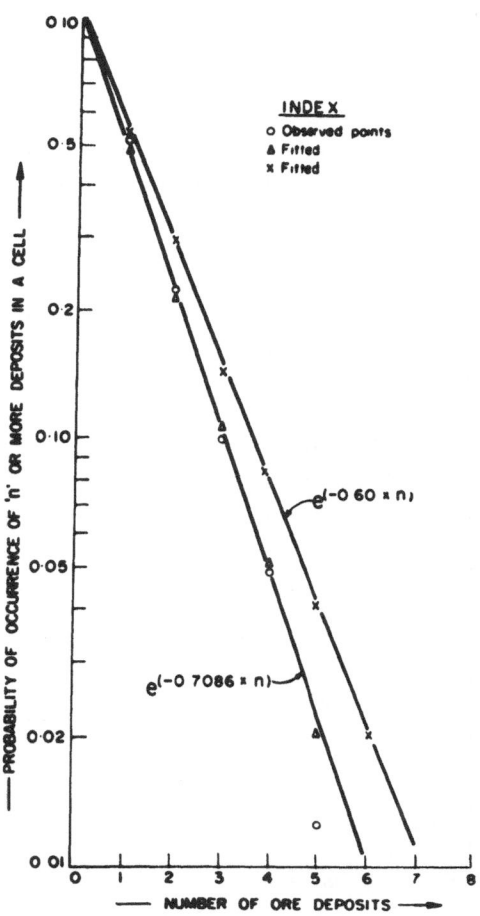

Figure 2.11. Fitting of a negative exponential model to the areal distribution of Cu deposits of Central India (cell size 50 x 50 km). From Sarma, D. D. Mineralium Deposita (47) by permission K. Springer Verlag, Berlin & New York.

If the mean/variance ratio equals 1, we have a random distribution of occurrences represented by the Poisson model. If the mean/variance is less than 1, there will be some degree of clustering, indicating a contagious process which can be modelled by a geometric distribution of the generalized negative binomial or Neyman A type (section 3, 40). A mean/variance ratio is greater than 1 signifies a more regular pattern, possibly indicating a competitive process with a normal model.

2.6.2 Poisson and Negative Exponential models

2.6.2.1. Introduction

As stated above, there is a functional relationship between the Poisson and exponential models as illustrated in Figure 2.8. If there are on average (m) deposits per cell in the Poisson model, then the mean area for the occurrence of one deposit is the fraction of a cell, 1/m, which is the mean of the exponential areal distribution.

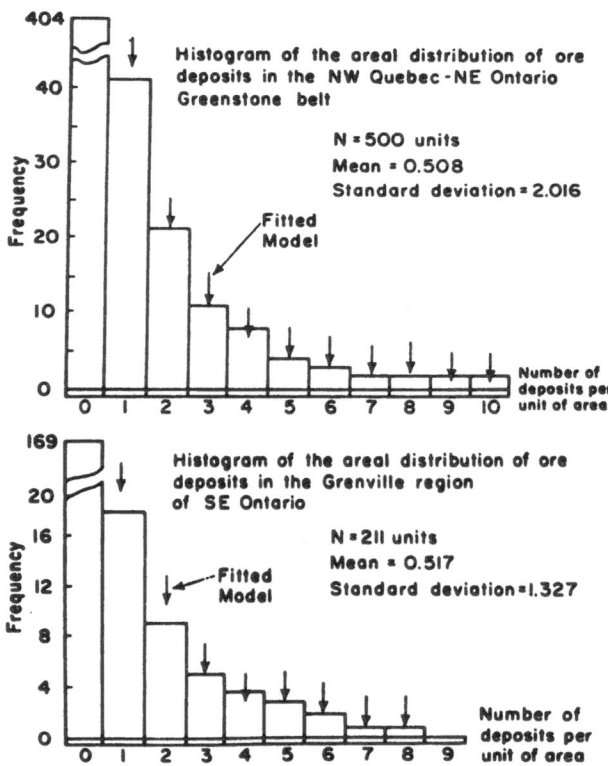

Figure 2.12. Fitting of negative binomial models to the areal distribution of ore deposits of all types in the N. W. Quebec - N. E. Ontario section of the Canadian Shield (top), and of the Grenville region of S. E. Ontario (bottom).

2.6.2.2. General properties

The Poisson model is defined by

$$Pr\{x \text{ ore deposits in a cell}\} = (m^x/x!)e^{-m},$$

where $x = 0, 1, 2, \ldots$
The following argument will indicate the manner in which a Poisson discrete occurrence model with mean m leads to an exponential continuous model with mean 1/m. A mean of (m) deposits per cell yields a mean of mt in the next t cells, and applying the Poisson model with mean (mt) instead of (m) yields the following for $x = 0$:

$$Pr\{0 \text{ deposits in the next t cells}) = e^{-mt}, \text{ using the Poisson}$$
distribution formula.

From this fact, the exponential distribution function is defined by:

$F(y)$ = Probability{a deposit will occur in the next y cells or less}
$$= 1 - \text{Probability}\{0\} = 1 - e^{-my},$$

using the Poisson result above. As a result, the exponential distribution is formulated direct from the Poisson and given by

$$F(x) = 1 - e^{-mx},$$

and the mean of which is 1/m, with the standard deviation also 1/m.

Readers should note that the mean of the Poisson distribution equals the variance, while the mean of the exponential distribution equals the standard deviation, that is the square root of the variance.

2.6.2.3. Applications of the Poisson and exponential models

The first worker to apply the Poisson model to the areal distribution of ore deposits was Allais (section 3, 1), who used it to predict the number of commercial ore deposits expected to occur within a large portion of the Sahara Desert territory of North Africa, based upon several control sampling areas of Europe and North America.

Another field study based on the negative exponential and Poisson models of occurrence of ore deposits is that of Sarma (section 1, 18). A portion of central India was gridded by means of 50 km x 50 km cells leading to a total number of 642 cells, not including un-mineralized areas. Figure 2.11 shows the fitted exponential distribution model, and the observed data, plotted on probability graph paper by Sarma. However, re-calculating the mean from the data yields m = 1.07 per cell as the success rate.

The Poisson model is therefore
$$Pr\{x \text{ successes in a cell}\} = (1.07^x/x!)e^{-1.07}, \quad x = 0, 1, 2, \ldots$$

and the spatial distribution measured in cell-units is the exponential distribution

$$Pr\{\text{at least 1 success in x cells}\} = 1 - e^{-1.07x},$$

with a mean of 1/1.07 = 0.93 cells occurring between deposits.

Another interesting example of the application of the Poisson model is that used to analyze the field results of a geochemical gas-bubble sniffing survey over the North Sea in the mid-1960's. The technique employed was to count the number of gas bubbles rising to the surface from the ocean-floor in each quadrat of 1 km x 1 km. The field data were then statistically processed by the authors using the POPMIX program (see Appendix 7) to separate the anomalous set from the background set, and then calculating the two Poisson parameters. The next step was to assign cells to the anomalous and background sets, using the Poisson classification function formulated in Chapter 6. The result was a map of the anomalous areas of the North Sea, which now are mainly proving to be the North Sea oil and gas fields.

Other examples of the application of the exponential model are shown further down in Tables 2.24 and 2.26 covering the areal distributions of various types of ore deposits in the North American Cordillera, Mid-western America, Western Australia and India.

In conclusion, it was found that the Poisson and associated models are generally satisfactory for the analysis of very large regions by the quadrat method using rather large cells. However, in the case of more detailed studies of smaller regions using relatively small cells, these models prove unsatisfactory because they do not take into account the natural geological processes which are reflected by the spatial aggregation of deposits into nests or clusters. As mentioned in the introduction, if there is clustering or "contagion", the variance will increase and m/s becomes less than 1, which indicates that a new model must be considered. We will see that the negative binomial model can be used to obviate this difficulty.

2.6.3. Negative Binomial model

2.6.3.1. General properties

As indicated by Bliss (section 3, 8) and Williamson (section 3, 50), the generalized negative binomial model has two parameters k and p, and the probability function, P, is defined by

$$P(x) = Pr\{x \text{ ore deposits in a cell}\}$$
$$= (k + x - 1)!*R^x/x!/(k - 1)!/q^k \text{ , for } x = 0, 1, \ldots$$

where

$$p = m/k,$$
$$q = 1 + p,$$
$$R = p/q,$$

and k & m are the maximum likelihood estimators considered in De Geoffroy & Wu (section 3, 19), and formulated from first principles in this section. However, the probability function, P, can be given in recurrence form, which is far more convenient for computing purposes (see computer program 3.2 in Appendix 3). P is given by:

$$P(0) \quad = 1/q^k$$

and

$$P(x + 1) = (x + k)*R*P(x)/x, \text{ for } x = 0, 1, 2, \ldots$$

The parameter, (k), is estimated from the mean, (m) and the variance, (s^2) of the frequency distribution formed from the sample size N of ore deposits counted within each cell of the gridded area as follows:

(k_1), the first estimate for k is determined from the first and second central moments estimators, m & s^2, by

$$k_1 = m^2/(s^2 - m).$$

The second estimate, k_2, is found by iteration, based on maximum likelihood principles and Taylor's theorem. The analysis for determining the maximum likelihood estimator, k_2, for the generalized negative binomial parameter, k, is formulated as follows:

first we define the function, y, by

$$y(k_2) = \Sigma_{i=0}^{n} A_i/(k_2 + i) - N*Log(k_2 + m) + N*Log(k_2),$$

where n is the number of classes with non-zero observed frequencies, N is the sample size, and A_i = cumulative observed frequency of X > i,

e.g. A_0 = 150 - 70 = 80, in example 1;

$k_2 = k_1$ initially;

then k_2 is determined as follows by Taylor's theorem:

$$y(k_2 + h) = y(k_2) + h*y'(k_2) + \ldots,$$

where

$$y'(k_2) = N/k_2 - N/(k_2 + m) - \Sigma_{i=1}^{n} A(i)/(k_2+i-1)/(k_2+i-1)$$

$$h = -y(k_2)/[y'(k_2)]$$

let: $k_2 = k_2 + h$,

using BASIC computer programming terminology.
The process is repeated iteratively until the absolute value of $y(k_2)$ approaches zero, i.e. in the computer program:

if the absolute value of

y(k) < .0001,
then
$k = k_2$,
otherwise repeat the algorithm until it is.
if y → 0, then k = k_2,
and

p = m/k.

The computer program for the negative binomial fitting to oberved quadrat frequency distributions is given in Appendix 3, program 3.2.

In many examples of the application of the generalized negative binomial model, the authors have not used the maximum likelihood estimator for the parameter, k; and it is interesting to compare their results with those given by our computer program, 3.2. (appendix 3). Our first three examples are taken from Bliss and Fisher (section 3, 8), who did use the maximum likelihood estimator, but not the Taylor Theorem application, which affected the accuracy of their results. In all examples the observed and expected frequencies given are for the negative binomial fit on the domain: 0, 1, ..., n. The comparative results are:

Example 1:

Calculated parameters: m = 1.146667, k = 1.02459

Observed frequency	Expected frequency (program 3.2.)	Bliss & Fisher expected frequency
70	69.49	69.49
38	37.60	37.60
17	20.10	20.10
10	10.70	10.70
9	5.69	5.69
3	3.02	3.02
2	1.60	1.60
1	0.85	0.85

Chi-Sq. value = 2.69 < Chi-sq.(4, 0.05) = 9.49,

which implies the negative binomial fit is significant.

Example 2:

Calculated parameters: m = 2.398, k = 3.898

Observed frequency	Expected frequency (program 3.2.)	Bliss & Fisher expected frequency
56	61.55	64.2
104	91.38	90.3
80	85.25	82.7
62	63.84	62.1
42	41.94	41.6
27	25.24	25.8
9	14.26	15.1
9	7.68	8.5
5	3.99	4.7
3	2.01	2.5
2	0.99	1.3

Chi-Sq. value = 7.00 < Chi-sq.(7, 0.05) = 14.07,

which implies the negative binomial fit is significant.

Example 3:

Calculated parameters: m = 0.6825, k = 3.5858

Observed frequency	Expected frequency (program 3.2.)	Bliss & Fisher expected frequency
213	214.15	214.2
128	122.79	122.8
37	45.01	45.0
18	13.40	13.4
3	3.53	3.5
1	0.86	0.9

Chi-Sq. value = 3.35 < Chi-sq.(2, 0.05) = 5.99,

which implies the negative binomial fit is significant. It will be observed that the output from the program 3.2. exactly replicates the Bliss & Fisher result, and the maximum-likelihood estimates of the parameters are therefore accurate and reliable. However in examples 2 and 3, Bliss and Fisher's results lacked the accuracy of example 1.

2.6.3.2. Examples of applications to mineral exploration

The first example is from Uhler & Bradley (22, section 2). They applied both the Poisson model and the negative binomial model to the analysis of the areal distribution of petroleum reservoirs occurring in Alberta; the cell size chosen for the study was 5 miles x 5 miles. The Poisson model yielded a fair result, but the negative binomial provided a much closer fit. The comparative results for the Negative Binomial fit are:

Example 1:

Calculated parameters: m = 0.0356, k = 0.0473

Observed frequency	Expected frequency (program 3.2.)	Uhler & Bradley expected frequency
8586	8586.0	8584.3
176	174.4	176.8
35	39.2	39.1
13	11.5	11.3
6	3.8	3.6
1	1.3	1.2

Chi-Sq. value = 2.01 < Chi-sq.(3, 0.05) = 5.99,

which implies the negative binomial fit is significant.

The second example is from De Geoffroy & Wu (section 3, 19), who applied the negative binomial distribution to the areal distribution of ore deposits in the Canadian Shield, using the maximum likelihood estimator for k with excellent results, which are summarized in the top portion of Figure 2.12., and in Table 2.22. It will be noted that there is a printing error: p is in fact 3.63, instead of 2.63 as shown. The comparative results are:

Table 2.22. Fitting of a negative binomial model and derivation
of probability distribution based on the observed
areal distribution of ore deposits in the Northwest
Quebec-Northeast Ontario section of the Canadian Shield.

Number of Commercial deposits per cell	Observed frequency	Observed relative frequency	Expected frequency (Neg. Binomial)	Expected relative frequency (Neg. Binomial)	Probability number of Commercial deposits per cell is > = class value
0	404	0.813	404	0.811	1.000
1	41	0.081	44	0.088	0.189
2	21	0.040	21	0.039	0.101
3	11	0.020	11	0.022	0.062
4	8	0.016	7	0.013	0.040
5	4	0.008	4	0.009	0.027
6	3	0.006	3	0.006	0.018
7	2	0.004	2	0.005	0.012
8	2	0.004	2	0.003	0.007
9	2	0.004	1	0.002	0.004
10	2	0.004	1	0.002	0.002
Total	500	1.000	500	1.000	

Results of chi-squared significance test at the ∂ - 0.05 significance level:
Chi-squared - 1.02 < Chi-sq. (0.05, 5 d.f.) - 11.07, which implies the distribution
is accepted as negative binomial at the 95% confidence level

Example 2:

Calculated parameters: $m = 0.508$, $k = 0.1398$

Observed frequency	Expected frequency (program 3.2.)	De Geoffroy & Wignall expected frequency
404	403.5	404
41	44.2	44
21	19.8	21
11	11.1	11
8	6.8	7
4	4.4	4
3	3.0	3
2	2.0	2
2	1.4	2
2	1	1
2	1	1

Table 2.23. Fitting of a negative binomial to the observed areal distribution of ore deposits in Southeast Ontario.

Number of Commercial deposits per cell	Observed frequency	Observed relative frequency	Expected relative frequency Neg. Binomial	Expected frequency Neg. Binomial	Probability number of Commercial deposits per cell is > = class value
0	168	0.797	0.783	165	1.000
1	18	0.089	0.100	20	0.217
2	9	0.042	0.052	11	0.117
3	5	0.024	0.026	5	0.065
4	4	0.019	0.016	3	0.039
5	3	0.010	0.009	3	0.023
6	2	0.009	0.007	2	0.014
7	1	0.005	0.004	1	0.007
8	1	0.005	0.003	1	0.003
Total	211	1.000	1.000	211	

Results of chi-squared significance test at the ∂ = 0.05 significance level:

Chi-squared = 0.97 < Chi-sq. (0.05, 5 d.f.) = 11.07, which implies the distribution is accepted as negative binomial at the 95% confidence level

The Chi-Sq. value = 4.69 < Chi-sq.(7, 0.05) = 14.07,

which implies the negative binomial fit is significant.

The third example is from De Geoffroy and Wignall (section 3, 21), who describe the successful fitting of a negative binomial model to the areal distribution of commercial deposits in a portion of the Grenville Province of the Canadian Shield, as shown by the histogram in the bottom portion of Figure 2.12 and in Table 2.23.

Table 2.24. Fitting a negative binomial and exponential models to the observed areal distribution of ore deposits to three regions of North America.

(1) North American Cordilleran Belt

(a) British Columbia & Yukon Territories, Canada	Probability Distribution based on Fitted Model	
	Number of deposits in 2500 sq. ml. cell	Probability
Area size: 625,000 sq. miles	at least 1 deposit	0.5066
Type of deposits: Porphyry-Cu-Mo, tactite, Massive Sulfides	at least 2 deposits	0.3362
Number of commercial & sub-economic deposits counted: 240	at least 3 deposits	0.2080
	at least 4 deposits	0.1340
Cell size: 50 miles x 50 miles	at least 5 deposits	0.0714
Number of cells: 250	at least 6 deposits	0.0513
Mean of Areal Distribution: 240/250 = 0.960	at least 7 deposits	0.0327
Fitted Model: Negative Exponential	at least 8 deposits	0.0149

(b) Basin & Range Area, U.S.A.	Probability Distribution based on Fitted Model	
	Number of deposits in 1000 sq. ml. cell	Probability
Area size: 357,000 sq. miles		
Type of deposits: Commercial base metal deposits	at least 1 deposit	0.3001
Number of commercial deposits counted: 154	at least 2 deposits	0.0910
	at least 3 deposits	0.0276
Cell size: 1000 square miles	at least 4 deposits	0.0084
Number of cells: 357	at least 5 deposits	0.0025
Mean of Areal Distribution: 154/357 = 0.431	at least 6 deposits	0.0008
Fitted Model: Negative Binomial		

(2) Appalachian Belt, U.S.A. & Canada

	Probability Distribution based on Fitted Model	
	Number of deposits in 3600 sq. ml. cell	Probability
Area size: 432,000 sq. miles		
Type of deposits: Massive sulfides	at least 1 deposit	0.3251
Number of commercial & sub-economic deposits counted: 96	at least 2 deposits	0.2250
	at least 3 deposits	0.1167
Cell size: 60 miles x 60 miles	at least 4 deposits	0.0667
Number of cells: 120	at least 5 deposits	0.0500
Mean of Areal Distribution: 96/120 = 0.80	at least 6 deposits	0.0167
Fitted Model: Negative Exponential		

(3) North American Precambrian Shield

N.E. Ontario & N.W. Quebec, Canada	Probability Distribution based on Fitted Model	
	Number of occurrences in 100 sq. ml. cell	Probability
	at least 1 deposit	0.189
	at least 2 deposits	0.101
	at least 3 deposits	0.062
Area size: 50,000 sq. miles	at least 4 deposits	0.040
Types of deposits: All types	at least 5 deposits	0.027
Number of commercial deposits counted: 254	at least 6 deposits	0.018
Cell size: 10 miles x 10 miles	at least 7 deposits	0.012
Number of cells: 500	at least 8 deposits	0.007
Mean of Areal Distribution: 254/500 = 0.508	at least 9 deposits	0.004
Fitted Model: Negative Binomial	at least 10 deposits	0.002

N.B.   1 mile = 1.6 km,   1 sq. ml. = 2.6 sq. km.

References: (1) (a) partly based on Annis & Cranstone, 1978
                      (b) Brant, 1968
              (2)   Heyl & Bozion, 1971, & McAllister, 1960
              (3)   De Geoffroy & Wu, 1970

Table 2.25. Fitting a negative binomial models to the observed
areal distribution of ore deposits in the Canadian
Shield and in the Paleozoic Platform of the U. K.

North American Precambrian Shield	

(1) Archean Superior Province, Canada	Probability Distribution based on Fitted Model	
Area size: 84,000 sq. miles	Number of deposits in 36 sq. ml. cell	Probability
Type of deposits: Ni-Cu mafic & Cu-Zn Volc. Mass. Sulfides	at least 1 deposit	0.164
Number of commercial deposits counted: 162	at least 2 deposits	0.083
	at least 3 deposits	0.031
Cell size: 6 miles x 6 miles	at least 4 deposits	0.012
Number of cells: 2300	at least 5 deposits	0.004
Mean of Areal Distribution: 162/2300 = 0.07	at least 6 deposits	0.001
Fitted Model: Negative Binomial		

(2) Grenville Province, Canada	Probability Distribution based on Fitted Model	
Area size: 21,000 sq. miles	Number of deposits in 100 sq. ml. cell	Probability
Types of deposits: All types	at least 1 deposit	0.217
Number of commercial deposits counted: 109	at least 2 deposits	0.117
	at least 3 deposits	0.065
Cell size: 10 miles x 10 miles	at least 4 deposits	0.039
Number of cells: 211	at least 5 deposits	0.023
Mean of Areal Distribution: 109/211 = 0.517	at least 6 deposits	0.014
Fitted Model: Negative Binomial	at least 7 deposits	0.007
	at least 8 deposits	0.003

(3) North-eastern Ontario, Canada	Probability Distribution based on Fitted Model	
	Number of occurrences in 64 sq. ml. cell	Probability
	at least 1 occurrence	0.4617
	at least 2 occurrences	0.3461
Area size: 8960 sq. miles	at least 3 occurrences	0.2754
Types of deposits: Vein-gold	at least 4 occurrences	0.2240
	at least 5 occurrences	0.1840
Number of commercial & non-commercial deposits counted: 572	at least 6 occurrences	0.1512
	at least 7 occurrences	0.1241
Cell size: 8 miles x 8 miles	at least 8 occurrences	0.1006
Number of cells: 140	at least 9 occurrences	0.0806
Mean of Areal Distribution: 573/140 = 4.08	at least 10 occurrences	0.0628
Fitted Model: Negative Binomial	at least 11 occurrences	0.0471
	at least 12 occurrences	0.0335

Paleozoic Platform	

(4) Pennine Region, U.K.	Probability Distribution based on Fitted Model	
	Number of deposits in 4 sq. km. cell	Probability
Area size: 936 sq. miles	at least 1 deposit	0.1397
Type of deposits: Mississippi Valley-type Pb-Zn	at least 2 deposits	0.0488
Number of commercial deposits counted: 140	at least 3 deposits	0.0195
	at least 4 deposits	0.0080
Cell size: 2 km x 2 km	at least 5 deposits	0.0032
Number of cells: 616	at least 6 deposits	0.0016
Mean of Areal Distribution: 140/616 = 0.227		
Fitted Model: Negative Binomial		

N.B.   1 mile = 1.6 km,   1 sq. ml. = 2.6 sq. km.
References: (1) private sources
               (2) De Geoffroy & Wignall, 1971
               (3) Agterberg, 1974
               (4) Bozdar, 1972

Table 2.26. Fitting a negative exponential models to the observed areal distribution of ore deposits Paleozoic Platform of the U. S. A., and the Precambrian Shields of Australia and India.

(1) Paleozoic Platforms		
Upper Mississippi Valley Zn-Pb District (U.S.A) (Wisconsin, Illinois & Iowa) Area size: 5400 sq. miles Type of deposits: Mississippi Valley-type Zn-Pb Number of commercial deposits counted: 141 Cell size: 6 miles x 6 miles Mean of Areal Distribution: 141/150 = 0.940 Fitted Model: Negative Exponential	**Probability Distribution based on Fitted Model**	
	Number of deposits in 36 sq. mi. cell	Probability
	at least 1 deposit	0.293
	at least 2 deposits	0.204
	at least 3 deposits	0.149
	at least 4 deposits	0.102
	at least 5 deposits	0.065
	at least 6 deposits	0.041
	at least 7 deposits	0.020
	at least 8 deposits	0.014
	at least 9 deposits	0.006
	at least 10 deposits	0.001
(2) Australian Precambrian Shield		
Archean Belts of Western Australia Area size: 1 million sq. miles Types of deposits: Ni-Cu mafic & Cu-Zn Volcanogenic sulfides Number of commercial deposits counted: 86 Cell size: 100 miles x 100 miles Mean of Areal Distribution: 86/100 = 0.860 Fitted Model: Negative Exponential	**Probability Distribution based on Fitted Model**	
	Number of deposits in 10,000 sq. mi. cell	Probability
	at least 1 deposit	0.461
	at least 2 deposits	0.192
	at least 3 deposits	0.086
	at least 4 deposits	0.035
	at least 5 deposits	0.011
(3) Precambrian Shield of India		
Archean Belts of Southern & Central India Area size: 642,000 sq. miles Types of deposits: Cu deposits of all types Number of commercial deposits counted: 585 Cell size: 1000 sq. miles Mean of Areal Distribution: 585/642 = 0.911 Fitted Model: Negative Exponential	**Probability Distribution based on Fitted Model**	
	Number of deposits in 1000 sq. mi. cell	Probability
	at least 1 deposit	0.4922
	at least 2 deposits	0.2352
	at least 3 deposits	0.1122
	at least 4 deposits	0.0515
	at least 5 deposits	0.0219
	at least 6 deposits	0.0071

N.B. 1 mile = 1.6 km, 1 sq. mi. = 2.6 sq. km.

References: (1) Heyl & Agnew, 1959
            (2) Australian Institute of M. & M., 1967
            (3) Sarma, 1979

Other examples of the application of the negative binomial models to the areal distributions of various types of deposits in the Western U. S. A., Appalachian of the Eastern U. S. A., and Canada, the Western and Eastern parts of the Superior Province of the Canadian Shield, and the Pennine region of the U. K. are listed in Tables 2.24 and 2.25.

The negative binomial model is the one to use whenever m/s is significantly smaller than 1, which will almost certainly mean that the deposits cluster within the region. In such circumstances, the Poisson model yields much poorer results than the negative binomial model. However, as pointed out by Agterberg (section3, 28), a main drawback of the quadrat method is that the results are not comparable when the cell sizes are modified. This may be caused by a characteristic of the negative binomial distribution, which is: the negative bimomial random variable is the sum of Poisson distributed random variables.

For this reason, some workers prefer to use the "nearest neighbour" model, which is based on mutual distances between occurrences, rather than density within cells of arbitrary size (section 3, 16 and 17). Mather (section 3, 33) provides the listings of a computer program which can handle the analysis of point patterns by the nearest neighbour technique. Miller and Kahn (section 3, 35) provide a useful account of the method and some geological examples, which could be of interest to some readers.

E2 Exercises

Based on computer program MODFIT in Appendix 3, in exercises
E2.1. & 2.2. use program 3.2. to derive the expected
generalized Negative Binomial frequencies, and the Chi-sqared
value to test the goodness of fit.
In exercises E2.3. to E2.9., using program 3.1., derive a
frequency distribution using the given number of classes on the
logarithm (base 10) of the data set referred to for each of the
parameters: (i) Length, & (ii) B/L ratio. Give a 95% confidence
interval for the geometric mean in each case, and test the
frequency distribution for Lognormal goodness of fit.

E2.1. Sum the frequencies in Examples 1 & 3: e.g. f(0) =
       70 + 213 = 283: apply program 3.2. of Appendix 3

E2.2. Sum the frequencies in examples 5 & 6

E2.3. Contact Metasomatic (Cu-Fe-Au) data set 8.1.10 (1),using
       8 classes with lowest limit in (i) 2, and (ii) -2, and
       class width 0.3.

E2.4. Volcanogenic Sulfides (Iberia) data set 8.1.8 (8),using
       8 classes with lowest limit in (i) 2, and (ii) -2, and
       class width 0.3.

E2.5. Vein Gold (N. American Shield) data set 8.1.4 (1),using
       10 classes with lowest limit in (i) 2, and (ii) -3.6, and
       class width 0.3.

E2.6. Ni-Cu Ultramafic (N. America) data set 8.1.6 (1),using
       8 classes with lowest limit in (i) 1.4, and (ii) -2.6,
       and class width 0.3.

E2.7. Vein Gold (W. Australian Shield) data set 8.1.4 (2),using
       10 classes with lowest limit in (i) 2, and (ii) -3.6, and
       class width 0.3.

E2.8. Ni-Cu Ultramafic (W. Australia) data set 8.1.6 (2),using
       8 classes with lowest limit in (i) 1.4, and (ii) -2.6,
       and class width 0.3.

E2.9. Vein Gold (N. American Shield) data set 8.1.3 (1),using
       8 classes with lowest limit in (i) 0.3, (ii) 0.0,
       and (iii) 0.3, and class width 0.3.

BIBLIOGRAPHY FOR CHAPTER 2

Section 1: Deterministic modelling

1. DICKINSON, S.B., 1965, Mineral exploration costs, in 8th
    Commonw. M. & M. Congress, V.2; Austral. Inst. of Min.
    and Met., pp.283-288.

2. EMERSON, D.W., 1977, Australian exploration and development:
    Comments and Costs; Bull. Aust. Soc. Expl.
    Geophysics, V.8, No.4, pp.91-94.

3. GRISWOLD, W.T. and WALLACE, W.K., 1967, Organization and costs
    for mineral exploration in Australia; Paper presented
    at Pacific Southwest Mineral Industry Conference
    (A.I.M.E.), Pacific Grove, California, May 8, 1967.

4. PERRY, A.J., 1968, Organisation and costs for mineral
    exploration in the Southwest U.S.A.; Pac. S.W. Min.
    Indust. Conf., AIME, May.

5. PETERS, W.C., 1959, Cost of exploration for mineral raw
    materials; Cost Engineering, V.4, No.3, July.

6. PETERS, W.C., 1967, Cost and value of drill hole information;
    Mining Congress Journal, V.53, No.1, Jauary, pp.56-59.

7. SHALLEY, M.J., 1971, Cost-probability approach to induced
    polarization in Australia; Bull. Austr., Soc. Expl.
    Geophys., V.2., No.4, pp.25-33.

8. ZIMMERMAN, O.T., 1968, Elements of capital cost estimation;
    Cost. Eng., V.13, No.4, pp.4-18.

Section 2: Stochastic modelling: economic and geometric modelling

1. AGTERBERG, F.P., 1974, Geomathematics (Chpt. 7, 11 & 15);
    Elsevier Publ., Amsterdam. Holland.

2. AGTERBERG, F.P., 1980, Lognormal models for several metals in
    selected areas of Canada; Reprint, Memoire du
    B.R.G.M., No.106, pp.83-90.

3. AGTERBERG, F.P. and DIVI, S.R., 1978, A statistical model for
    the distribution of copper, lead and zinc in the
    Canadian Appalachian region; Econ. Geol., V.73,
    pp.230-245.

4. AITCHINSON, J. and BROWN, J.A.C., 1966, The lognormal
    distribution with special references to its use in
    Economics; Cambridge Univ. Press, U.K.

5. ALLAIS, M., 1957, Methods of appraising economic prospects of
    mining exploration over large territories; Manag. Sc.,
    V.3, pp.285-347.

6. BOLDY, J., (Un)certain exploration facts and figures; Can.
    Inst. M.& M. Bull., May, pp.86-95.

7.   DAVIS, J.C., 1973, Statistics and data analysis in Geology; J.Wiley, N.Y.

8.   DIVI, S.R., 1980, Deposit modeling and resource estimation of stratiform massive sulphide deposits in Canada; Comput. & GeoSc., V.6, No.2, pp.163-174.

9.   GIBBONS, J.G., OLKIN, I. and FOBEL, M., 1977, Selecting and ordering populations; J. Wiley, New York.

10. GRIFFITHS, J.C., 1962, Frequency distributions of some natural resources materials; Bull. No.63, Min. Ind. Expmt. Stat., University of Pennsylvania, pp.174-198.

11. JOHNSON, N.I. and KOTZ, S., 1970, Continuous univariate distributions (V.21); Houghton Mifflin, Boston, Mass.

12. JONES, M.P. and BEAVEN, C.H.J., 1971, Sampling of non-gaussian mineralogical distributions; Trans. Inst. M.& M., V.80, pp.B316-B322.

13. KAUFMAN, G., 1964, Size and distribution of oil and gas fields, Bull. Am Assoc., Petrol. Gelogists, V.48, p.534.

14. KOCH, G.S. and LINK, R.F., 1970, Statistical analysis of geological data, (Chpt. 6, 12, 13 and 16), J.Wiley, New York, U.S.A.

15. KRUMBEIN, W.C. and GRAYBILL, F.A., 1965, Introduction to statistical models in Geology; (Chpts 5, 6, 7, 8) McGraw Hill, New York.

16. PRETORIUS, D.A. and HEMPKINS, W.B., 1968, Statistical moments of the frequency distribution or Rhodesian gold mineralization in time and space; a preliminary analysis; Trans. Geol. Soc. S. Africa, V.71, pp.9-19.

17. SANGSTER, D.F., 1980, Quantitative characteristics of volcanogenic massive sulfide deposits; Can. Inst. M.& M., Bull., February, pp.74-81.

18. SARMA, D.D., 1969, A preliminary statistical study on the basic mine valuation problems at Kolar Goldfields, Mysore State, India; Geo-expl., V.7, No.2, pp.97-105.

19. SARMA, D.D., 1979, An exploration strategy for prospecting with a case study on copper prospects at Ingladahl, India; Miner. Deposits, V.14, pp.263-279.

20. SHARP, W.E., 1976, A log-normal distribution of alluvial diamonds with an economic cutoff; Econ. Geology, V.71, pp.648-655.

21. SINCLAIR, A.J., 1974, Probability graphs of ore tonnages in mining camps, a guide to exploration; Can. Inst. M.& M., Bull., Oct., pp.71-75.

22.   UHLER, R.S. and BRADLEY, P., 1970, Stochastic model for
            determining the economic prospects of petroleum
            exploration over large regions; Jour. Am. Stat.
            Assoc., V.65, No.330, pp.623-630.

23.   WALPOLE, R.E., 1974, Introduction to Statistics; Macmillan, New
            York.

Section 3: Stochastic modelling: ore occurrence

1.   ALLAIS, M., 1957, Method of appraising economic prospects of
            mining exploration over large territories; Management
            Sci., V.3, pp.285-347.

2.   ANNIS, R.C., CRANSTONE, D.A. and VALLEE, M., 1978, A survey of
            known deposits in Canada that are not being mined;
            Bull. M.R.181, Dept. of Mines, Energy and Resources,
            Ottawa, Canada.

3.   ANSCOMBE, F.J., 1949, The statistical analysis of insect counts
            based on the negative binomial distribution;
            Biometrics, V.5, pp.165-173.

4.   ANSCOMBE, F.J., 1950, Sampling theory of the negative binomial
            and logarithmic series distributions; Biometrika,
            V.37, pp.358-382.

5.   AUSTRALASIAN INSTITUTE OF MINING AND METALLURGY, 1975, Economic
            geology of Australia and Papua- New Guinea;
            Melbourne, Australia.

6.   BARTLETT, M.S., 1960, An introducton to stochastic processes,
            Cambridge, U. K.

7.   BARTLETT, M.S., 1960, Stochastic population models in ecology
            and epidemiology, Methuen, London

8.   BLISS, C., 1953, Fitting the negative binomial distribution to
            biological data; Biometrics, V.2, pp.176-200.

9.   BOZDAR, L.B. and KITCHENHAM, B.A., 1972, Statistical appraisal
            of the occurrence of leadmines in the Northern
            Pennines; Trans. Inst. Min. & Metall., V.81, pp.B183-
            B187.

10.   BRANT, A.A., 1968, The pre-evaluation of the possible
            profitability of exploration prospects; Mineral.
            Deposita, V.3, pp.1-17.

11.   BROWN, D. and ROTHERY, P., 1978, Randomness and local
            regularity of points in a plane; Biometrika, V.65,
            pp.115-122.

12.   CLARK, P.J., 1955, On some aspects of spatial patterns in
            biological populations; Science, V.121, pp.397-398.

13.   CLARK, P.J., 1956, Grouping in spatial distributions, Science,
            V.123, pp.123-125.

14. CLARK, P.J. and EVANS, F.C., 1954, Distance to nearest neighbour as a measure of spatial relationships in populations; Ecology, V.35, pp.445-453.

15. CONOVER, W.J., BEMENT, T.R. and IMAN, R.L., 1979, On a method for detecting clusters of possible Uranium deposits; Technom., V.21, No.3, pp.277-282.

16. DACEY, M.F., 1960, A note on the derivation of nearest neighbour distances; J. of Reg. Sc., V.2, pp.81-87.

17. DACEY, M.F., 1962, Analysis of central place and point patterns by a nearest neighbour method; Land studies in Geography, V.24, pp.55-75.

18. DACEY, M.F., 1964, Two-dimensional random point patterns: a review and an interpretation; Papers, Regional Science Association, V.13, pp.41-55.

19. DE GEOFFROY, J.G. and WU, S.M., 1970, A statistical study of ore occurrences in the Greenstone Belts of the Canadian Shield; Econ. Geol., V.65, pp.496-504.

20. DE GEOFFROY, J. and WIGNALL, T.K., 1970, Application of statistical decision techniques to the selection of prospecting areas and drilling targets in regional exploration; Can. Inst. M.& M., Bull., August, pp.893-899.

21. DE GEOFFROY, J.G. and WIGNALL, T.K., 1971, A probabilistic appraisal of mineral resources in a portion of the Grenville Province of the Canadian Shield; Econ. Geol., V.66, pp.466-479.

22. DOUGLAS, J.B., 1955, Fitting the Neyman type A (two parameter) distribution; Biometrics, V.11, pp.149-173.

23. DREW, L.J. and GRIFFITHS, J.C., 1965, Size, shape and arrangements of some oilfields in the U.S.A.; short course and symposium on Computers and Computer Applications in Mining and Exploration, University of Arizona, V.3, pp.FF1-FF31.

24. DREW, M.W., 1974, A deposit distribution model for uranium; Programmes Analysis Unit Report 10/74, London, U.K. Atomic Energy Authority.

25. EBDON, D., 1977, Statistics in Geography; a practical approach, (Chpt 7); B. Blackwell, Oxford, U.K.

26. EVANS, D.A., 1953, Experimental evidence concerning contagious distributions in ecology; Biometrika, V.40, pp.186-211.

27. GRIFFITHS, J.C., 1962, Frequency distributions of some natural resources materials; Bull. No.63, Min. Ind. Expmt. Stat., University of Pennsylvania, pp.174-198.

28. HARRIS, D.P. and AGTERBERG, F.P., 1981, The appraisal of mineral resources; Econ. Geol., V.75, pp.897-938.

29. HARVEY, D.W., 1966, Geographical processes and the analysis of point patterns; Trans. Inst. Br. Geogr., V.49, pp.81-95.

30. HEYL, A.V. and AGNEW, A.F., 1959, The Geology of the Upper Mississippi Valley Zinc-Lead District; U.S.G.S. Prof. Paper No.309.

31. JOHNSTON, N.I. and KOTZ, S., 1970, Discrete univariate distributions (V.1); Houghton Mifflin, Boston, Mass., U.S.A.

32. MacDOUGAL, E.B., 1976, Computer programming for spatial problems; E. Arnold Publ., London.

33. MATHER, P.M., 1976, Computers in Geography: a practical approach, (Chpt 4); also computer program for nearest neighbour analysis, B. Blackwell, Oxford, U.K., pp.52-54.

34. McCONNELL, M., 1966, Quadrat methods in map analysis, Discussion Paper No.3; Dept. of Geography, University of Iowa.

35. MILLER, R.L. and KAHN, J.S., 1962, Statistical analysis in the Geological Sciences, (Chpt 16); J. Wiley, New York, U.S.A.

36. MORISITA, M., 1954, Estimation of population density by spacing method; Memoirs, Faculty of Science, Kyushu University, Series E, No.1, pp.187-197.

37. MORISITA, M., 1959, Measuring the dispersion of individuals and analysis of the distributional patterns; Memoirs, Faculty of Science, Kyushu University, Series E, No.2, pp.215-233.

38. NEFT, D.S., 1966, Statistical analysis for areal distribution, Monograph 2; Regional Sc. Res. Inst., Philadelphia, Penna. U.S.A.

39. NEYMAN, J., 1939, On a new class of 'contagious' distributions, applicable in Entomology and Bacteriology; Annals of Mathematical Statistics, V.10, pp.35-37.

40. NEYMAN, J. and SCOTT, E.L., 1957, On a mathematical theory of populations conceived as conglomerations of clusters; Cold Spring Harbor Symposia on Quantitative Biology, V.22, pp.109-120.

41. SHENTON, L.R., 1949, On the efficiency of the method of moments and Neyman's Type A distribution; Biometrika, V.36, pp.450-454.

42. SKELLAM, J.G., 1951, Random dispersal in theoretical populations; Biometrika, V.38, pp.196-218.

43. SKELLAM, J.G., 1953, Studies in statistical Ecology -
     1. spatial pattern; Biometrika, V.39, pp.346-362.

44. SKELLAM, J.G., 1958, On the derivation and applicablity of
     Neyman's Type A distribution; Biometrika, V.45, pp.32-
     36.

45. SLICHTER, L.B., 1960, The need for a new philosophy of
     prospecting; Mining Eng., June, pp.570-576.

46. ROGERS, A., 1974, Statistical Analysis of Spatial Dispersion;
     the Quadrat Method, (Chpt 1); Pion Ltd., London, U.K.

47. SARMA, D.D., 1979, An exploration strategy for prospecting with
     a case study on copper prospects at Ingladahl, India;
     Mineral. Deposita, V.14, pp.263-279.

48. TAKACS, L., 1960, Stochastic processes, Methuen, London.

49. THOMAS, M., 1949, A generalisation of Poisson's binomial limit
     for use in Ecology; Biometrika, V.36, pp.18-25.

50. WILLIAMSON, E. and BRETHERTON, M.H., 1963, Tables of the
     negative binomial probability distribution; John
     Wiley, New York, U.S.A.

Section 4: Evaluation of the probability of ore occurrence: subjective approach

1. BARRY, G.S. and FREYMAN, A.J., 1970, Mineral endowment of the
    Canadian Northwest: a subjective probability
    assessment; Can. Inst. M.& M. Bull., September,
    pp.1031-1038.

2. ELLIS, J., HARRIS, D.P. and VAN WIE, N., 1975, A subjective
    probability appraisal of uranium resources in the
    State of New Mexico; U.S. Energy Research and
    Development Administration, Grand Junction, Colorado,
    USA.

3. GOLABI, K. and LAMONT, A., 1981, A probabilistic approach to
    the assessment of uranium resources; Jour. Math. Geol.
    V.13, No.6.

4. HARRIS, D.P., 1973, A subjective appraisal of metal endowment
    of Northern Sonora, Mexico; Econ. Geol., V.68, No.2,
    pp.222-242.

CHAPTER THREE

STATISTICAL POOLING OF REGIONS TO ASSIST THE OPTIMIZATION OF
WORLD-WIDE ORE SEARCH PROGRAMS

3.1. GENERAL STATEMENT

3.1.1. Rationale of Statistical Pooling of Regions

The purpose of statistical pooling is two-fold. One aspect is
practical: to reduce the work-load in the statistical processing leading to
the optimization of ore search programs. The other one is theoretical: to
increase the reliability of predictions made on detection and economic
parameters which are required for the optimal selection of ore deposit types
and regions of search.

The pooling of regions of occurrence of a specific type of ore deposit
on the basis of economic and/or geometric parameters results in a single
optimized function being applicable throughout the pool, either for
detection purposes or for economic strategies, instead of the application of
a different function to each component of the pool. The savings in tedious
and costly computerized data processing which is required to formulate and
apply the optimizing functions can be considerable when dealing with large
data bases such as is the case in our project. The reader will remember from
Chapter 1 that we are dealing with six types and six sub-types of ore
deposits occurring in twenty-five regions included in five continents of the
World. The optimization of search programs for the various types and sub-
types of ore deposits for each of these numerous regions would prove an
extremely cumbersome and time-consuming undertaking.

An additional benefit resulting from the pooling of regions in the
search for a specific type of ore deposit is the increase in sample size
resulting from dealing with pools rather than individual regions.
Consequently, the standard error of the mean parameters used for the
detection ore deposits or the selection of regions will decrease; the width
of the confidence intervals for the expected values of the parameters will
obviously therefore decrease, leading to a welcome gain in the reliability
of statistical predictions and significance tests.

3.1.2. Summary of statistical pooling methodology

3.1.2.1. Introduction

The method of finding out in an objective manner if a group of regions
of occurrence of a specific type of ore deposit can be pooled for ore
detection or selection purposes relies on the statistical testing of
population commonality. The use of multiple comparison tests run in a
sequential manner is at the core of the methodology. If significant

differences are established among a group of regions, then the Studentized multiple range test may be used for testing the commonality of subsets.

3.1.2.2. Methodology

The statistical pooling methodology can conveniently be broken into the three following steps:

Step 1: Lognormal Transformation

Since nearly every parametric test involved in the pooling methodology is based upon normal distribution theory, all quantitative variates should be normalized before analysis. In Chapter 2, it was established that almost all the economic and geometric data connected with ore deposits are distributed lognormally. Therefore, the lognormal transformation was found to be appropriate for most geometric and economic parameters of ore bodies.

Step 2: Commonality testing

A number of individual regions grouped on the basis of some common attribute such as the occurrence of a specific type of ore deposit are considered for statistical testing of commonality based on some common type of parameter, either geometric or economic, (univariate situations), or set of parameters (multivariate situations). The Multiple Comparison Test is then used to test the null hypothesis that there is no difference between the regions. If the null hypothesis cannot be rejected at say a 5% level of significance, then we should pool the group, unless there are valid scientific grounds for not doing so.

Step 3: Multiple Range Test

If the null hypothesis is rejected in step 2, and if there are more than two regions, then we should determine whether there are any subsets that are not significantly different at the 5% significance level.

3.2. TESTING OF POPULATION COMMONALITY

3.2.1. Introduction: Dual nature of population commonality testing

Most statistical populations are numerically described and summarized by means of two statistics, namely the variance and mean (univariate populations), or the variance-covariance matrix and vector mean (multivariate populations). Therefore, if we wish to test the commonality of two populations, we will have to test both the variances and the means for statistical equality. Since all statistical tests on the equality of means require homoscedascity (equality of variances), the test on variances is always run first. If the populations are heteroscedastic, i.e. have unequal variances, then Agterberg (5, Chapter 5) indicates a way out which is to take logarithmic transforms of the data.

3.2.2. Testing univariate population commonality

3.2.2.1. General methodology

The flow-diagram for univariate commonality testing is given in Figure 3.1, with equal variance tests being followed by tests on the means of samples. For k populations, k ≥ 2, the general procedure is to implement the path via Bartlett's test for equal variances, followed by ANOVA to test for

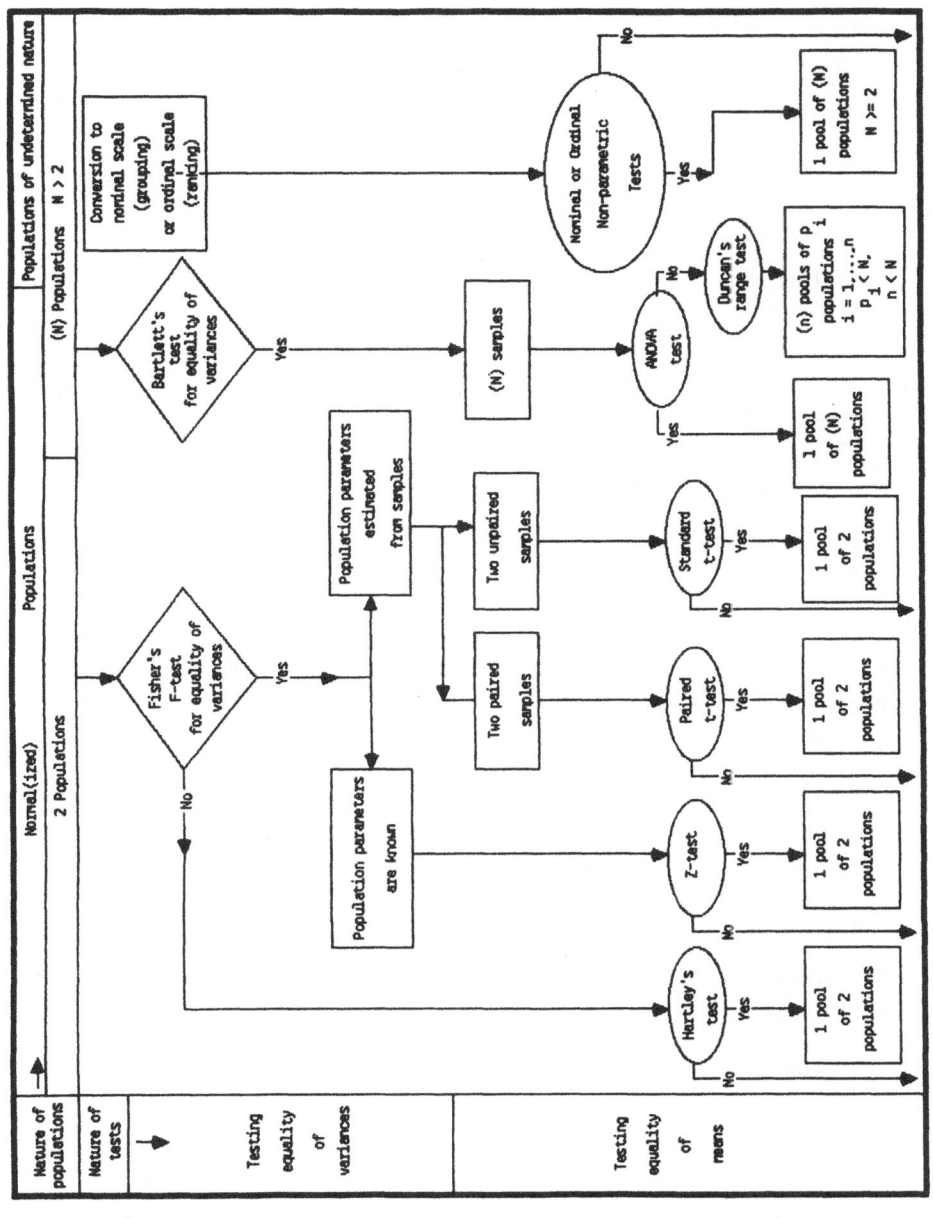

Figure 3.1. Flow diagram for commonality testing for univariate populations.

99

equal means. If H_0 is not rejected, then the k regions may be pooled. If it is rejected, then a Studentized multiple range test, such as Duncan's, which is particularly suitable for samples of unequal sizes, will find which regions are different, and subsets of homogeneous regions may then be formed for those that group together.

3.2.2.2. ANOVA testing

The ANOVA test is designed to test the null hypothesis,

$$H_0: \mu_1 = \mu_2 = \ldots = \mu_k,$$

at a given level of significance; if the test procedure fails to reject the hypothesis, then pooling of the k regions is valid.

The test-procedure is as follows:

step 1: calculate the variance between the means of the k samples, often called the "treatment" variance.

step 2: calculate the sum of square deviations within each sample about the sample mean. Sum the results over the k samples, and divide by the (total sample size - k). The result is the "within" or "error" variance.

step 3: Calculate the statistic F = (step 1 result)/(step 2 result), with k - 1, (total sample size - k) degrees of freedom.

step 4: Are the three following assumptions of the model satisfied?
   (i)   Normal distribution(s),
   (ii) Equal variances
   (iii) Equal means

step 5: If so, then the F-statistic has the F-distribution, and H_0 is rejected at the α significance level (one-tail test) only if $F > F_\alpha$, with the degrees of freedom as given in step 3.

For greater computational ease, the method is split into six steps as follows:

step 1: Calculate the total sum of squares, SST, defined by:

$SST = \Sigma_{i=1}^{k} \Sigma_{j=1}^{n_k} x_{ij}^2 - T^2/N$, where T is the grand total of the x_{ij}, and where $N = \Sigma_{j=1}^{k} n_j$, where n_j is the sample size of the jth sample, j = 1, 2, ..., k.

step 2: Calculate the treatment sum of squares, SSTr, defined by:

$SSTr = \Sigma_{i=1}^{k} T_i^2/n_i - T^2/N$, where T is the grand total of the x_{ij},

and where $N = \Sigma_{i=1}^{k} n_i$, where n_i is the sample size of the ith

sample, and $T_i = \Sigma_{i=1}^{n_i} x_i$, is the total of the ith sample,

$$i = 1, 2, \ldots, k.$$

step 3: Calculate the within or error sum of squares, SSE, defined by:
$$SSE = SST - SSTr$$

step 4: Calculate the variance estimates, the mean squares, MSTr and MSE, defined by

$$MSTr = SSTr/(k - 1),$$

$$MSE = SSE/[\Sigma_{i=1}^{k}(n_i - 1)] = SSE/(\Sigma_{i=1}^{k}n_i - k) = SSE/(N - k)$$

step 5: Calculate the statistic F, which is $F_{k-1, \, N-k}$ distributed, defined by

$$F = MSTr/MSE.$$

step 6: Under the null hypothesis: H_0: $\mu_1 = \mu_2 = \ldots = \mu_k$, the rejection region for the significance level, α, is given by: reject H_0 if

$$F > F_{\alpha, \, k-1, \, N-k} \text{ (one-tail test)}.$$

If H_0 is rejected then a Multiple Range Test is applied to see which means or groups of means differ.

3.2.2.3. Multiple range test

Multiple Range Tests determine which sets of means are different, when H_0 is rejected at the given significance level, α. Duncan's test (8) and (11) involves the use of the Studentized range statistic, which has a Q-distribution. The test is carried out in the following manner:

step 1: Arrange the sample means in ascending order, m_1, m_2, ..., m_k; and calculate the differences between the end ones of each set, in 2's, 3's, ..., k - 1.

step 2: Find $Q = Q_{\alpha, \, d, \, N-k}$, from a Q-table, where the d entry is the number of means in the set being tested for homogeneity.

step 3: Calculate

$$w = (MSE/n_b)\hat{\,}0.5*(Q),$$

for d = 2, ..., k - 1, where n_b is the mean sample size of the set. Then w is the least range for the differences between the end means of the set to be homogeneous; that is they differ significantly if the difference is greater than w.

step 4: Compare each difference calculated in step 1, with the appropriate w, and underline the set only if the difference is less than w: which indicates that they do not differ significantly.

3.2.3. Testing multivariate population commonality: the MANOVA test

The MANOVA test is used to test the null hypothesis,

$$H_0: \underline{\mu}_1 = \underline{\mu}_2 = \cdots = \underline{\mu}_k,$$

where $\underline{\mu}_i$ is a p-variate vector, for i = 1, 2, ..., k,

The assumptions for the validity of the test are that each of the k distributions is multivariate normal, and that all have equal variance-covariance matrices. The test procedure is formulated by incorporating a specific design matrix in the general linear model presented in Chapter 5.

3.3. SEQUENTIAL TESTING OF POPULATION COMMONALITY

3.3.1. Introduction

In the previous section, we briefly described the three main component blocks of the statistical methodology of population commonality testing, namely ANOVA, Multiple Range Test and MANOVA, in order of increasing statistical complexity for explanatory purposes and didactic reasons. However, we shall present now the actual structure of the testing methodology which is essentially sequential in nature. Most readers hardly need to be reminded of the power and efficiency of the sequential approach which justifies its frequent use in many statistical methodologies as illustrated in the following Chapters 4, 5, and 6.

3.3.2. Structure of the sequence

The sequence of the population commonality testing is as follows:

(1) MANOVA: required to deal with actual situations which are generally multivariate in nature,
(2) ANOVA on each variate, and finally
(3) Multiple Range Test on each variate to delve into the comparison of populations in a univariate context, if required. The procedure is computerized by means of the SEQPOOL program whose listings are shown in Appendix 3.

The structure of the sequence is as follows:

step 1: MANOVA tests $H_0: \underline{\mu}_1 = \underline{\mu}_2 = \cdots = \underline{\mu}_k$.

If H_0 is rejected then go to step 2; if it is not rejected then all k regions may be pooled and the sequence ends.

step 2: If H_0 is rejected, then apply ANOVA, to each variate$_i$, i = 1, 2, ..., p.

If $H_0: \mu_{1i} = \mu_{2i} = \cdots = \mu_{pi}$, for i = 1, 2, ..., p, is not rejected on specific variates, then the k regions may be pooled on each of those variates. Then go to step 3 on the variates for which H_0 is rejected.

step 3: Apply Duncan's multiple range test on each variate, for which the
null hypothesis in step 2 was rejected. The result will be to split
the k populations on each variate into homogeneous subsets.

The first two statements of the SEQPOOL program are for the Honeywell
version of the SPSS; they therefore need to be changed to systems statements
appropriate to the type of computer in use. The accompanying flow diagram
shown in Figure 3.2 illustrates the application of the sequential procedure

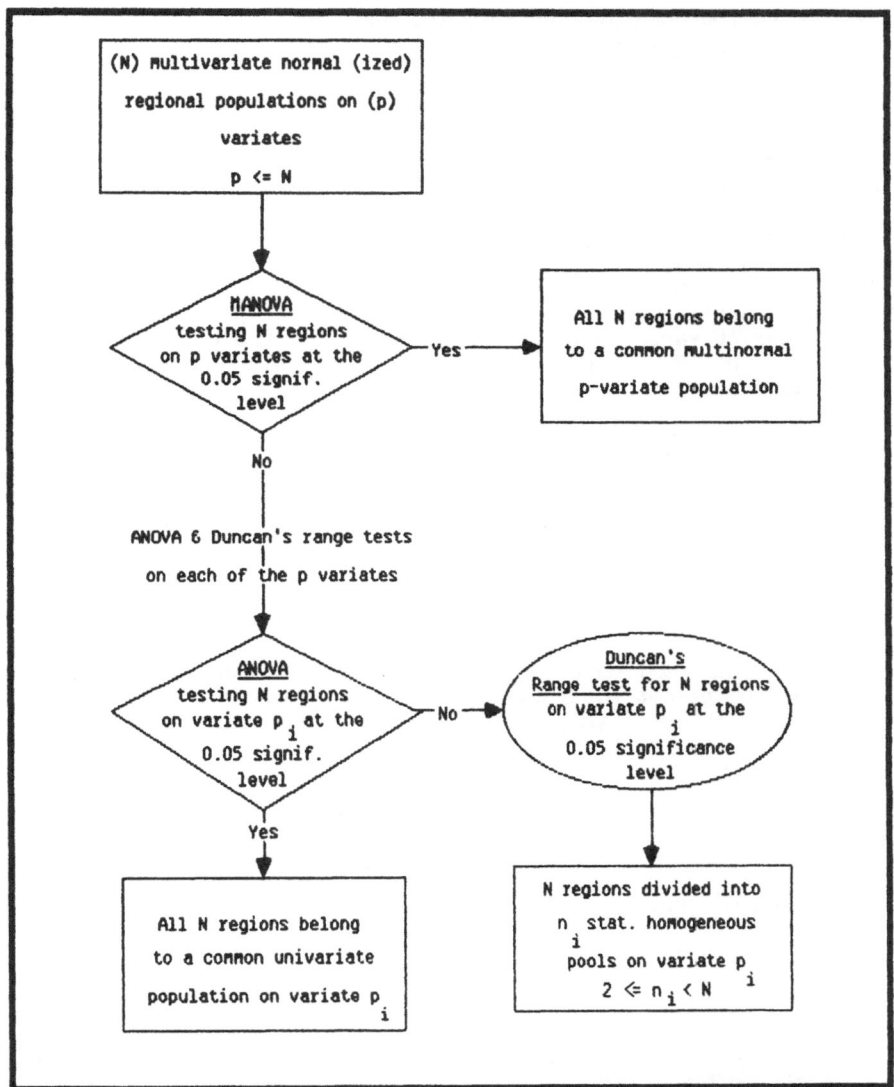

Figure 3.2. Flow diagram of sequential testing for multivariate
population commonality for optimized world-wide ore search

to a practical case of the multivariate population commonality testing which is used to assist in the planning of a world-wide search program for a specific type of ore deposit. We first consider (k) search regions located in various parts of the world to be compared on the basis of (p) variates either geometric or economic in nature, in order to assist detection or economic planning, respectively. We end up with the following alternative outcomes:

(1) all (k) regions belong to a common multivariate population on all (p) variates, or
(2) all (k) regions belong to a common univariate population on specific variate (p_i), or
(3) the (k) regions are divided into (n) statistically homogeneous pools of regions on specific variate (p_i).

For a large planning effort of this kind, the benefit in labour-saving brought by the application of the sequential testing methodology is quite obvious. Instead of repeating (k) x (p) times the complex statistical procedures required by the optimization of the search programs for each region on each variate, we could end up with only one round, if we are lucky, as in case (1). Even in the worst case (3), we face a much-reduced work-load because the number of pools (n) is usually substantially smaller than (k), and the number (q) of variates on which the ANOVA test results in rejection of the null hypothesis is also substantially smaller than the original number of variates (p). This is well illustrated by the following application of the SEQPOOL procedure to our world-wide data-base covering many important types of ore deposits occurring in many regions over five continents.

3.4. APPLICATION OF THE SEQUENTIAL TESTING PROCEDURE TO ASSIST THE WORLD-WIDE SEARCH FOR SIX TYPES OF ORE DEPOSITS

3.4.1. Introduction

The SEQPOOL procedure was applied to assist optimization of the search for six genetic types and six sub-types of base and precious metal deposits in twenty-five regions located in five continents of the World. Because of lack of space, we cannot possibly describe the results of the whole study; we have to restrict our coverage to a series of carefully chosen examples, grouped into three categories: namely:

(1) examples of very good work-load reduction,
(2) examples of satisfactory reduction, and
(3) lack of reduction.

3.4.2. Examples of very good pooling

We are providing below two good examples of very satisfactory pooling of regions which results in a considerable reduction of statistical work-load. One concerns the porphyry-Cu-Mo type of deposits for four regions of the World, and the second one deals with the Volcanogenic sulfide deposits for eleven regions.The first example is the application to the geometric parameters of porphyry-Cu-Mo deposits. In Chapter 2, the chi-squared test for goodness-of-fit was applied to the porphyry-Cu-Mo deposits in the regions of the North & South American Cordillera, South Pacific Island Arc, and Australasia. In every case, the geometric parameters were all shown to conform to a lognormal distribution. As a result, a logarithmic

Table 3.1. Pooling of four regions in three continents based on the geometric parameters of porphyry-Cu-Mo deposits.

Type of deposit: Porphyry Cu-Mo   Regions: N. & S. American Cordillera, B.C. Cordillera,
Sample size   : 102   Tasman Geosyncline, and S. Pacific Island Arc

(1) Testing the significance of the difference between regions

Parameter(s)	Test procedure	F-value or equivalent F-value	degrees of freedom	Significance level of F-value	Conclusion on differences between regions
all	MANOVA	2.82	8, 194	0.015	Not significant
length,l	ANOVA	3.36	3, 99	0.045	significant
breadth,b	ANOVA	7.11	3, 99	0.002	significant
b/l ratio	ANOVA	3.51	3, 99	0.039	significant

(2) Results of Duncan's multiple range tests at the ∂ = 0.05 significance level

Parameter	Region	Geometric mean	Rank
length in metres	Tasman Geosyncline	750	4
	B.C. Cordillera	820	3
	North & South American Cordillera	990	2
	South Pacific Island Arc	1180	1

Conclusion: The four regions form one homogeneous group on length parameter

Parameter	Region	Geometric mean	Rank
breadth/length ratio	B.C. Cordillera	0.25	4
	North & South American Cordillera	0.46	3
	South Pacific Island Arc	0.47	2
	Tasman Geosyncline	0.48	1

Conclusion: The four regions form one homogeneous group on shape ratio parameter

transformation was made on each variate to carry out the sequential testing which requires normality of distributions. This implies that the commonality test is being made on the geometric mean parameters of the regions.

It should be noted that the geometric parameters involved in our examples of testing are only those required for the calculation of the probability of detection and for the optimization of search grids. They include the length of the horizontal section of the portion of the deposit which lies within the average range of detection (100 meters below the surface), and the ratio (B/L) of the breadth of the section over the length. The results of the test are shown in Table 3.1 for porphyry-Cu-Mo deposits.

Table 3.2. Pooling of eleven regions in four continents based on the economic parameters of volcanogenic sulfide deposits.

(1) Testing the significance of the difference between regions (Continued)

Parameter(s)	Test procedure	F-value or equivalent F-value	degrees of freedom	Significance level of F-value	Conclusion on differences between regions
tonnage (millions)	ANOVA	16.93	10, 356	0.0001	significant
gross value ($mil.)	ANOVA	18.59	10, 356	0.0001	significant

(2) Results of Duncan's multiple range tests at the ∂ = 0.05 significance level

Parameter	Region	Geometric Mean	Rank
tonnage in millions of tonnes	W. Australian Shield	1.3	11
	Appalachian Belt	2.2	10
	N. American Shield	2.3	9
	Tasman Geosyncline	2.8	8
	Scandinavian Caledonides	3.4	7
	N. American Cordillera	4.3	6
	Kuroko Belt of Japan	5.2	5
	Scandinavian Shield	6.0	4
	Eastern Mediterranean	8.1	3
	Iberian Peninsula Pyrite Belt	28.2	2
	N. American Cordillera (exhalative)	29.5	1

Conclusion: Groups of geologic regions which do not differ significantly on on tonnage parameter

Group 1: W. Australian Shield, Tasman Geosyncline, Appalachian Belt, North American Shield, Scandinavian Caledonides

Group 2: Scandinavian Shield, E. Mediterranean , Kuroko Belt of Japan, North American Cordillera,

Group 3: Iberian Peninsula Pyrite Belt, North American Cordillera (exhalative)

In Table 3.1, the first line gives the result of the first SEQPOOL-SPSS program:

MANOVA: F = 2.82.

Comparing this with the $F_{8, 194}$ distribution, the probability of such a result is 0.015, so we do not reject the null hypothesis of equal means; and the regions are pooled. It will be noted that the next step in the analysis, the three ANOVA's, was still carried out. However, the multivariate test over-rides these, so that they do not add further information. Pooling

Table 3.2. (Continued) Pooling of eleven regions in four continents based on the economic parameters of volcanogenic sulfide deposits.

Type of deposit: volcanogenic sulfides

Sample Size: 367

Regions: N. American Shield, W. Australian Shield, Appalachian Belt, Scandinavian Caledonides, Scandinavian Shield, Eastern Mediterranean Iberian Peninsula Pyrite Belt, Tasman Geosyncline, N. American Cordillera, N. American Cordillera, (exhal.), Kuroko Belt of Japan

(1) Testing the significance of the difference between regions

Parameter(s)	Test procedure	F-value or equivalent F-value	degrees of freedom	Significance level of F-value	Conclusion on differences between regions
all	MANOVA	9.08	30, 1068	0.0001	significant
grade in $/tonne	ANOVA	6.37	10, 356	0.0001	significant
tonnage (millions)	ANOVA	16.93	10, 356	0.0001	significant
gross value ($mil.)	ANOVA	18.59	10, 356	0.0001	significant

(2) Results of Duncan's multiple range tests at the ∂ - 0.05 significance level

Parameter	Region	Geometric Mean	Rank
grade in US $ / tonne	Scandinavian Caledonides	36	11
	Eastern Mediterranean	42	10
	Scandinavian Shield	43	9
	N. American Shield	50	7
	Iberian Peninsula Pyrite Belt	50	7
	Appalachian Belt	54	6
	W. Australian Shield	72	5
	Tasman Geosyncline	86	4
	N. American Cordillera	91	3
	Kuroko Belt of Japan	101	2
	N. American Cordillera (exhalative)	109	1

Conclusion: Groups of geologic regions which do not differ significantly on on $ grade parameter

Group 1: Scandinavian Caledonides, Scandinavian Shield, E. Mediterranean, North American Shield, Iberian Peninsula Pyrite Belt, Appalachian Belt

Group 2: W. Australian Shield, Tasman Geosyncline, N. American Cordillera, Kuroko Belt of Japan, North American Cordillera (exhalative)

was therefore implemented, and the geometric mean length parameter for detection purposes was re-calculated to be 950 meters, with the ratio for the pool, R = B/L = 0.37.

The results of the sequential procedure to test the commonality of eleven regions of occurrence of volcanogenic sulfide deposits based on economic parameters are shown in Table 3.2. The economic parameters being considered are: dollar grade, tonnage and gross dollar value. Again, these parameters were tested on all variates and in each region in Chapter 2, and found to be

Table 3.2. (Continued) Pooling of eleven regions in four continents based on the economic parameters of volcanogenic sulfide deposits.

(1) Testing the significance of the difference between regions (Continued)

Parameter(s)	Test procedure	F-value or equivalent F-value	degrees of freedom	Significance level of F-value	Conclusion on differences between regions
gross value ($mil.)	ANOVA	18.59	10, 356	0.0001	significant

(2) Results of Duncan's multiple range tests at the α = 0.05 significance level

Parameter	Region	Geometric Mean	Rank
gross value in U.S. $ millions	W. Australian Shield	92	11
	N. American Shield	111	10
	Appalachian Belt	119	9
	Scandinavian Caledonides	120	8
	Tasman Geosyncline	239	7
	Scandinavian Shield	271	6
	Eastern Mediterranean	328	5
	N. American Cordillera	386	4
	Kuroko Belt of Japan	531	3
	Iberian Peninsula Pyrite Belt	1393	2
	N. American Cordillera (exhalative)	3207	1

Conclusion: Groups of geologic regions which do not differ significantly on on gross value parameter

Group 1: W. Australian Shield, Appalachian Belt, North American Shield, Scandinavian Caledonides

Group 2: Tasman Geosyncline, Scandinavian Shield, E. Mediterranean, , Kuroko Belt of Japan, North American Cordillera,

Group 3: Iberian Peninsula Pyrite Belt, North American Cordillera (exhalative)

lognormal, so the log-tranformation was taken. The results are given in Table 3.2. and the first line of the table gives the result of the first SEQPOOL-SPSS program:

MANOVA: F = 9.08.

Comparing this with the $F_{30,1068}$ distribution, the probability of such a result is less than 0.0001, so H_0, the null hypothesis of equal means, is rejected, the F value being highly significant, and the sequence therefore continues with:

Table 3.3. Pooling of three regions in three continents based on the economic parameters of Ni-Cu ultramafic deposits.

Type of deposit: Ni-Cu Ultramafic   Regions: N. American Shield, W. Australian Shield,
Sample Size: 65   Scandinavian Shield.

(1) Testing the significance of the difference between regions

Parameter(s)	Test procedure	F-value or equivalent F-value	degrees of freedom	Significance level of F-value	Conclusion on differences between regions
all	MANOVA	5.35	6, 120	0.0001	significant
grade in $/tonne	ANOVA	8.81	2, 62	0.0001	significant
tonnage (millions)	ANOVA	2.20	2, 62	0.120	not significant
gross value ($mil.)	ANOVA	1.17	2, 62	0.316	not significant

(2) Results of Duncan's multiple range tests at the $\partial = 0.05$ significance level

Parameter	Region	Geometric Mean	Rank
grade in US $ / tonne	Scandinavian Shield	47	1
	N. American Shield	92	2
	W. Australian Shield	100	3

Conclusion: Groups of geologic regions which do not differ significantly on grade:
Group 1: Scandinavian Shield.
Group 2: W. Australian Shield, North American Shield.

Parameter	Region	Geometric Mean	Rank
tonnage in millions of tonnes	Scandinavian Shield	1.3	1
	N. American Shield	2.1	2
	W. Australian Shield	3.9	3

Conclusion: The three regions form one homogeneous group on tonnage parameter

Parameter	Region	Geometric Mean	Rank
gross value in U.S. $ millions	Scandinavian Shield	63	1
	N. American Shield	197	2
	W. Australian Shield	394	3

Conclusion: Groups of geologic regions which do not differ significantly on gross value:
Group 1: Scandinavian Shield.
Group 2: W. Australian Shield, North American Shield.

ANOVAs: $F_1 = 6.37$, $F_2 = 16.93$ and $F_3 = 18.59$.

Again all results are highly significant at the 0.0001 level, compared with $F_{10, 356}$, so that H_0, the null hypothesis of equal means, is rejected on all three variates. Therefore, Duncan's Multiple Range Test, the next step in the sequential testing is now invoked.

Table 3.4. Pooling of three regions in three continents
based on the geometric parameters of Ni-Cu ultramafic
deposits.

Type of deposit: Ni-Cu Ultramafic   Regions: North American Shield, Scandinavian Shield,
Sample size   : 61   Western Australian Shield

(1) Testing the significance of the difference between regions

Parameter(s)	Test procedure	F-value or equivalent F-value	degrees of freedom	Significance level of F-value	Conclusion on differences between regions
all	MANOVA	5.01	6, 112	0.001	significant
length,l	ANOVA	7.98	2, 58	0.001	significant
breadth,b	ANOVA	4.86	2, 58	0.011	significant
b/l ratio	ANOVA	3.84	2, 58	0.027	significant

(2) Results of Duncan's multiple range tests at the ∂ = 0.05 significance level

Parameter	Region	Geometric mean	Rank
length in metres	Scandinavian Shield	180	1
	North American Shield	330	2
	Western Australian Shield	550	3

Conclusion: Each region differs significantly from the other two
on the length parameter

Parameter	Region	Geometric mean	Rank
breadth/length ratio	North American Shield	0.06	1
	Western Australian Shield	0.07	2
	Scandinavian Shield	0.17	3

Conclusion: Groups of geologic regions which do not differ significantly on shape
ratio parameter
Group 1: North American Shield, Western Australian Shield
Group 2: Scandinavian Shield

Duncan's multiple range test results

(1) grade $/tonne: the geometric means were ranked in order from $37/tonne
for the Scandinavian Caledonides to $109/tonne for the North American
Cordillera Exhalative deposits; the resulting grouping was very clear,
with $54 being the cut-off for group 1, with six regions in the group.
The top five were grouped together as the other group.

(2) tonnage (millions): the 11 regions were again put in rank order and
ranged from 1.3 (Western Australia) to 29.5 million tonnes (North

110

Table 3.5. Pooling of three regions in two continents based on the economic parameters of contact metasomatic deposits.

Type of deposit: Contact Metasomatic   Regions: N. American Cordillera: (Cu-Fe-Au),
                                                   (Pb-Zn-Cu-Ag), (W-Mo), (Cu-Au-Mo)
Sample Size: 136   East Asia: (W-Mo)

(1) Testing the significance of the difference between regions

Parameter(s)	Test procedure	F-value or equivalent F-value	degrees of freedom	Significance level of F-value	Conclusion on differences between regions
all	MANOVA	31.53	12, 341	0.0001	significant
grade in $/tonne	ANOVA	52.18	3, 131	0.0001	significant
tonnage (millions)	ANOVA	21.56	3, 131	0.0001	significant
gross value ($mil.)	ANOVA	36.55	3, 131	0.0001	significant

(2) Results of Duncan's multiple range tests at the ∂ = 0.05 significance level

Parameter	Region	Geometric Mean	Rank
grade in US $ / tonne	N. American Cordillera: (Cu-Au-Mo)	29	1
	N. American Cordillera: (Cu-Fe-Au)	46	2
	N. American Cordillera: (W-Mo)	70	3
	East Asia: (W-Mo)	95	4
	N. American Cordillera: (Pb-Zn-Cu)	169	5

Conclusion: Groups of geologic regions which do not differ significantly on grade parameter

Each of the four types differs significantly from the other three on the $ grade parameter, so that the only grouping is N. American Cordillera (W-Mo) and East Asia (W-Mo).

American Cordillera Exhalative). The test assigned them to three groups.

(3) gross value ($millions): Ordering the 11 regions, the geometric mean gross values ranged from $92 millions (Western Australia) to $3207 millions (North American Cordillera Exhalative).

The test result was again highly satisfactory, assigning the 11 regions to three groups.

We can see that the commonality testing was highly successful in the case of the geometric parameters of the porphyry-Cu-Mo deposits, reducing the application of complex detection calculations to only one homogeneous pool of regions instead of the original four. Similarly, in the case of the economic parameters of the volcanogenic sulfide deposits, we now have to deal with only three statistically homogeneous pools of regions instead of the original eleven regions if we are considering the tonnage and gross value parameters for economic planning purposes. However, only two pools are to be dealt with if we are interested in the dollar grade parameter.

Table 3.5. (Continued) Pooling of three regions in two continents based on the economic parameters of contact metasomatic deposits.

(1) Testing the significance of the difference between regions (Continued)

Parameter(s)	Test procedure	F-value or equivalent F-value	degrees of freedom	Significance level of F-value	Conclusion on differences between regions
tonnage (millions)	ANOVA	21.56	3, 131	0.0001	significant
gross value ($mil.)	ANOVA	36.55	3, 131	0.0001	significant

(2) Results of Duncan's multiple range tests at the ∂ - 0.05 significance level

Parameter	Region	Geometric Mean	Rank
tonnage in millions of tonnes	N. American Cordillera: (W-Mo)	0.6	1
	East Asia: (W-Mo)	0.9	2
	N. American Cordillera: (Pb-Zn-Cu)	1.2	3
	N. American Cordillera: (Cu-Fe-Au)	6.0	4
	N. American Cordillera: (Cu-Au-Mo)	64.0	5

Conclusion: Groups of geologic regions which do not differ significantly on tonnage parameter

Group 1: N. American Cordillera: (W-Mo), East Asia: (W-Mo)
N. American Cordillera: (Pb-Zn-Cu)

Group 2: N. American Cordillera: (Cu-Fe-Au)

Group 3: N. American Cordillera: (Cu-Au-Mo)

Parameter	Region	Geometric Mean	Rank
gross value in U.S. $ millions	N. American Cordillera: (W-Mo)	41	1
	East Asia: (W-Mo)	86	2
	N. American Cordillera: (Pb-Zn-Cu)	208	3
	N. American Cordillera: (Cu-Fe-Au)	265	4
	N. American Cordillera: (Cu-Au-Mo)	1865	5

Conclusion: Groups of geologic regions which do not differ significantly on gross value parameter

Group 1: N. American Cordillera: (W-Mo), East Asia: (W-Mo), N. American Cordillera: (Pb-Zn-Cu)
Group 2: N. American Cordillera: (Cu-Fe-Au),
Group 3: N. American Cordillera: (Cu-Au-Mo)

3.4.3. Example of satisfactory reduction of statistical work-load by pooling

In Tables 3.3 and 3.4, the Ni-Cu Ultramafic regions of the North American Shield, Western Australian Shield, and Scandinavian Shield were compared first by their economic parameters, and secondly on the geometric ones.

Table 3.6. Pooling of three regions in one continent based on
the geometric parameters of Mississipi Valley-type
Pb-Zn deposits.

Type of deposit: Mississipi Valley type Pb-Zn   Regions: Upper Mississipi Valley, Missouri - Tri-State,

Sample size: 99   North American Arctic

(1) Testing the significance of the difference between regions

Parameter(s)	Test procedure	F-value or equivalent F-value	degrees of freedom	Significance level of F-value	Conclusion on differences between regions
all	MANOVA	35.14	6, 188	0.00001	significant
length,l	ANOVA	46.65	2, 96	0.00001	significant
breadth,b	ANOVA	93.52	2, 96	0.00001	significant
b/l ratio	ANOVA	29.50	2, 96	0.00001	significant

(2) Results of Duncan's multiple range tests at the $\delta = 0.05$ significance level

Parameter	Region	Geometric	Rank
length in metres	Upper Mississipi Valley	340	3
	North American Arctic	410	2
	Missouri - Tri-State	2860	1

Conclusion: Groups of geologic regions which do not differ significantly
on length parameter

Group 1: Upper Mississipi Valley, North American Arctic

Group 2: Missouri - Tri-State

Parameter	Region	Geometric	Rank
breadth/length ratio	Upper Mississipi Valley	0.10	3
	Missouri - Tri-State	0.19	2
	North American Arctic	0.40	1

Conclusion: Each of the regions differs significantly from the other two
on shape ratio parameter

The results were as follows:

MANOVA: H_0 was rejected in both cases.

ANOVAS: When dealing with the economic parameters, H_0 was rejected in
the case of the grade parameter, but when considering the
geometric parameters, H_0 was rejected on all three parameters.

Table 3.7. Results of application of sequential commonality to the whole data-base.

Purpose of pooling of regions ➡			Computation of detection probabilities & optimization of field programs	Selection of ore deposit type (s) & region (s) of investigation
Nature & number of ore deposit variates ➡			3 to 5 geometric variates	3 economic variates
Type & sub-type of ore deposit covered by data-base ⬇	Number of regions covered by data-base	Number of Units (region - deposit type)	Number of pools & units used for further statistical processing ⬇	
Porphyry-Cu-Mo	5	5	1 pool of 5 units	1 pool of 4 units and 1 unit
Contact-metasomatic 4 sub-types Cu-Fe-Au Pb-Zn-Cu-Ag Cu-Mo-Au W-Mo	2	5	5 units	1 pool of 2 units and 3 units
Ni-Cu-Ultramafic	3	3	1 pool of 2 units and 1 unit	1 pool of 2 units and 1 unit
Volcanogenic sulfides 2 sub-types: volcanogenic massive sulfides and exhalative sulfides	10	11	1 pool of 3 units 1 pool of 6 units 1 pool of 2 units	1 pool of 3 units 3 pool of 5 units 1 pool of 3 units
Mississippi Valley type Pb-Zn	3	3	3 units	1 pool of 2 units and 1 unit
Vein-Gold	2	2	2 units	1 pool of 2 units
Total: 6 types & sub-types	Total: 25 regions	Total: 29 units	Total: 5 pools and 11 units	Total: 8 pools and 6 units

Duncan's multiple range tests

On the grade parameter, the Scandinavian Shield was significantly different from the other two which were pooled.

The test failed to group any of the three regions on the length parameter; but on the shape ratio, the North American Shield was grouped with the Western Australian Shield, while the Scandinavian Shield was assigned on its own to the other group.

We can conclude that the application of the SEQPOOL procedure to the Ni-Cu ultramafic deposits of three important nickel producing regions of the world was not as spectacularly successful as in the previous case of the porphyry-Cu-Mo and Volcanogenic sulfide deposits, with the exception of the testing on the tonnage parameter which led to the pooling of the three regions into one group for economic planning purposes. The procedure was moderately successful for the testing based on the grade and gross value parameters for economic planning purposes, as well as on the geometric shape ratio parameter for detection planning, leading to the pooling of two of the three regions. Finally, no pooling could be statistically justified, when dealing with the length parameter of the Ni-Cu deposits.

3.4.4. Examples of the unsuccessful application of the SEQPOOL procedure

Four sub-types of contact-metasomatic deposits were tested on their three economic parameters. They included the Cu-Fe-Au, Pb-Zn-Cu-Ag, Cu-Mo-Au and W-Mo sub-types occurring in two regions of the World, namely the North American Cordillera and East Asia (South Korea and Japan). The results of the testing are as follows, based on Table 3.5:

MANOVA: H_0 was rejected at the 0.0001 significance level, with

$F = 31.53$;

ANOVAs: H_0 was rejected on all 3 parameters at the 0.0001 significance level.

Duncan's multiple range tests: resulted in some grouping on each of the three parameters.

Among the four sub-types of contact metasomatic deposits, the SEQPOOL procedure was successful in pooling the two regions of occurrence based on each of the three parameters only for the W-Mo sub-type. The testing of the other three sub-types on the three economic parameters proved totally unsuccessful.

Finally, Table 3.6 shows the results of the testing of commonality between three regions of occurrence of the Mississipi Valley Pb-Zn - type deposits in the North American Continent, namely Missouri-Tri-State, the Upper Mississipi Valley and the Paleozoic Arctic regions, based on the geometric parameters required for optimal detection planning. Only one grouping of two regions was justified on the length parameter, but none was obtained on the B/L shape ratio parameter.

3.4.5. Summary of the results of the application of the SEQPOOL procedure to the whole data-base

The results of the application of the sequential population commonality testing based upon economic and geometric parameters to the whole world-wide data-base of 900 deposits are summarized in Table 3.7. Starting from a rather unwieldy total of 29 individual units combining types of deposits and regions of occurrence, the application of the SEQPOOL procedure led to a reduction of the total to a much more manageable set of 16 pools and units based on the geometric parameters, and to an even smaller set of only 14 pools and units based on economic parameters. The pooling system thus reduced the number of sets of airborne and ground detection tables by nearly 45%, and economic planning tables by 52%.

Table 3.8. Pooling of five region-deposit types in Europe based
on the economic parameters of ore deposits (grade).

Sample Size: 121	Types of deposits	Geologic Region
	Ni-Cu Ultramafic, Volc. Sulf.	Scandinavian Shield
	Volc. Sulf.	Scandinavian Caledonides
	Volc. sulf.	Iberian Peninsula Pyrite Belt
	Volc. sulf.	Eastern Mediterranean

(1) Testing the significance of the difference between regions

Parameter(s)	Test procedure	F-value or equivalent F-value	degrees of freedom	Significance level of F-value	Conclusion on differences between regions
all	MANOVA	5.34	12, 348	0.0001	significant
grade in $/tonne	ANOVA	2.59	4, 116	0.041	not significant
tonnage (millions)	ANOVA	10.65	4, 116	0.0001	significant
gross value ($mil.)	ANOVA	12.24	4, 116	0.0001	significant

(2) Results of Duncan's multiple range tests at the ∂ = 0.05 significance level

Parameter	Region & Deposit type	Geometric Mean	Rank
grade in US $ / tonne	Scandinavian Caledonides (Volc. sulf.)	36	5
	Eastern Mediterranean (Volc. sulf.)	42	4
	Scandinavian Shield (Volc. sulf.)	43	3
	Scandinavian Shield (Ni-Cu Ultramaf.)	47	2
	Iberian Penins.Pyr. Belt (Volc. sulf.)	50	1

Conclusion: The five region - deposit types form one homogeneous group on
on $ grade parameter

3.5. APPLICATION OF SEQUENTIAL TESTING TO ASSIST CONTINENT-WIDE SEARCH PROGRAMS FOR SIX TYPES OF ORE DEPOSITS

3.5.1. Introduction

In the previous section, 3.4, the sequential testing of population
commonality based on both economic and geometric parameters was applied to
various groups of regions located throughout the World for each of six
specific genetic types of ore deposits which are commonly chosen as targets
for exploration programs. Although this may appear to be a rather
theoretical approach, it proved very useful as a labour-saving exercise,
particularly when dealing with detection planning. Furthermore, the testing

Table 3.8. (Continued) Pooling of five region-deposit types in Europe based on the economic parameters of ore deposits (tonnage & gross value).

(2) **Results of Duncan's multiple range tests at the δ = 0.05 significance level (Continued)**

Parameter	Region & Deposit type	Geometric Mean	Rank
tonnage in millions of tonnes	Scandinavian Shield (Ni-Cu Ultramaf.)	1.8	5
	Scandinavian Caledonides (Volc. sulf.)	3.4	4
	Scandinavian Shield (Volc. sulf.)	6.0	3
	Eastern Mediterranean (Volc. sulf.)	8.1	2
	Iberian Penins.Pyr. Belt (Volc. sulf.)	28.2	1

Conclusion: Groups of region - deposit types which do not differ significantly on on tonnage parameter

Group 1: Scandinavian Shield (Ni-Cu Ultramafic), Scandinavian Caledonides (Volc. sulf.)

Group 2: Scandinavian Shield (Volc. sulf.), Eastern Mediterranean (Volc. sulf.)

Group 3: Iberian Peninsula Pyrite Belt (Volc. sulf.)

Parameter	Region & Deposit type	Geometric Mean	Rank
gross value in US $ millions	Scandinavian Shield (Ni-Cu Ultramaf.)	63	5
	Scandinavian Caledonides (Volc. sulf.)	120	4
	Scandinavian Shield (Volc. sulf.)	271	3
	Eastern Mediterranean (Volc. sulf.)	328	2
	Iberian Penins.Pyr. Belt (Volc. sulf.)	1393	1

Conclusion: Groups of region - deposit types which do not differ significantly on on gross value parameter

Group 1: Scandinavian Shield (Ni-Cu Ultramafic), Scandinavian Caledonides (Volc. sulf.)

Group 2: Scandinavian Shield (Volc. sulf.), Eastern Mediterranean (Volc. sulf.)

Group 3: Iberian Peninsula Pyrite Belt (Volc. sulf.)

statistically confirmed some important and long-held views of many economic geologists regarding the remarkable constancy of geologic and economic characteristics of some deposit types such as porphyry-Cu and volcanogenic sulfides, through geologic time and space.

We are now considering a more practical approach directly aimed at assisting the economic and detection planning work of exploration staff. Since many mining companies search for mining deposits on a geographical basis rather than specializing in a particular type of deposit, the next step is to apply the pooling on a continent by continent basis. We are choosing the European Continent and Australasia & East Asia as examples for the application of the sequential testing procedure to assist continent-wide ore search programs.

3.5.2. Application of the SEQPOOL procedure to Europe

We are considering two types of ore deposits, namely the Ni-Cu ultramafic and volcanogenic sulfides, occurring in two regions of Northern Europe and two regions of the Mediterranean, which are combined into five units. The results of the commonality testing based on three economic parameters are shown in Table 3.8.

MANOVA: Reject H_0 at the 0.0001 significance level.

ANOVAs:

Grade:	no significant differences: pool all 5,
Tonnage:	reject H_0,
Gross value:	reject H_0.

Duncan's multiple range tests

Tonnage:	2 groups of two, and 1 unit.
Gross value:	2 groups of two, and 1 unit.

As a result of the testing, we find that the five populations are statistically similar on the basis of the dollar grade variate. However, when it comes to the tonnage and gross dollar value parameters which are of paramount importance for economic planning, pooling is justified for the Ni-Cu ultramafic deposits of the Scandinavian Shield and volcanogenic sulfide deposits of the Caledonides on the one hand, and for the volcanogenic sulfide deposits of the Scandinavian Shield and the East Mediterranean on the other hand. Please note that the very large volcanogenic pyritic deposits of the Iberian Peninsula stand on their own on the basis of tonnage and gross value.

3.5.3. Application of the SEQPOOL procedure to Australasia and East Asia

We are considering five types of ore deposits occurring in three regions of Australasia and two regions of East Asia which are listed at the head of Table 3.9., and are grouped into eight units (region-ore deposit type) for testing purposes. The commonality test results on the economic parameters for the eight units read as follows:

MANOVA

Reject H_0 at the 0.0001 significance level

ANOVAs

Reject H_0 at the 0.0001 significance level on all three parameters

Duncan's multiple range tests

Grade:	Group 1: Tasman Geosyncline & S. Pacific Island Arc (porphyry-Cu-Au).
	Group 2: the remaining six region-deposit types
Tonnage:	2 groups of three, and 1 group of two
Gross value:	2 groups of three, and 1 group of two

Table 3.9. Pooling of eight region-deposit types in
Australasia and East Asia based on the economic
parameters of ore deposits (grade).

	Types of deposits	Geologic Region
	Vein Gold, Ni-Cu Ultramafic, Volc. Sulf.	Western Australian Shield
	Porphyry Cu-Mo, Volc. Sulf.	Tasman Geosyncline
Sample Size: 181	Porphyry Cu-Mo	South Pacific Island Arc
	Contact metasomatic (W - Mo)	East Asia
	Volc. sulf.	Kuroko Belt of Japan

(1) Testing the significance of the difference between regions

Parameter(s)	Test procedure	F-value or equivalent F-value	degrees of freedom	Significance level of F-value	Conclusion on differences between regions
all	MANOVA	15.51	21, 519	0.0001	significant
grade in $/tonne	ANOVA	62.63	7, 173	0.0001	significant
tonnage (millions)	ANOVA	11.24	7, 173	0.0001	significant
gross value ($mil.)	ANOVA	9.24	7, 173	0.0001	significant

(2) Results of Duncan's multiple range tests at the ∂ = 0.05 significance level

Parameter	Region & Deposit type	Geometric Mean	Rank
grade in US $ / tonne	Tasman Geosyncline (P-Cu-Mo)	11	8
	South Pacific Island Arc (P-Cu-Mo)	11	7
	W. Australian Shield (Volc. sulf.)	72	6
	Tasman Geosyncline (Volc. sulf.)	86	5
	East Asia (Contact metas: W-Mo)	95	4
	W. Australian Shield (Ni-Cu Ultramaf.)	100	3
	Kuroko Belt of Japan	101	2
	W. Australian Shield (Vein Gold)	156	1

Conclusion: Groups of region - deposit types which do not differ significantly on
on $ grade parameter

Group 1: Tasman Geosyncline (Porphyry Cu-Mo), South Pacific Island Arc (Porphyry Cu-Mo)
Group 2: comprises the other six region - deposit types.

Grade-wise, the statistical testing indicates one grouping for
porphyry-Cu-Mo of two regions of Australasia, and a pool of the remaining
four types of ore deposits occurring in Australasia and East Asia. Tonnage-
wise, the grouping of the two regions of occurrence of porphyry-Cu-Mo is
again justified statistically, while the other regions of occurrence of the
four other types of deposits in Australasia and in East Asia are split into
two pools. Finally, if we consider the gross value parameter, one grouping
comprising the East Asian W-Mo and Australasian vein gold and volcanogenic
sulfide deposits coincides with one of the pools indicated by the testing

Table 3.9. (Continued) Pooling of eight region-deposit types in
Australasia and East Asia based on the economic
parameters of ore deposits (tonnage & gross value).

(2) Results of Duncan's multiple range tests at the ∂ – 0.05 significance level (Continued)

Parameter	Region & Deposit type	Geometric Mean	Rank
tonnage in millions of tonnes	East Asia (Contact metas: W-Mo)	0.9	8
	W. Australian Shield (Vein Gold)	1.0	7
	W. Australian Shield (Volc. sulf.)	1.3	6
	Tasman Geosyncline (Volc. sulf.)	2.8	5
	W. Australian Shield (Ni-Cu Ultramaf.)	3.9	4
	Kuroko Belt of Japan	5.2	3
	Tasman Geosyncline (P-Cu-Mo)	39	2
	South Pacific Island Arc (P-Cu-Mo)	96	1

Conclusion: Groups of region – deposit types which do not differ significantly on
on tonnage parameter

Group 1: East Asia (Cont. metas. W-Mo), W. Australian Shield (Vein Gold), W. Australian Shield (Volc. sulf.)
Group 2: Tasman Geosyncline (Volc. sulf.), W. Australian Shield (Ni-Cu Ult.), Kuroko Belt of Japan (Volc. sulf.)
Group 3: Tasman Geosyncline (Porphyry Cu-Mo), South Pacific Island Arc (Porphyry Cu-Mo)

Parameter	Region & Deposit type	Geometric Mean	Rank
gross value in U.S. $ millions	East Asia (Contact metas: W-Mo)	86	8
	W. Australian Shield (Volc. sulf)	92	7
	W. Australian Shield (Vein Gold)	132	6
	Tasman Geosyncline (Volc. sulf.)	239	5
	W. Australian Shield (Ni-Cu Ultramaf.)	394	4
	Tasman Geosyncline (P-Cu-Mo)	409	3
	Kuroko Belt of Japan	531	2
	South Pacific Island Arc (P-Cu-Mo)	1092	1

Conclusion: Groups of region – deposit types which do not differ significantly on
on gross value parameter

Group 1: East Asia (Cont. metas. W-Mo), W. Australian Shield (Vein Gold), W. Australian Shield (Volc. sulf.)
Group 2: Tasman Geosyncline (Volc. sulf.), W. Australian Shield (Ni-Cu Ult.), Tasman Geosyncline (Porphyry Cu-Mo)
Group 3: Kuroko Belt of Japan (Volc. sulf.), South Pacific Island Arc (Porphyry Cu-Mo)

based on the tonnage parameter. The other two groupings differ from those
obtained for the tonnage parameter.

3.5.4. Application of the results of commonality testing to economic planning

3.5.4.1. Introduction

One of the principal tasks of economic planning for mineral exploration
programs is the selection of ore deposit types and regions of search on the
basis of economic criteria which reflect corporate goals. The principal

Table 3.10. Ranking of pooled region-deposit types in Europe
based on the geometric means and dispersion
coefficients of three parameters of economic worth.

Ranking of pooled types of ore deposits & regions of Europe based on gross value parameter in US $Millions				
Pooled regions & types of ore deposits	Geometric Mean	Rank	Dispersion coefficient	Rank
Iberian Peninsula Pyr. Belt (Volc. sulfides)	1393	1	1.1	1
Scandinavian Shield (Volc. sulfides), & Eastern Mediterranean (Volc. sulfides) }	298	2	1.3	2
Scandinavian Shield (Ni-Cu Ultramafic), & Scandinavian Caledonides (Volc. sulfides) }	87	3	1.4	3
Ranking based on tonnage parameter in Millions of tonnes				
Iberian Peninsula Pyr. Belt (Volc. sulfides)	28.2	1	0.7	1
Scandinavian Shield (Volc. sulfides), & Eastern Mediterranean (Volc. sulfides) }	7.0	2	1.3	2
Scandinavian Shield (Ni-Cu Ultramafic), & Scandinavian Caledonides (Volc. sulfides) }	2.4	3	1.4	3

criterion on which the comparative merits of various types of ore deposits
and regions of occurrence may be judged is the gross dollar value as defined
in Chapter 1. However, as the gross dollar value is a composite parameter
obtained by multiplication of the tonnage by the average dollar grade, a
high gross value can result either from low tonnage and very high dollar
grade, or from a large tonnage of low grade material. Thus, the ranking of
deposits based solely on gross value leads to some ambiguity in the
selection which has to be resolved by consideration of additional rankings
based first on the tonnage parameter ans finally on the dollar grade
parameter.

Table 3.11. Ranking of pooled region-deposit types in Australasia and East Asia based on the geometric means and dispersion coefficients of three parameters of economic worth.

Ranking of pooled types of ore deposits & regions of Australasia based on gross value parameter in US $Millions				
Pooled regions & types of ore deposits	Geometric Mean	Rank	Dispersion coefficient	Rank
Kuroko Belt of Japan (Volc. sulfides), & South Pacific Island Arc (Porphyry-Cu-Mo)	760	1	1.5	1
Tasman Geosyncline (Volc. sulfides), W. Australian Shield (Ni-Cu Ultramafic), & Tasman Geosyncline (Porphyry-Cu-Mo)	330	2	1.6	2
East Asia (Contact metasomatic: W-Mo), W. Australian Shield (Volc. Sulfides), & W. Australian Shield (Vein gold)	100	3	2.0	3
Ranking based on tonnage parameter in Millions of tonnes				
Tasman Geosyncline (Porphyry-Cu-Mo), & South Pacific Island Arc (Porphyry-Cu-Mo)	61.0	1	1.1	1
Tasman Geosyncline (Volc. sulfides), W. Australian Shield (Ni-Cu Ultramafic), & Kuroko Belt of Japan (Volc. sulfides)	3.9	2	1.5	2
East Asia (Contact metasomatic: W-Mo), W. Australian Shield (Volc. sulfides), & W. Australian Shield (Vein gold)	1.0	3	2.3	3
Ranking based on grade parameter in US $/tonne				
W. Australian Shield (Volc. sulfides), Tasman Geosyncline (Volc. sulfides), East Asia (Contact metasomatic: W-Mo), W. Australian Shield (Ni-Cu Ultramafic), Kuroko Belt of Japan (Volc. sulfides), & W. Australian Shield (Vein gold)	101	1	0.8	2
Tasman Geosyncline (Porphyry-Cu-Mo), & South Pacific Island Arc (Porphyry-Cu-Mo)	11	2	0.7	1

This ranking exercise has been carried out previously in Chapter 2, based on the two statistics for individual populations of regions of occurrence for each of six types of ore deposits. However, this type of ranking is fraught with difficulties, because we are considering the variability of parameters for individual regions only, rather than taking into account their variabilities in relation to one another. This is where the usefulness of the SEQPOOL procedure becomes evident. For now we can compare the statistics of pools of regions and ore deposit types which were determined by the SEQPOOL process. This has been done for Europe and for Australasia and East Asia; the results are shown below.

3.5.4.2. Application of SEQPOOL to the selection of region-ore deposit types in Europe

The results are given in Table 3.10. The geometric means and dispersion coefficients were calculated for the gross value and tonnage parameters of each of the pools of regions. The ranking based on the dollar grade parameter is not shown because we found in Section 3.5.2 that all units could be pooled into a single group with geometric mean value equal to $43 per tonne. We note that rankings based upon gross value and tonnage are similar, whether considering the geometric mean or the dispersion coefficient as criterion. The pyritic volcanogenic deposits of the Iberian Peninsula obviously should be prime exploration prizes with the volcanogenic sulfide deposits of the Scandinavian Shield and East Mediterranean coming in second position.

3.5.4.3. Application of SEQPOOL to the selection of region-ore deposit types in Australasia and East Asia

The geometric means and dispersion coefficients were calculated for each parameter (gross value, tonnage, and grade) for each pool of regions to enable the ranking of eight units in Australasia and East Asia. The results are given in Table 3.11.

The ranking order is the same whether we use the geometric mean or the dispersion coefficient criterion when we consider the gross value parameter. This applies also to the ranking based upon tonnage. However, the ranking of the two pools obtained for the dollar grade parameter shows that the porphyry-Cu-Mo group of Australasia comes a poor second to the first pool based on the geometric mean, while there is scarcely any difference if we consider the dispersion coefficient instead.

3.5.5. Conclusion

We hope this chapter has convincingly demonstrated the considerable merits of the SEQPOOL procedure for the purpose of objectively grouping ore deposit types and regions of occurrence on the basis of geometric and economic parameters. When dealing with the optimal designing of ore search programs which requires much tedious and complex statistical processing, the saving in costly computer time is quite substantial, amounting to as much as 45%. When dealing with economic planning of exploration programs, the merits of the method are of two kinds. One is a startling reduction in computer time, by as much as 52%. The second one is a noticeable clarification of the information on which the ranking of the ore deposit type-region units can be achieved, which leads to a much more reliable selection for planning purposes.

This is demonstrated by the comparison of the ranking of European ore deposit type-region units before application of the SEQPOOL procedure (see Table 2.20) with that obtained after statistical pooling (Table 3.9). A comparison of the results for Australasia and East Asia in Table 3.10 with the results given in Table 2.21, presents an even clearer picture with the larger groupings. For example, when considering the grade parameter, the reduction is from 8 individual region-ore deposit types to only two pools of region-ore deposit types, and the choice is clearly between small rich deposits and large lower-grade ones.

CHAPTER 3: Exercises

E3 Exercises

Using the computer program SEQPOOL in Appendix 3, test the region-deposit types for differences on the given parameters using the logarithm (base 10) of the data set referred to.

E3.1. Vein Gold: Economic: data set 8.1.3.

E3.2. Vein Gold: Geometric: data set 8.1.4.

E3.3. Porphyry Cu-Mo: Geometric: data set 8.1.2.

E3.4. Ni-Cu Ultramafic: Economic: data set 8.1.5.

E3.5. Contact Metasomatic: Geometric: data set 8.1.10.

E3.6. Porphyry Cu-Mo: Economic: data set 8.1.1.

E3.7. Mississippi Valley-type Pb-Zn: Economic: data set 8.1.11.

E3.8. Volcanogenic sulfides: Geometric: data set 8.1.8.

BIBLIOGRAPHY FOR CHAPTER 3

1. ANDERSON, T.W., 1958, Introduction to multivariate statistical analysis; Wiley, New York.

2. BARTLETT, M.S., 1947, The use of transformations; Biometrics, V.3, pp.39.

3. BOOT, J.C.G. and COX, E.B., 1970, Statistical analysis for managerial decisions; McGraw Hill, New York.

4. COCHRAN, W.G. and COX, G.M., 1957, Experimental design; J. Wiley, New York.

5. COX, D.R., 1958, Planning of experiments; Wiley, New York.

6. DANIEL, W.W., 1974, Biostatistics: a foundation for analysis in the Health Sciences; John Wiley & Sons, New York.

7. DAVIES, O.L., (Ed.), 1960, The design and analysis of experiments; Hafner, New York.

8. DUNCAN, D.B., 1949, Significance tests for differences between ranked variates drawn from normal populations; Ph.D. Thesis, Iowa State College.

9. DUNCAN, D.B., 1951, A significance test for differences between ranked treatments in an Analysis of Variance; Virginia Journal of Science, V.2, pp.171-189.

10. DUNCAN, D.B., 1952, On the properties of the multiple comparisons test; Virginia Journal of Science, V.3, pp.50-67.

11. DUNCAN, D.B., 1955, Multiple range and multiple-F tests; Biometrics, V.11, pp.1-42.

12. EISENHART, C., 1947, The assumptions underlying the Analysis of Variance; Biometrics, V.3, pp.1-21.

13. FINNEY, D.J., 1955, Experimental design and its statistical basis; the University of Chicago Press, Chicago.

14. FISHER, R.A., 1966, The design of experiments, 8th Ed.; Oliver and Boyd, Edinburgh.

15. KEMPTHORNE, O., 1952, The design and analysis of experiments; Wiley, New York.

16. KEULS, M., 1952, The use of the studentized range in connection with the Analysis of Variance; Euphytica, V.1, pp.112-122.

17. KIRK, R., 1968, Experimental design: procedures for the Behavioral Sciences; Belmont, Calif.; Brooks.

18. KOCH, G.S. and LINK, R.F., 1970, Statistical analysis of geological data, V.1; J. Wiley, New York.

20. KRAMER, C.Y., 1956, Extension of multiple range tests to group means with unequal numbers of replications; Biometrics, V.12, pp.307-310.

21. KRUMBEIN, W.C. and GRAYBILL, F.A., 1965, An introduction to
        statistical models in Geology (Chpt 9); McGraw-Hill,
        New York.

22. LIGHT, R.J. and MARGOLIN, B.H., 1971, An Analysis of Variance
        for categorical data; Journal of the American
        Statistical Association, V.66, pp.534-544.

23. MAXWELL, A.E., 1961, Analyzing qualitative data; John Wiley &
        Sons, Inc., New York.

24. MILLER, R.L. and KAHN, J.S., 1962, Statistical analysis in the
        geological sciences; J. Wiley, New York.

25. MORRISON, D.F., 1967, Multivariate statistical methods;
        McGraw-Hill, N.Y.

26. NEWMAN, D., 1939, The distribution of the range in samples
        from a normal population in terms of an independent
        estimate of standard deviation; Biometrika, V.31,
        pp.20-30.

27. SCHEFFE, H.A., 1959, The Analysis of Variance; New York,
        Wiley.

28. TUKEY, J.W., 1949, Comparing individual means in the Analysis
        of Variance; Biometrics, V.5, pp.99-114.

30. WALPOLE, R.E., 1968, Introduction to statistics; The Macmillan
        Company, N.Y.

31. WINER, B.J., 1971, Statistical principles in experimental
        design; 2d ed., McGraw-Hill, New York.

CHAPTER FOUR

PROBABILISTIC MODELS FOR THE OPTIMAL DETECTION OF ORE DEPOSITS
BY AIRBORNE AND GROUND EXPLORATION PROGRAMS

4.1. GENERAL STATEMENT

4.1.1. The ore detection process

The principal aim of field programs is to collect information which
will lead to the detection of mineral deposits. There are two approaches to
the detection problem: one is direct and the other one indirect; they are
used simultaneously or sequentially to best advantage.

The direct approach is a search for the mineral deposit by direct
observation of geological features. The observation is based either on
visual recognition either unaided, or aided by photogeology carried out from
aeroplane platforms or by satellite imagery obtained by remote sensing
techniques, or by mechanical probes such as drills in three-dimensional
investigations.

The indirect approach deals mainly with geophysical and geochemical
features and searches for mineral deposits or associated features such as
halos in an indirect manner through local changes called anomalies, by
sampling the continuum of the regional geoenvironment. There are two main
types of procedures involved: (1) continuous sampling, and (2) discrete
sampling.

Continuous geophysical sampling surveys are feasible only from the
air; the data are recorded as graphs, or on photographic films, or on
magnetic tapes, while sweeping across the area along regularly spaced flight
lines. The continuously recorded survey data are then sampled at regular
intervals making sure to include all high and low points. This digitization
is carried out to obtain representative discrete point data which are much
easier to process statistically than continuous data. Discrete sampling is
carried out on the ground at various points related to the topography
(outcrops usually occur at or near topographic highs) in reconnaissance
surveys, or at points indicated by the processing of the airborne survey
data, or at control grid points, which is the case in all systematic surveys
and drilling programs.

The indirect approach is employed to detect either the targeted
mineral deposits themselves through their own geophysical or geochemical
signatures or the halos associated with the deposits, which are larger and
thus more easily detected than the deposits. Following the processing of the
data gathered, the halos are delineated, and a second round of the search
proceeds to detect the mineral deposits themselves within the halos, either
indirectly with further geo-surveys or directly by systematic drilling.

4.1.2. Rationale of the introduction of geometric probability theory as a foundation for the construction of probabilistic detection models in the search for ore deposits

The parameters required for the construction of models for the detection of ore bodies or halos by any field program, airborne or ground, require the use of the expected target length, L_t, and the shape ratio, ($R_t = B_t/L_t$), the true dip, plunge and strike direction. In addition, the spacing of sampling controls and the orientation of the control grid are particularly important. While the parameters of the geometry of the search grid, including the attitude of the detector, are under the exploration planner's control and so deterministic in nature, the geometric parameter of the targeted prize (size, shape, attitude, and location) are all unknown. They are stochastic in nature, and their expected values may be derived by the statistical modelling of known ore deposits in the region of the same genetic type as the expected prizes. They are expressed within confidence intervals which are derived from fitted probability distributions. Therefore, it seems quite appropriate to apply the geometric probability theory as the foundation for the computation of ore-detection probabilities.

Geometric probability theory deals with geometric random variables such as point-nets, lines and convex planar figures or sets of a combination of these. A principal theme of geometric probability theory is the relationship between various geometric objects as expressed by intersection or coverage. Kendall & Moran (13), Savinskii (18), and Singer & Wickham (22) have made substantial contributions to the theory of coverage and intersection. This work laid the foundation for the computation of the probabilities of detection of ore deposits.

4.1.3. Continuous sampling models for airborne detection

Over the past thirty years, since most of the easily detected ore deposits had been found by ground prospecting, one of the main strategies for the initial search for hidden deposits in regional exploration programs is the use of geophysical airborne surveys, with the aircraft carrying continuously recording sensors of a variety of kinds, and more recently satellites using remote sensors. If an airborne recording device registers the signature of an ore deposit as one "blip", that is if the flight line intersects the deposit or target at least once, then it is said to be detected. If it intersects the target at least twice, then the detection is said to be confirmed. The calculation of the probabilities of detection or confirmed detection will be dealt with in Section 4.2.

4.1.4. Discrete sampling models for ground detection

Systematic discrete sampling on ground grids generally orthogonal in design was used before airborne surveys. Most of the work done on the application of geometric probability theory to ground exploration programs since the 1950's was mainly concerned with drilling rather than geochemical or geophysical surveys, because of the high cost of systematic drilling programs. Savinskii (18) and Singer (22, 23, 24, 25) were amongst the earlier workers in the field. In our study, we have taken 100 meters as the maximum vertical depth of drilling and as the maximum depth of ore detection by geophysical instruments in the present state of the art.

4.1.5. Detection of ore deposits based on Optimal Search theory

From Stone (29), the theory of Optimal Search is the maximization of the probability of target detection, P_d, subject to budgetary and resource constraints, $f(x_1, x_2)$. The idea is to employ the traditional Lagrangian

Table 4.1. Expressions of the probability of detection and confirmed detection of randomly oriented targets of various shapes by airborne surveys on various types of grid designs.

Probability of at least One Intersection of Randomly Oriented Targets of various shapes by Parallel or Orthogonal grids of lines			
Grid design Characteristics	Target/Grid ratio Geometry	Probability of intersection	
		Linear Shape	Circular Shape
Parallel Spacing (S)	$0 < U <= 1$ $U > 1$	$P_I = 2U / \pi$ $P_I = (2U/\pi)[1 - \sqrt{(1 - 1/U^2)} + (1/U)Acs(1/U)]$	$P_c = U$ $P_c = 1$
Square (S X S)	$0 < U <= 1$	$P_I = (U / \pi)(4 - U)$	$P_c = U(2 - U)$
Rectangular (S X T)	$W = S/T <= 1$	$P_I = (U / \pi)[2(W + 1) - UW]$	$P_c = U[(W + 1) - UW]$

Probability of at least two intersections of Randomly Oriented Targets of various shapes by Parallel grid			
Grid design Characteristics	Target/Grid ratio Geometry	Probability of at least two intersections	
		Linear Shape	Circular Shape
Parallel Spacing (S)	$U > 1$	$P_I = (2/\pi)[Cotan(a) + (a - \pi/2)]$, where $a = Arcsin(1/U)$	$P_c = 1$

for elliptical targets with shape ratio $R = B/L$
$$P_e = (1 - R)P_I + RP_c$$

theory to make the problem into an unconstrained one. However, in the detection of ore deposits, the method is unsatisfactory due mainly to the complex nature of the geometry functions, which are diverse functions on each subdomain. Therefore, in our work, the optimization of mineral is achieved by Dynamic Programming using the method of stages.

4.2. DETECTION OF RANDOMLY ORIENTED ORE DEPOSITS BY AIRBORNE SURVEYS ON GRIDS OF VARIOUS DESIGNS

4.2.1. Introduction

Airborne geophysical surveys consist of pattern-grid searches for sub-adjacent deposits conducted from aircraft platforms equipped with a set of continuously recording sensors of various kinds. In this work, we consider parallel flight lines with equal spacing, s meters, and orthogonal square grids with spacings s meters x s meters. The problem of airborne detection probabilities involves the intersection of convex planar figures (targets) by either equally spaced straight lines or orthogonal grid lines (search patterns).

Agocs (1) originally formulated without proof the expressions for the airborne detection probabilities of targets of various shapes by parallel equally spaced flight-lines. McCammon (15), using conditional probabilities, supplied the proofs. McCammon also considered the use of orthogonal grids of lines for the search for elliptical targets.

Table 4.2. Airborne detection probability tables for porphyry-Cu-Mo
deposits of the North & South American Cordillera and
Australasia.

Probability of detection of Porphyry Cu-Mo deposits by airborne geophysical surveys			
Pool of regions: N. & S. American Cordillera, B.C. Cordillera, South Pacific Island Arc, Tasman Geosyncline		Grid design: Parallel flight-lines with spacing S meters	
Randomly orientated elliptical targets with expected major axis = L meters in the confidence interval: l.c.l. < geometric mean < u.c.l. R is defined as the ratio B/L with geometric mean = 0.37			
Grid spacing S meters	l.c.l. = 775 m	geom. mean = 950 m	u.c.l. = 1180 m
200	1.000	1.000	1.000
400	0.894	0.914	1.000
600	0.835	0.869	0.896
800	0.745	0.818	0.858
1000	0.596	0.733	0.816
1200	0.497	0.610	0.758
1400	0.426	0.523	0.650
1600	0.373	0.458	0.569
2000	0.298	0.366	0.455
2500	0.238	0.293	0.364
3000	0.199	0.244	0.303

4.2.2. Geometric probability models for the vertical detection of randomly
oriented targets of various shapes by airborne surveys on parallel
and orthogonal grids

The parameters of the study are as follows:

(1) target geometry parameters are length, L_t, and shape ratio,
$R_t = B_t/L_t$.

(2) target-detector configuration: the vertical distance between the
detector and target is made up of two components:

(a) flight elevation above the ground and:

(b) depth of the center of the target below the ground.

Since the targeted deposits are assumed to be randomly oriented, the
target-grid configuration is expressed only by the dimensional parameter,
which is the ratio, U_t, of the longest dimension of the expected target, L_t,
over the grid spacing, S:

$$U_t = L_t/S.$$

Table 4.2. (Continued) Airborne detection probability tables for porphyry-Cu-Mo deposits of the North & South American Cordillera and Australasia.

Probability of confirmed detection of Porphyry Cu-Mo deposits by airborne geophysical surveys			
Pool of regions: N. & S. American Cordillera, B.C. Cordillera South Pacific Island Arc, Tasman Geosyncline		Grid design: Parallel flight-lines with spacing S meters	
Grid spacing S meters	l.c.l. = 775 m	geom. mean = 950 m	u.c.l. = 1180 m
200	1.000	1.000	1.000
400	0.621	0.778	0.992
600	0.422	0.507	0.633
800	0.000	0.398	0.474
1000	0.000	0.000	0.397
1200	0.000	0.000	0.000

Determination of grid-size for specified probability levels of detection of Porphyry Cu-Mo deposits by airborne geophysical surveys		
Specified detection probability level	Required grid spacing S in meters	Corresponding cost in $ per km square
0.10	6200	$43.62
0.20	3400	$47.75
0.30	2300	$52.12
0.40	1800	$55.87
0.50	1500	$59.32
0.60	1200	$64.50
0.70	1050	$68.20
0.80	800	$77.45
0.90	400	$116.28
0.95	200	$193.95

The expressions describing the probabilistic models used for the detection of randomly oriented targets by continuous-sampling airborne surveys are listed in Table 4.1. The table covers the cases of detection and confirmed detection for targets with linear shapes, $R_t < 0.05$, and circular shapes, $R_t = 1$. For elliptical targets, with $0.05 < R_t < 1$, the probability of detection is the weighted sum of the probabilities for linear shape, with weighting $1 - R_t$, and circular shape, with weighting R_t. While the weighting, R_t, is naturally in the 0 to 1 interval, the ratio, U_t, is taken in the interval $0 < U_t \leq 2$ for all practical field situations.

4.2.3. Application of probabilistic models to the airborne search for specific types of ore deposits in various regions of the World

The models listed in Table 4.1 are used to calculate the probabilities of detection and confirmed detection of porphyry-Cu-Mo deposits and their

Table 4.3. Airborne detection probability table for porphyry-Cu-Mo pyritic halos of the B.C. Cordillera of North America.

Probability of detection of Porphyry Cu-Mo Pyritic halos by airborne geophysical surveys			
Region: B.C. Cordillera Grid design: Square grid with spacings S x S meters			
Randomly orientateded elliptical targets with expected major axis = L meters in the confidence interval: l.c.l. < geometric mean < u.c.l. R is defined as the ratio B/L with geometric mean = 0.62			
Grid spacing S meters	l.c.l. = 1050 m	geom. mean = 1500 m	u.c.l. = 2100 m
1000	1.000	1.000	1.000
1500	0.749	1.000	1.000
2000	0.596	0.789	1.000
2500	0.493	0.665	0.855
3000	0.420	0.572	0.749
3500	0.366	0.502	0.665
4000	0.324	0.447	0.596
4500	0.290	0.402	0.540
5000	0.263	0.366	0.493

Table 4.4. Airborne detection probability tables for volcanogenic sulfide deposits of the East Mediterranean region and the Kuroko Belt of Japan (parallel grid).

Probability of detection of Volcanogenic sulfide deposits by airborne geophysical surveys			
Pool of regions: Kuroko Belt of Japan, Eastern Mediterranean region Grid design: Parallel flight-lines with spacing S meters			
Randomly orientated elliptical targets with expected major axis = L meters in the confidence interval l.c.l. < geometric mean < u.c.l. R is defined as the ratio B/L with geometric mean = 0.40			
Grid spacing S meters	l.c.l. = 305 m	geom. mean = 372 m	u.c.l. = 450 m
100	1.000	1.000	1.000
150	0.903	0.921	1.000
200	0.868	0.893	0.913
300	0.785	0.833	0.866
400	0.596	0.725	0.813
500	0.477	0.580	0.705
600	0.397	0.483	0.588
700	0.341	0.414	0.504
800	0.298	0.362	0.441
1000	0.238	0.290	0.353

Table 4.4. (Continued) Airborne detection probability tables for volcanogenic sulfide deposits of the East Mediterranean region and the Kuroko Belt of Japan (parallel grid).

Probability of detection of Volcanogenic sulfide deposits by airborne geophysical surveys

Pool of regions: Kuroko Belt of Japan, Eastern Mediterranean region

Grid design: Parallel flight-lines with spacing S meters

Grid spacing S meters	l.c.l. = 305 m	geom. mean = 373 m	u.c.l. = 450 m
100	1.000	1.000	1.000
150	0.676	0.827	1.000
200	0.515	0.617	0.751
300	0.401	0.439	0.509
400	0.000	0.000	0.416

Determination of grid size for specified levels of detection probabilities

Specified detection probability level	Required grid spacing S in meters	Corresponding cost in $ per km square
0.10	2450	$51.29
0.20	1325	$62.06
0.30	925	$72.20
0.40	700	$82.99
0.50	575	$92.64
0.60	475	$104.02
0.70	425	$111.71
0.80	325	$134.21
0.90	175	$216.15
0.95	75	$452.86

associated pyritic halos, (Tables 4.2 & 4.3), and of volcanogenic sulfide deposits, (Tables 4.4 & 4.5), by means of airborne surveys run on parallel or square grids for increasing grid spacings within a realistic range. The calculations are based upon the geometric mean length, L_t, and limits of the associated 95% confidence interval, and the geometric mean shape ratio, R_t, which were all re-calculated for each pool of regions resulting from the sequential commonality testing study of Chapter 3.

In Table 4.2, the airborne detection probabilities are given for porphyry-Cu-Mo deposits using parallel flight lines with spacings S meters apart, where the domain for S is

200 m < S < 3000 m.

The parameters for the 95% confidence interval for the expected target lengths are the parameters re-calculated for the pool of regions including the North & South American Cordillera, B.C. Cordillera, South Pacific Island Arc and the Tasman Geosyncline. There is a rapid decline in probability as

Table 4.5. Airborne detection probability table for volcanogenic
sulfide deposits of the East Mediterranean region and
the Kuroko Belt of Japan (square grid).

Probability of detection of Volcanogenic sulfide deposits by airborne geophysical surveys			
Pool of regions: Kuroko Belt of Japan, Eastern Mediterranean region		Grid design: Square grid with spacings S x S meters	
Randomly orientated elliptical targets with expected major axis = L meters in the confidence interval l.c.l. < geometric mean < u.c.l. R is defined as the ratio B/L with geometric mean = 0.40			
Grid spacing S meters	l.c.l. = 305 m	geom. mean = 372 m	u.c.l. = 450 m
250	1.000	1.000	1.000
300	0.948	1.000	1.000
400	0.809	0.904	1.000
500	0.694	0.792	0.887
600	0.605	0.698	0.795
800	0.478	0.559	0.649
1000	0.394	0.465	0.544
1200	0.335	0.397	0.467
1500	0.273	0.325	0.385
2000	0.209	0.249	0.297

the spacing, S, is increased. The confirmed detection results are given in
the second part of Table 4.2, where it will be noted how dramatically the
confirmed detection probabilities fall as S increases. The third portion of
Table 4.2 is a very useful one for exploration planning purposes, although
it required a great deal of computer time to construct. The cost
calculations are based on an average cost of US$50 per km line. The table
enables explorationists to budget the cost of coverage of an area of given
extent and to design the grid in order to ensure a pre-specified detection
probability level.

In Table 4.3, the airborne detection probabilities are given for the
pyritic halos associated with porphyry-Cu-Mo deposits using parallel flight
lines with spacings S meters apart, where the domain for S is

1000 m < S < 5000 m.

In Table 4.4, the airborne detection probabilities are given for
Volcanogenic sulfide deposits using parallel flight lines with spacings S
meters apart, where the domain for S is

100 m < S < 1000 m.

The deposits belong to a pool of two regions, the Kuroko Belt of Japan
and the Eastern Mediterranean, which was indicated by the statistical

testing theory. The parameters and the 95% confidence interval limits were re-calculated for the pool of regions. The confirmed detection results are given in the second part of Table 4.4 and the third section of the table displays the "inverse" results, which are useful for finding the grid spacing and cost of detection given a specific probability goal. For example, to be 95% certain of intersecting a target within the pool of regions, the grid spacing required is 75 m, and the cost is $452.86 per km square.

In Table 4.5, the airborne detection probabilities are given for volcanogenic sulfide deposits using square grid design with spacings S meters x S meters apart, where the domain for S is:

250 m < S < 2000 m.

A comparison of the first section of Table 4.4 with Table 4.5 underscores the considerable improvement in detection capability gained by using a square grid design instead of a parallel one which has to be measured against a substantial increase in cost. For instance, certainty of detection with a probability = 1.0 requires a 100 m spacing for parallel design, while only a 300 m spacing is necessary when using the square design.

4.3. DETECTION OF RANDOMLY ORIENTED ORE DEPOSITS BY GROUND SURVEYS OR VERTICAL DRILLING PROGRAMS ON SQUARE GRIDS

4.3.1. Introduction

Ground grids for discrete sampling are point nets at the nodes of regular adjoining polygons. The sampling is done at the nodes of the net. A great deal of research work has been done on the shape of the polygon for the drilling grid to determine the most cost-effective design (4), (6), (7), (10), (22), (23), (24), (25). However, only orthogonal grids, and particularly square grids have been considered in actual field work. In this study, all results will be for square grids with spacing S x S meters.

The determination of the proper drill-hole spacing has attracted a great deal of attention because of the cost involved in systematic drilling, and spacing requirements are dependent upon the nature of the program and the target. The topic of adjusting the spacing to yield maximum efficiency is formulated in section 4.5. The deterministic model describing quantitatively the vertical detection situation was previously considered in section 2.1.1. of Chapter 2 and illustrated in Figure 2.1.

The ratio, $A_t = L_t/(2S)$, of the target semi-length over the grid spacing, is a very important parameter in the calculation of detection probabilities based on geometric probability theory.

4.3.2. Methodology: Introduction of Simpson's rule

For discrete sampling, the formulation of the detection probability function is far more complex than in the case of continuous line sampling. Singer & Wickham (22) & Singer (23) followed up the early work of Savinskii (18) on the problem of detection probabilities of any type of elliptical target by point-net sampling. However, in order to calculate the detection probabilities for specific types of targets, the authors formulated and wrote the OPTGRID program package given in Appendix 5 which covers the cases of oriented as well as randomly oriented targets. The simplified flow diagram for OPTGRID is presented in Figure 4.1.

For the detection of randomly oriented elliptical targets, there is no preferred or expected strike orientation, so the optimum grid orientation is not considered. Because of the symmetry introduced by the use of a square grid design, the domain for the detection probability function is 0° to 45°, or 0 to π/4 radians. For randomly oriented targets, the probability function for a specific target length and any grid spacing, S, P(S), is integrated over the whole domain and P(S) is

$$P(S) = 4/\pi * \int_0^{\pi/4} P(x)\ dx.$$

In Figure 4.1, the modification consists of replacing the 10 degree

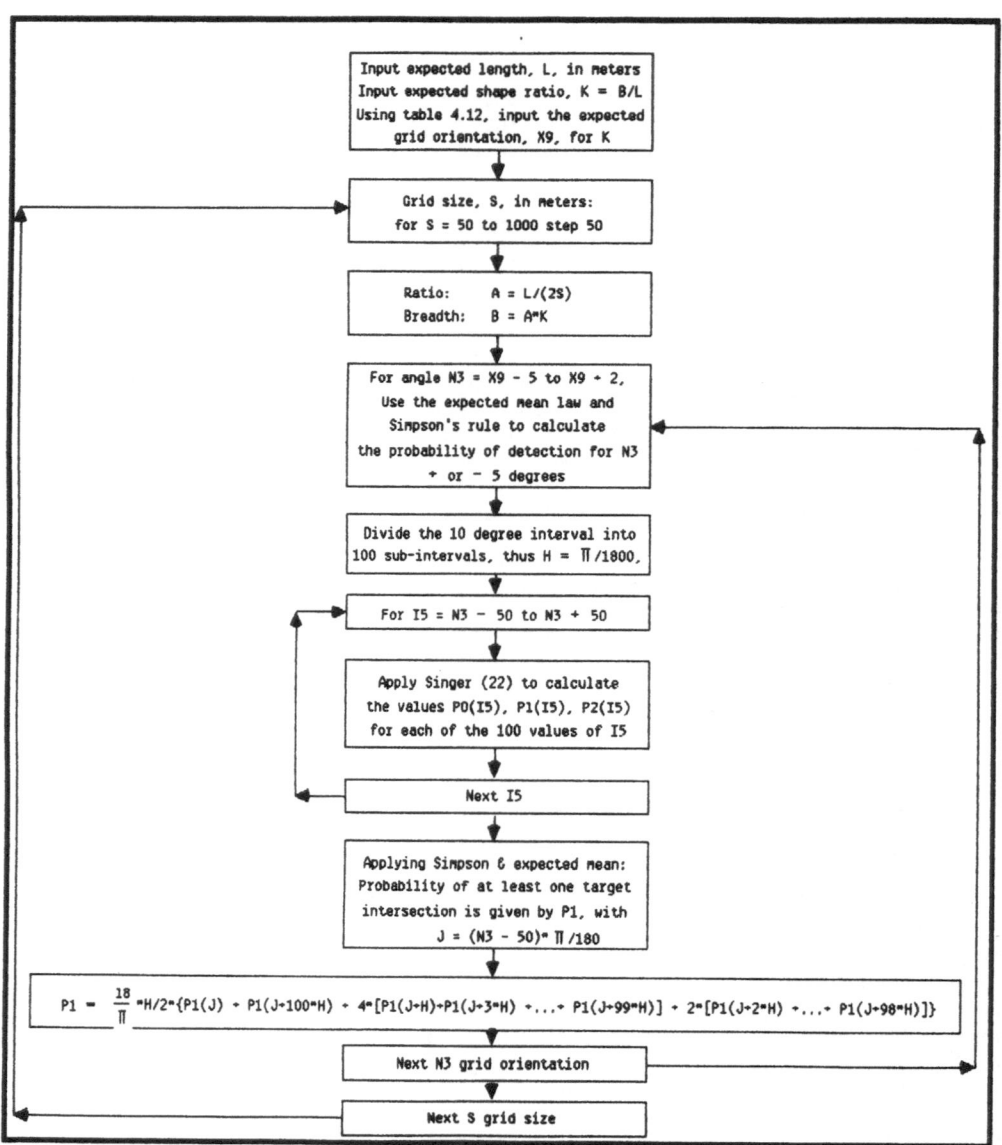

Figure 4.1. Simplified flow-diagram of the OPTGRID program for calculation of the probability of detection of elliptical targets by discrete sampling.

interval by a 45° interval, thus:
$$H = \pi/400.$$
There is no modification required however in Appendix 5, since program 5.7 is designed especially for randomly oriented targets.

Since the probability of detection, P(S), is based upon conditional probabilities, with too many diversities and complexities to do the mathematical integration, we obviously have to be satisfied with the approximate integration using Simpson's Rule.
The actual formula used for P(S) is given by

$$P(S) = 4/\pi * \pi/400 * \{P(a) + P(\pi/4) + 4 * P(a+H) + 2 * P(a+2H)+$$
$$4 * P(a+3H) + ...\}, \text{ where the summation is over the 101 terms.}$$

4.3.3. Application to specific types of ore deposits

4.3.3.1. Introduction

The calculation of ground detection probabilities for randomly oriented targets, (see Appendix 4, section 4.7.), requires the following parameters:

(1) The shape ratio: $R_t = B_t/L_t$,

(2) The lower confidence interval limit for the expected length of the target (or pool): l.c.l. = lower confidence limit in the same units as S, the spacing,

(3) The geometric mean of the expected target or pool of targets,

(4) The upper confidence limit, u.c.l. of the expected target or pool of regions.

(5) The domain for S, using the same units as for the above 3 parameters. Statements 31 to 36 of program 4.7 are the limits for the spacing. In the listing, the statements imply:

31: Lowest limit =   50 meters, with statement 35 implying an increment of 25 meters, until the spacing reaches:
32: Upper limit =   100 meters, then with statement 36 implying an increment of 50 meters, until the grid spacing reaches:
34: Highest limit = 1000 meters.

The program 4.7, Appendix 4, is designed for some interactive input: The first message reads:

"INPUT EXPECTED LENGTH OF TARGET IN METERS"

The reply is the L.c.l. on the first run: the program then prints out the probabilities for the specified range given above, using the shape ratio of the expected target, in reply to the second message:

"EXPECTED SHAPE RATIO: $R_t = B_t / L_t$"

On the second and third runs respectively, the mean and U.c.l are input.

4.3.3.2. Application to the ground detection of porphyry-Cu-Mo deposits

Table 4.6 gives the ground detection probabilities for porphyry Cu-Mo deposits of the pool of regions including North and South American Cordillera, B. C. Cordillera, South Pacific Island Arc and the Tasman

Table 4.6. Probability table for the detection of porpyry-Cu-Mo
deposits of the N. and S. American Cordillera and
Australasia by ground geophysical surveys or vertical
drilling programs to a depth of 100 meters.

Probability of detection of Porphyry Cu-Mo deposits

by ground geophysical surveys or drilling programs

Pool of regions: N. & S. American Cordillera,   Grid design: Square grid
B.C. Cordillera, South Pacific Island Arc,   with spacing S x S meters
Tasman Geosyncline

Randomly orientated elliptical targets with expected major axis = L meters in the confidence interval
l.c.l. < geometric mean < u.c.l.
R is defined as the ratio B/L with geometric mean = 0.37

Grid spacing S meters	l.c.l. = 775 m	geom. mean = 950 m	u.c.l. = 1180 m
250	1.000	1.000	1.000
300	0.990	1.000	1.000
400	0.846	0.967	1.000
500	0.636	0.831	0.965
600	0.469	0.658	0.861
700	0.353	0.513	0.720
800	0.272	0.404	0.591
900	0.215	0.323	0.484
1000	0.174	0.262	0.400
1200	0.121	0.182	0.281
1400	0.089	0.134	0.207
1500	0.077	0.117	0.180
2000	0.043	0.066	0.101

Geosyncline of Australasia. The parameters of these deposits were re-
calculated on the pooled lognormal distribution, with the expected length
parameters in meters calculated to be

l. c. l.   = 775 m,
geometric mean = 950 m,
u. c. l.   = 1180 m.

For the spacing domain, S, in meters, statements 31 to 36 were modified to:

31: lowest limit = 250 meters, with statement 35 implying an increment of
     50 meters, until the spacing reaches:

32: upper limit = 300 meters,
     then with statement 36 implying an increment of 100 meters, until the
     grid spacing reaches:

34: highest limit = 2000 meters.

     The results demonstrate how quickly the probabilities decrease from a
probability ≈ 1 at a grid spacing, S = 250 meters, down to a probability ≈
0, when the spacing is 2000 meters. Table 4.7 gives the results for the
probability of confirmed detection, i.e. the probability of two or more
target hits, given the spacing, S, for the pool of porphyry Cu-Mo deposits.
It can be seen from a comparison of the first section of Table 4.7 with

Table 4.7. Probability tables for the detection of porpyry-Cu-Mo deposits of the N. and S. American Cordillera and Australasia by ground geophysical surveys or vertical drilling programs to a depth of 100 meters.

Probability of confirmed detection of Porphyry Cu-Mo deposits by ground geophysical surveys or drilling programs			
Pool of regions: N. & S. American Cordillera, B.C. Cordillera, South Pacific Island Arc, Tasman Geosyncline		Grid design: Square grid with spacing S x S meters	
Grid spacing S meters	l.c.l. = 775 m	geom. mean = 950 m	u.c.l. = 1180 m
250	1.000	1.000	1.000
300	0.937	1.000	1.000
400	0.240	0.860	1.000
500	0.059	0.218	0.853
600	0.014	0.070	0.265
800	0.000	0.006	0.027

Determination of grid-size for specified probability levels of detection of Porphyry Cu-Mo deposits by vertical drilling programs			
Specified detection probability level	Required grid spacing S in meters	Corresponding cost in $1000's per km square percussion	diamond
0.10	1600	2.26	6.77
0.20	1200	2.87	8.61
0.30	1000	3.41	10.24
0.40	800	4.31	12.94
0.50	725	4.82	14.46
0.60	650	5.48	16.45
0.70	575	6.38	19.15
0.80	525	7.17	21.52
0.90	475	8.20	24.59
0.95	425	9.55	28.65

Table 4.6 that the confirmed detection probabilities cut off much more quickly than those of single detection, as could be expected, for when the spacing exceeds the length of the deposit, the probability of two intersections is zero.

Finally, the "inverse" problem needs to be solved: given a specified detection probability, what spacing is required for the detection of targets of a specific expected length and shape ratio, R_t. The resulting program is given in Appendix 4, program 4.8, and the output is listed in the second

Table 4.8. Probability table for the detection of volcanogenic
sulfide deposits of the Eastern Mediterranean region
and the Kuroko Belt of Japan by ground geophysical
surveys or vertical drilling programs to a depth of
one hundred meters.

Probability of detection of Volcanogenic sulfide deposits by ground discrete point geophysical surveys or drilling programs			
Pool of regions: Kuroko Belt of Japan, Eastern Mediterranean region		Grid design: Square grid with spacings S x S meters	
Randomly orientated elliptical targets with expected major axis = L meters in the confidence interval l.c.l. < geometric mean < u.c.l. R is defined as the ratio B/L with geometric mean = 0.40			
Grid spacing S meters	l.c.l. = 305 m	geom. mean = 370 m	u.c.l. = 450 m
100	1.000	1.000	1.000
150	0.923	0.995	1.000
200	0.668	0.855	0.971
250	0.459	0.648	0.830
300	0.326	0.470	0.653
400	0.183	0.271	0.397
500	0.117	0.173	0.256
600	0.081	0.120	0.178
800	0.046	0.068	0.100
1000	0.029	0.043	0.064
1200	0.020	0.030	0.044
1500	0.013	0.019	0.028

section of Table 4.7. For example, if a detection probability of 0.90 is
specified, then the required grid spacing is 475 meters, and the cost of the
drilling program in $1000's per km square is: 9.55 for percussion drilling,
and 28.65 for diamond drilling.

Table 4.8 gives the ground detection probabilities for Volcanogenic
sulfide deposits of the pool of regions of Kuroko Belt of Japan and the
Eastern Mediterranean. The parameters of these deposits were calculated on
the pooled lognormal distribution, with the expected length parameters in
meters calculated to be: geometric mean = 370, and its 95% confidence
limits are: l. c. l. = 305, u. c. l. = 450, and the geometric mean shape
ratio parameter, R_t = 0.40.

For the spacing domain, S, in meters, statements 31 to 36 were
modified as follows:

31: Lowest limit = 100 meters, with statement 35 implying an increment of
50 meters, until the spacing reaches an upper limit = 200 meters (32),
then with statement 36 implying an increment of 100 meters, until the
grid spacing reaches the highest limit = 1500 meters (34).

Table 4.9. Probability tables for the detection of volcanogenic
sulfide deposits of the East Mediterranean region and
the Kuroko Belt of Japan by ground geophysical surveys
or vertical drilling programs to a depth of 100 meters.

Probability of confirmed detection of Volcanogenic sulfide deposits by ground discrete point geophysical surveys or drilling programs			
Pool of regions: Kuroko Belt of Japan, Eastern Mediterranean region		Grid design: Square grid with spacing S x S meters	
Grid spacing S meters	l.c.l. = 305 m	geom. mean = 370 m	u.c.l. = 450 m
100	1.000	1.000	1.000
150	0.385	0.990	1.000
200	0.066	0.232	0.819
250	0.010	0.054	0.198
300	0.000	0.012	0.059
400	0.000	0.000	0.003

Determination of grid-size for specified probability levels of detection of Volcanogenic sulfide deposits by vertical drilling programs			
Specified detection probability level	Required grid spacing S in meters	Corresponding cost in $1000's per km square	
		percussion	diamond
0.10	650	5.48	16.45
0.20	500	7.65	22.95
0.30	400	10.40	31.21
0.40	350	12.63	37.89
0.50	290	16.79	50.36
0.60	260	19.92	59.75
0.70	240	22.63	67.89
0.80	220	26.06	78.19
0.90	190	33.23	99.70

A comparison of table 4.8 with the first section of Table 4.7 shows the
very strong influence of the mean target length on the probabilities of
detection when the shape ratios are nearly the same. If we consider the 300
meter spacing for instance, the probability of detection of the volcanogenic
sulfide deposits whose mean length is 370m is only 0.326 as compared to 1.00
for the porphyry-Cu-Mo deposits whose mean length is 950m.

The first section of Table 4.9 lists the probabilities of confirmed
detection (the probability of two or more target hits) given the spacing, S,
for the pool of Volcanogenic sulfide deposits. Finally, we may wish to
determine the spacing and associated coverage cost per unit area which is
required in order to obtain a pre-specified level of probability of
detection of the volcanogenic sulfide deposits within the two regions of the
pool. The resulting program is given in Appendix 4, program 4.8, and the
output is shown in Table 4.9. For example, if a detection probability of
0.80 is required then the required grid spacing is 220 meters, and the cost

Table 4.10. Application of Dynamic Programming to the maximization of the probability of detection of oriented ore deposits.

Optimum grid orientation to expected strike line for specific shape ratios, R.

Survey or program: Ground geophysical at discrete points or drilling programs

Grid design: Square grid with spacings S x S meters

Orientated elliptical targets with expected major axis = L meters in the confidence interval
l.c.l. < geometric mean < u.c.l.
R is defined as the ratio R = B/L

Ratio R = B/L	Optimum Grid Orientation (degrees)
0.05	14
0.10	19
0.20	28
0.30	29
0.40	36
>= 0.50	45

**Optimum grid orientation to expected strike line:
Dynamic program method for example with R = 0.3**

STAGE 1:

Grid Orientation (degrees):	10	15	20	25	30 *	35	40	45
Ratio A = L/2S	Probability of detection (at least one intersection)							
0.70	0.420	0.447	0.460	0.462	0.462	0.462	0.462	0.462
0.80	0.499	0.544	0.583	0.601	0.603	0.600	0.588	0.579
0.90	0.575	0.633	0.694	0.741	0.756	0.732	0.697	0.681
1.00	0.649	0.720	0.796	0.864	0.886	0.849	0.799	0.779
1.10	0.723	0.804	0.894	0.958	0.969	0.940	0.896	0.873
1.20	0.795	0.887	0.968	0.999	0.999	0.988	0.969	0.960
1.30	0.867	0.953	0.997	1.000	1.000	1.000	0.999	0.999

* Optimal grid orientation

of the drilling program in $1000's per km square is: 28.06 for percussion drilling, (unit cost: $2190 per 100 meter hole), and 99.70 for diamond drilling, (unit cost: $6560 per 100 meter hole).

4.4. DETECTION OF ORIENTED ORE DEPOSITS BY AIRBORNE AND GROUND SURVEYS & DRILLING PROGRAMS. APPLICATION OF DYNAMIC PROGRAMMING

4.4.1. Introducing the Dynamic Programming approach

Mathematical Programming models are designed with industrial applications in mind. They are used in situations where an objective function, which quantifies a goal, is defined subject to constraints of a budgetary, or geometric, or geographic nature, etc. These models are useful in optimization problems which are beyond the scope of Calculus. The best-known are Dynamic Programming, Linear Programming, and Network models.

Table 4.10. (Continued) Application of Dynamic Programming to the maximization of the probability of detection of oriented ore deposits.

Optimum grid orientation to expected strike line: Dynamic program method for example with R = 0.3 (Continued) STAGE 2:								
Grid Orientation (degrees):	28	29 *	30	31	32	33	34	35
Ratio A = L/2S	Probability of detection (at least one intersection)							
0.70	0.462	0.462	0.462	0.462	0.462	0.462	0.462	0.462
0.80	0.603	0.603	0.603	0.603	0.603	0.602	0.601	0.600
0.90	0.755	0.756	0.756	0.754	0.750	0.745	0.739	0.732
1.00	0.884	0.887	0.886	0.882	0.877	0.869	0.860	0.849
1.10	0.971	0.971	0.969	0.965	0.960	0.955	0.949	0.940
1.20	1.000	1.000	0.999	0.998	0.996	0.994	0.991	0.988
* Optimal grid orientation								

Optimum grid orientation to expected strike line: Dynamic program method for a specific ore deposit								
Region & Type of deposit: W. Australian Shield, Vein Gold Expected Length: 595 meters Expected shape ratio, R = B/L = 0.09								
Grid Orientation (degrees):	16	17	18 *	19	20	21	22	23
Spacing S meters	Probability of detection (at least one intersection)							
150	0.928	0.944	0.950	0.945	0.932	0.913	0.894	0.876
200	0.615	0.621	0.623	0.621	0.615	0.607	0.599	0.590
250	0.400	0.400	0.400	0.400	0.400	0.400	0.399	0.399
300	0.278	0.278	0.278	0.278	0.278	0.278	0.278	0.278
500	0.100	0.100	0.100	0.100	0.100	0.100	0.100	0.100
* Optimal grid orientation								

Linear Program and Network models will be fully dealt with in Chapter 5, and illustrated by several applications to mineral exploration.

Dynamic Program models handle a wide range of Decision theoretic problems, many of which can be represented by a Bayesian-type decision tree, that is to say the decisions at each succeeding stage are conditional on the decisions taken at the preceding stage. The algorithm is to lop off any branch of the tree whenever its outcome is dominated by another outcome which uses the same amount of resources at any stage. The process ends at the optimal outcome of the final stage, and the path which led to this optimal outcome is traced back to the beginning root of the tree and thus the optimal decision at each stage is determined. This is the essence of Dynamic Programming, which together with the new generation of computers can take over most of the work of optimization since it can operate on every problem of optimization that the Calculus could handle, and also solve those that Calculus could not. The Dynamic Programming approach is particularly well-suited for application to most mineral exploration situations because of their multi-stage sequential structure.

Table 4.11. Probability tables for the detection of volcanogenic
sulfide deposits of the North American Shield, the
Appalachian Belt and the Tasman Geosyncline by
ground geophysical surveys and drilling programs
to a vertical depth of 100 m.

Probability of detection of Volcanogenic sulfide deposits by ground geophysical surveys or drilling programs			
Pool of regions: N. American Shield, Appalachian Belt, Tasman Geosyncline		Grid design: Square grid with spacing S x S meters	
Orientated elliptical targets with expected major axis = L meters in the confidence interval l.c.l. < geometric mean < u.c.l. R is defined as the ratio B/L with geometric mean = 0.19 Optimun grid orientation with expected strike line = 27°			
Grid spacing S meters	l.c.l. = 255 m	geom. mean = 300 m	u.c.l. = 370 m
50	1.000	1.000	1.000
100	0.913	0.999	1.000
150	0.431	0.597	0.877
200	0.243	0.336	0.511
250	0.155	0.215	0.327
300	0.108	0.149	0.227
400	0.061	0.084	0.128
500	0.039	0.054	0.082
600	0.027	0.037	0.057
800	0.015	0.021	0.032
1000	0.010	0.013	0.020
Probability of confirmed detection			
25	1.000	1.000	1.000
50	0.979	1.000	1.000
100	0.388	0.518	0.695
150	0.098	0.195	0.361
200	0.014	0.050	0.143
250	0.000	0.007	0.046

4.4.2. Application of Dynamic Programming to the maximization of the detection probability of oriented ore deposits

The problem posed was: for oriented targets with all their expected
parameters calculated from control samples, orient a specific grid with
respect to the expected direction of strike of targets to maximize the
probability of detection. In a word, the answer is OPTGRID (31), which is
illustrated by the flow-diagram given in Figure 4.1. When dealing with the
determination of the optimal orientation of parallel grids for airborne
geophysical surveys, the application of the dynamic programming method fully
confirmed the validity of the time-honoured practice of laying the control-
grid at right-angles to the expected direction of strike of the targets, or
at least, to the general direction of the geological grain of the region, as
a second best alternative.

Table 4.11. (Continued) Probability tables for the detection of
volcanogenic sulfide deposits of the North American
Shield, the Appalachian Belt and the Tasman Geosyncline
by ground geophysical surveys and drilling programs
to a vertical depth of 100 m.

	Probability of detection of Volcanogenic sulfide deposits 55° angled drilling programs		

Pool of regions: N. American Shield, Appalachian Belt, Grid design: Square grid

Tasman Geosyncline with spacing S x S meters

Orientated elliptical targets with expected major axis = L meters in the confidence interval

l.c.l. < geometric mean < u.c.l.

R is defined as the ratio B/L with geometric mean = 0.32

Optimum grid orientation with expected strike line = 30°

Grid spacing S meters	l.c.l. = 255 m	geom. mean = 300 m	u.c.l. = 370 m
100	1.000	1.000	1.000
150	0.723	0.914	1.000
200	0.409	0.565	0.832
250	0.261	0.362	0.551
300	0.155	0.251	0.382
400	0.102	0.141	0.215
500	0.065	0.090	0.114
600	0.045	0.063	0.096
800	0.026	0.035	0.054
1000	0.016	0.016	0.034
2000	0.004	0.006	0.009
Probability of confirmed detection			
75	1.000	1.000	1.000
100	0.658	1.000	1.000
150	0.197	0.370	0.621
200	0.037	0.109	0.278
250	0.000	0.021	0.102
300	0.000	0.000	0.028

Let us now deal with the optimal orientation of control grids in the
case of ground detection by geophysical surveys or drilling programs by
means of examples.The program 4.6 of Appendix 4 was modified to yield the
table of results given in Table 4.10 as follows:

(1) The optimal orientation of the grid, and

(2) The probabilities of detection associated with the optimal orientation
of the grid for oriented targets of a specific type, such as the
Western Australian Shield vein gold deposits.

Table 4.11. (Continued) Probability tables for the detection of volcanogenic sulfide deposits of the North American Shield, the Appalachian Belt and the Tasman Geosyncline by ground geophysical surveys and drilling programs to a vertical depth of 100 m.

Determination of grid-size for specified probability levels of detection of Volcanogenic sulfide deposits by vertical drilling programs			
Pool of regions: N. American Shield, Appalachian Belt, Tasman Geosyncline		Grid design: Square grid with spacing S x S meters	
Specified detection probability level	Required grid spacing S in meters	Corresponding cost in $1000's per km square percussion	diamond
0.10	350	12.63	37.89
0.20	275	18.23	54.69
0.30	225	25.12	75.37
0.40	200	30.50	91.51
0.50	175	38.18	114.55
0.60	150	49.77	149.30
0.70	145	42.79	158.38
0.80	135	59.83	179.50
0.90	125	68.56	205.67

Determination of grid-size for specified probability levels of detection of Volcanogenic sulfide deposits by 55° angled drilling programs			
Specified detection probability level	Required grid spacing S in meters	Corresponding cost in $1000's per km square percussion	diamond
0.10	500	18.42	55.26
0.20	350	32.40	97.20
0.30	275	48.51	145.53
0.40	250	57.14	171.41
0.50	225	68.64	205.92
0.60	200	84.50	253.50
0.70	190	92.59	277.76
0.80	175	107.31	321.93
0.90	170	113.07	339.22

In the example given in Table 4.10, the optimal grid orientation is 18°, with the best probabilities of detection given in that column. Note that even for a grid orientation off by as little as 5° the probabilities are significantly less, and of course if 45° had been chosen, then the detection probabilities would be **halved!**

$$R_t = B_t/L_t = 0.3,$$

and the expected target semi-length/grid spacing,

$$A_t = L_t/(2S),$$ varying regularly between 0.5 and 2.0, but here from 0.7 to 1.30.

Note that in stage 1, the search domain was 0° to 45°, with increments

Table 4.12. Probability tables for the detection of volcanogenic
sulfide deposits of the W. Australian Shield,
Scandinavian Shield, Iberian Peninsula Pyrite Belt, North
American Cordillera, and N. American Cordillera (exhal.)
by ground geophysical surveys and drilling programs
to a vertical depth of 100 m.

Probability of detection of Volcanogenic sulfide deposits			
by ground geophysical surveys or drilling programs			

Pool of regions: W. Australian Shield, Scandinavian Shield,
Iberian Peninsula Pyrite Belt,
N. American Cordillera, N. American Cord. (Exhal.)

Grid design: Square grid
with spacing S x S meters

Orientated elliptical targets with expected major axis = L meters in the confidence interval

l.c.l. < geometric mean < u.c.l.

R is defined as the ratio B/L with geometric mean = 0.19

Optimum grid orientation with expected strike line = 27°

Grid spacing S meters	l.c.l. = 420 m	geom. mean = 560 m	u.c.l. = 650 m
100	1.000	1.000	1.000
150	0.984	1.000	1.000
200	0.658	0.984	1.000
250	0.421	0.748	0.932
300	0.292	0.520	0.701
400	0.165	0.292	0.394
500	0.105	0.187	0.252
600	0.073	0.130	0.175
800	0.041	0.073	0.099
1000	0.026	0.047	0.063
2000	0.007	0.012	0.016
Probability of confirmed detection			
75	1.000	1.000	1.000
100	0.823	1.000	1.000
150	0.463	0.703	0.856
200	0.232	0.463	0.583
300	0.032	0.148	0.145
400	0.000	0.032	0.039

of 5°, leading to an optimum of 30°. All the other paths are dominated since
they have lower probabilities, so now the search centered upon 30°. In stage
2, the domain had narrowed to 28° to 35°, in increments of 1°, and 29° is
selected since it dominates all the other columns. If need be, at the third
stage the interval could be narrowed down to 28.1° to 29.9° in steps of
0.2°, if that degree of accuracy was thought to be desirable.

The results given in Table 4.10 may be used as a starting point for
finding the optimum grid orientation angle for specific dipping ore deposit
types, as the length of the deposit is also a factor influencing the
outcome. Program 4.6, OPTGRID, in Appendix 4, is designed to yield the

Table 4.12. (Continued) Probability tables for the detection of volcanogenic sulfide deposits of the W. Australian Shield, Scandinavian Shield, Iberian Peninsula Pyrite Belt, North American Cordillera, and N. American Cordillera (exhal.) by ground geophysical surveys and drilling programs to a vertical depth of 100 m.

Probability of detection of Volcanogenic sulfide deposits			
55° angled drilling programs			

Pool of regions: W. Australian Shield, Scandinavian Shield, Iberian Peninsula Pyrite Belt, N. American Cordillera, N. American Cord. (Exhal.)

Grid design: Square grid with spacing S x S meters

Orientated elliptical targets with expected major axis = L meters in the confidence interval

l.c.l. < geometric mean < u.c.l.

R is defined as the ratio B/L with geometric mean = 0.32

Optimum grid orientation with expected strike line = 30°

Grid spacing S meters	l.c.l. = 420 m	geom. mean = 560 m	u.c.l. = 650 m
150	1.000	1.000	1.000
200	0.952	1.000	1.000
250	0.708	0.989	1.000
300	0.493	0.842	0.972
400	0.277	0.493	0.664
500	0.177	0.315	0.425
600	0.123	0.219	0.295
800	0.069	0.123	0.106
1000	0.044	0.079	0.074
2000	0.011	0.020	0.027
Probability of confirmed detection			
125	1.000	1.000	1.000
150	0.764	1.000	1.000
200	0.433	0.764	1.000
250	0.187	0.512	0.680
300	0.074	0.287	0.472
400	0.001	0.074	0.162

optimum, as well as the associated maximum detection probabilities for these deposit types. It will be observed that when using this program, it is designed for inter-active input, where the orientation angle minus five degrees read from Table 4.10 is the first input in answer to the first question mark. This is followed by the expected length of the deposit type, and then by the shape ratio. The program should prove useful in those cases where the orientation of the sought-after deposits is either known or can be estimated.

4.4.3. Examples for specific types of oriented dipping ore deposits

We have chosen the following examples to illustrate the application of Dynamic Programming to the maximization of detection probabilities by ground surveys and drilling programs in the most general case, when the targets are

Table 4.12. (Continued) Probability tables for the detection of volcanogenic sulfide deposits of the W. Australian Shield, Scandinavian Shield, Iberian Peninsula Pyrite Belt, North American Cordillera, and N. American Cordillera (exhal.) by ground geophysical surveys and drilling programs to a vertical depth of 100 m.

Determination of grid-size for specified probability levels of detection of Volcanogenic sulfide deposits by vertical drilling programs			
Pool of regions: W. Australian Shield, Scandinavian Shield, Iberian Peninsula Pyrite Belt, N. American Cordillera, N. American Cord. (Exhal.)		Grid design: Square grid with spacing S x S meters	
Specified detection probability level	Required grid spacing S in meters	Corresponding cost in $1000's per km square percussion	diamond
0.10	675	5.24	15.73
0.20	475	8.20	24.59
0.30	400	10.40	31.21
0.40	350	12.63	37.89
0.50	325	14.11	42.32
0.60	300	15.93	47.79
0.70	275	18.23	54.69
0.80	250	21.20	63.59
0.90	225	25.12	75.37

Determination of grid-size for specified probability levels of detection of Volcanogenic sulfide deposits by 55° angled drilling programs			
Specified detection probability level	Required grid spacing S in meters	Corresponding cost in $1000's per km square percussion	diamond
0.10	850	8.68	26.05
0.20	650	12.52	37.55
0.30	550	15.96	47.88
0.40	450	21.67	65.01
0.50	400	26.10	78.30
0.60	375	28.95	86.86
0.70	350	32.40	97.20
0.80	325	36.62	109.85
0.90	300	41.86	125.58

oriented ore deposits with a mean dip of (α) degrees.

The detection probability results for our first example are given in Table 4.11: volcanogenic sulfide deposits of the pool including the North American Continent and the Tasman Geosyncline of Australasia. These are followed by the next pool of volcanogenic sulfide deposit regions in Table 4.12, including the North American Cordillera (two types), Europe and the Western Australian Shield.

Each table includes a section dealing with detection and confirmed detection by ground geophysical surveys and vertical drilling programs to a depth of 100 meters, followed by a section covering 55-degree angled drilling to a vertical depth of 100 m. In the case of vertical detection,

Table 4.13. Probability tables for the detection of Ni-Cu
ultramafic deposits of the North American Shield
and the Western Australian Shield by ground
geophysical surveys and drilling programs to
a vertical depth of 100 m.

Probability of detection Ni-Cu Ultramafic deposits by ground geophysical surveys or drilling programs			
Pool of regions: North American Shield, Western Australian Shield		Grid design: Square grid with spacing S x S meters	
Orientated elliptical targets with expected major axis = L meters in the confidence interval l.c.l. < geometric mean < u.c.l. R is defined as the ratio B/L with geometric mean = 0.12 Optimum grid orientation with expected strike line = 20°			
Grid spacing S meters	l.c.l. = 320 m	geom. mean = 440 m	u.c.l. = 605 m
75	1.000	1.000	1.000
100	0.984	1.000	1.000
150	0.529	0.779	1.000
200	0.235	0.450	0.770
250	0.152	0.293	0.555
300	0.108	0.254	0.385
400	0.053	0.115	0.220
500	0.031	0.073	0.150
600	0.026	0.049	0.095
1000	0.011	0.020	0.042
1500	0.003	0.011	0.025
Probability of confirmed detection			
25	1.000	1.000	1.000
50	0.929	1.000	1.000
100	0.400	0.634	0.897
150	0.141	0.338	0.566
250	0.007	0.061	0.211
400	0.000	0.001	0.025

the deterministic model described in section 2.1.1 and Figure 2.1 of Chapter
2 is used in conjunction with the geometric probabilistic model, while the
model illustrated by Figure 2.2 is used for the calculation of the
probabilities of detection by angled drilling. A third section of each table
covers the "inverse" problem, i. e. the determination of grid size and
associated cost required to obtain a specified level of probability of
detection by vertical and 55-degree angled drilling programs to a vertical
depth of 100 meters.

The following results were given in the output of Table 4.11 for the
vertical detection of volcanogenic sulfide deposits of the first pool:

(1) The optimum grid direction with the expected strike = 27°.

Table 4.13. (Continued) Probability tables for the detection
of Ni-Cu ultramafic deposits of the North American
Shield and the Western Australian Shield by
ground geophysical surveys and drilling programs
to a vertical depth of 100 m.

Probability of detection Ni-Cu Ultramafic deposits by 55° angled drilling programs			

Pool of regions: North American Shield, Western Australian Shield   Grid design: Square grid
with spacing S x S meters

Orientated elliptical targets with expected major axis = L meters in the confidence interval

l.c.l. < geometric mean < u.c.l.

R is defined as the ratio B/L with geometric mean = 0.26

Optimum grid orientation with expected strike line = 27°

Grid spacing S meters	l.c.l. = 320 m	geom. mean = 440 m	u.c.l. = 605 m
100	1.000	1.000	1.000
150	0.968	1.000	1.000
200	0.708	0.983	1.000
250	0.488	0.830	0.990
300	0.340	0.628	0.900
500	0.120	0.225	0.425
600	0.076	0.169	0.310
800	0.045	0.090	0.170
1000	0.030	0.059	0.104
1500	0.012	0.023	0.040
Probability of confirmed detection			
75	1.000	1.000	1.000
100	0.787	1.000	1.000
150	0.374	0.696	1.000
200	0.121	0.407	0.728
250	0.029	0.185	0.506
300	0.002	0.077	0.313
400	0.000	0.004	0.091

(2) The detection probabilities with the grid orientation 27° ± 5°, for the
expected length L_t = 255 m, 300 m, 370 m, the L.c.l., geometric mean, and
H.c.l. respectively.

(3) For 55°-angled drilling, the target shape, R_t increases to 0.30, and the
optimum grid direction with the expected strike = 30°.

(4) The detection probabilities with the grid orientation 30° ± 5°, for the
expected length L_t = 255 m, 300 m, 370 m, the L.c.l., geometric mean,
and H.c.l. respectively.

Table 4.12 gives similar results for the optimal orientation of ground
grids designed to detect volcanogenic sulfide deposits of the second pool.

Table 4.13. (Continued) Probability tables for the detection of Ni-Cu ultramafic deposits of the North American Shield and the Western Australian Shield by ground geophysical surveys and drilling programs to a vertical depth of 100 m.

Determination of grid-size for specified probability levels of detection of Ni-Cu Ultramafic deposits by vertical drilling programs			
Pool of regions: North American Shield, Western Australian Shield		Grid design: Square grid with spacing S x S meters	
Specified detection probability level	Required grid spacing S in meters	Corresponding cost in $1000's per km square	
		percussion	diamond
0.10	425	9.55	28.65
0.20	305	15.53	46.60
0.30	245	21.89	65.68
0.40	225	25.12	75.37
0.50	205	29.20	87.83
0.60	185	34.76	104.27
0.70	175	38.18	114.55
0.80	165	42.22	126.66
0.90	150	49.77	149.30

Determination of grid-size for specified probability levels of detection of Ni-Cu Ultramafic deposits by 55° angled drilling programs			
Specified detection probability level	Required grid spacing S in meters	Corresponding cost in $1000's per km square	
		percussion	diamond
0.10	600	14.04	42.13
0.20	450	21.67	65.01
0.30	375	28.95	86.86
0.40	325	36.62	109.85
0.50	300	41.86	125.58
0.60	275	48.51	145.53
0.70	250	57.14	171.41
0.80	225	68.64	205.92
0.90	210	77.50	232.50

The detection probabilities for the Ni-Cu ultramafic deposits are presented in Table 4.13 for the first pool of regions: the North American Shield and Western Australian Shield. The table shows that the optimal orientation of the grid for vertical detection is 20 degrees from the expected strike direction for Ni-Cu ultramafic deposits of the North American and Western Australian Shields. Regarding the 55-degree angled detection, the ratio R_t is inflated from 0.12 to 0.26, and the optimal grid orientation is at 27 degrees from the strike line of the targets.

However, when dealing with the "stubbier" Scandinavian Shield Ni-Cu deposits (Table 4.14), we find the optimal angle of grid orientation is 29 degrees in the case of vertical detection, and is inflated to 45 degrees in the case of angled drilling.

Table 4.14. Probability tables for the detection of Ni-Cu
ultramafic deposits of the Scandinavian Shield
by ground geophysical surveys and drilling programs
to a vertical depth of 100 m.

Probability of detection Ni-Cu Ultramafic deposits

by ground geophysical surveys or drilling programs

Region: Scandinavian Shield

Grid design: Square grid
with spacing S x S meters

Orientated elliptical targets with expected major axis = L meters in the confidence interval

l.c.l. < geometric mean < u.c.l.

R is defined as the ratio B/L with geometric mean = 0.31

Optimum grid orientation with expected strike line = 29°

Grid spacing S meters	l.c.l. = 125 m	geom. mean = 180 m	u.c.l. = 265 m
50	1.000	1.000	1.000
100	0.385	0.785	1.000
150	0.175	0.355	0.758
200	0.097	0.197	0.430
250	0.062	0.128	0.276
300	0.043	0.089	0.191
400	0.020	0.050	0.108
500	0.010	0.028	0.060
600	0.007	0.019	0.048
800	0.005	0.012	0.027
1000	0.003	0.008	0.017
Probability of confirmed detection			
25	1.000	1.000	1.000
50	0.616	1.000	1.000
75	0.174	0.572	1.000
100	0.030	0.240	0.680
250	0.000	0.000	0.002
300	0.000	0.000	0.000

Finally, when dealing with very narrow targets such as vein-gold
deposits of the Western Australian Shield (Table 4.15), the optimal grid
orientation with respect to the expected strike line is 19 degrees in the
case of vertical detection. In the case of angled detection, the target
shape ratio, R_t, increases from 0.09 to 0.19, and the optimal grid
orientation becomes 27 degrees.

Table 4.14. (Continued) Probability tables for the detection of
Ni-Cu ultramafic deposits of the Scandinavian Shield
by ground geophysical surveys and drilling programs
to a vertical depth of 100 m.

Probability of detection Ni-Cu Ultramafic deposits by 55° angled drilling programs			
Region: Scandinavian Shield		Grid design: Square grid with spacing S x S meters	
Orientated elliptical targets with expected major axis = L meters in the confidence interval l.c.l. < geometric mean < u.c.l. R is defined as the ratio B/L with geometric mean = 0.61 Optimum grid orientation with expected strike line = 45°			
Grid spacing S meters	l.c.l. = 125 m	geom. mean = 180 m	u.c.l. = 265 m
50	1.000	1.000	1.000
100	0.749	1.000	1.000
150	0.333	0.690	1.000
200	0.187	0.388	0.841
250	0.120	0.248	0.538
300	0.083	0.172	0.374
500	0.030	0.062	0.111
600	0.021	0.043	0.080
800	0.012	0.024	0.034
1000	0.007	0.016	0.023
Probability of confirmed detection			
50	1.000	1.000	1.000
75	0.378	1.000	1.000
100	0.074	0.607	1.000
150	0.000	0.052	0.519
200	0.000	0.000	0.114
250	0.000	0.000	0.007

4.5. OPTIMIZATION OF AIRBORNE & GROUND FIELD SURVEYS AND DRILLING PROGRAMS BASED ON DETECTION PROBABILITY MODELS

4.5.1. Introduction of the Detection Efficiency and Dynamic Programming models

4.5.1.1. Detection Efficiency Model

In the previous sections of the chapter, we were preoccupied mainly
with the calculation and maximization of detection probabilities based upon
deterministic detection models (see Figures 2.1 and 2.2 of Chapter 2) and on
probabilistic detection models derived from the theory of geometric
probabilities. However, because of the economic context in which mineral
exploration has to operate, it is obvious that we have to take into account
the amount of effort and resources required to acquire the expected prize.

Table 4.14. (Continued) Probability tables for the detection of
Ni-Cu ultramafic deposits of the Scandinavian Shield
by ground geophysical surveys and drilling programs
to a vertical depth of 100 m.

Determination of grid-size for specified probability levels of detection of Ni-Cu Ultramafic deposits by vertical drilling programs			
Region: Scandinavian Shield		Grid design: Square grid with spacing S x S meters	
Specified detection probability level	Required grid spacing S in meters	Corresponding cost in $1000's per km square	
		percussion	diamond
0.10	280	17.73	53.18
0.20	205	29.28	87.83
0.30	160	44.51	133.53
0.40	145	52.79	158.38
0.50	130	63.95	191.86
0.60	125	68.56	205.67
0.70	115	79.55	238.65
0.80	105	93.71	281.12
0.90	100	102.37	307.11

Determination of grid-size for specified probability levels of detection of Ni-Cu Ultramafic deposits by 55° angled drilling programs			
Specified detection probability level	Required grid spacing S in meters	Corresponding cost in $1000's per km square	
		percussion	diamond
0.10	250	57.14	171.41
0.20	200	84.50	253.50
0.30	155	133.71	401.14
0.40	135	172.26	516.77
0.50	115	231.92	695.77
0.60	105	274.95	824.86
0.70	95	331.94	995.83
0.80	85	409.74	229.23
0.90	75	520.04	1560.12

There has to be a balancing of the reward (detection) against the cost of
obtaining it. The necessity of a trade-off between the desire to maximize
detection and budgetary requirements is well described graphically by the
heuristic model shown in Figure 2.3 of Chapter 2, which relates the
probability of detection, P_d, of any type of mineral resource to the search
expenditure, x in dollars, for saturation coverage of a unit of area,
defined by the Exponential model as follows:

$$P_d = 1 - \exp(-kx),$$

where k is an empirically determined constant measuring the historical
success rate or assessment of it. It can be seen from the left-hand portion

Table 4.15. Probability tables for the detection of vein gold
deposits of the Western Australian Shield by
ground geophysical surveys and drilling programs
to a vertical depth of 100 m.

Probability of detection Vein gold deposits by ground geophysical surveys or drilling programs			
Region: Western Australian Shield		Grid design: Square grid with spacing S x S meters	
Orientated elliptical targets with expected major axis = L meters in the confidence interval l.c.l. < geometric mean < u.c.l. R is defined as the ratio B/L with geometric mean = 0.09 Optimum grid orientation with expected strike line = 19°			
Grid spacing S meters	l.c.l. = 430 m	geom. mean = 595 m	u.c.l. = 820 m
50	1.000	1.000	1.000
100	0.990	1.000	1.000
150	0.579	0.945	1.000
200	0.327	0.621	0.968
250	0.209	0.400	0.742
300	0.145	0.278	0.527
400	0.082	0.156	0.297
500	0.052	0.100	0.190
600	0.036	0.070	0.132
800	0.020	0.039	0.074
1000	0.013	0.025	0.048
Probability of confirmed detection			
25	1.000	1.000	1.000
50	0.930	1.000	1.000
100	0.458	0.718	0.911
150	0.193	0.400	0.644
250	0.026	0.109	0.271
400	0.000	0.011	0.061

of the graph shown in Figure 2.3 of Chapter 2, that the probability of
success rises rapidly to the 0.50 level or better for relatively small
increments in exploration expenditure. The right-hand portion of the graph,
however, illustrates a situation of diminishing returns: when we reach
probability levels of 0.80 or more, large expenditure increments bring about
ever-decreasing improvements in the success probability levels.

In practical terms, the question arises: what grid spacing is it best
to choose when the aim is cost-efficiency? A supplementary question also
arose: is there an optimal angle at which to drill in the case of dipping

Table 4.15. (Continued) Probability tables for the detection of vein gold deposits of the Western Australian Shield by ground geophysical surveys and drilling programs to a vertical depth of 100 m.

Probability of detection Vein gold deposits by 55° angled drilling programs			
Region: Western Australian Shield		Grid design: Square grid with spacing S x S meters	
Orientated elliptical targets with expected major axis = L meters in the confidence interval l.c.l. < geometric mean < u.c.l. R is defined as the ratio B/L with geometric mean = 0.19 Optimum grid orientation with expected strike line = 27°			
Grid spacing S meters	l.c.l. = 430 m	geom. mean = 595 m	u.c.l. = 820 m
100	1.000	1.000	1.000
150	0.992	1.000	1.000
200	0.690	0.998	1.000
250	0.441	0.833	1.000
300	0.307	0.587	0.971
400	0.172	0.330	0.627
500	0.110	0.211	0.401
600	0.077	0.147	0.279
800	0.043	0.083	0.157
1000	0.028	0.053	0.100
Probability of confirmed detection			
75	1.000	1.000	1.000
100	0.848	1.000	1.000
150	0.482	0.759	1.000
200	0.250	0.511	0.795
250	0.104	0.332	0.591
400	0.001	0.048	0.213

deposits, and if so, what is it, and how much does it save? To answer the questions, we used the Detection Efficiency model, which maximizes the probability of detection subject to the cost function constraint, in the context of the Dynamic Programming approach.

The reader will recall our brief description of a heuristic model labelled "Efficiency Function" which is decribed in section 2.2 of Chapter 2. The make-up of the function is illustrated by Figure 2.4, which shows how a probabilistic model (probability of detection) of a reward function is combined with a deterministic model of a cost function. The formal definition of the function, EFF, is as follows:

$$EFF(S) = k_1 P(S) - k_2 C(S),$$

Table 4.15. (Continued) Probability tables for the detection of
vein gold deposits of the Western Australian Shield
by ground geophysical surveys and drilling programs
to a vertical depth of 100 m.

Determination of grid-size for specified probability levels of detection of Vein gold deposits by vertical drilling programs			
Region: Western Australian Shield		Grid design: Square grid with spacing S x S meters	
Specified detection probability level	Required grid spacing S in meters	Corresponding cost in $1000's per km square	
		percussion	diamond
0.10	500	7.65	22.95
0.20	350	12.63	37.89
0.30	300	15.93	47.79
0.40	250	21.20	63.59
0.50	225	25.12	75.37
0.60	205	29.28	87.83
0.70	200	30.50	91.51
0.80	175	38.18	114.55
0.90	165	42.22	126.66

Determination of grid-size for specified probability levels of detection of Vein gold deposits by 55° angled drilling programs			
Specified detection probability level	Required grid spacing S in meters	Corresponding cost in $1000's per km square	
		percussion	diamond
0.10	850	8.68	26.05
0.20	600	14.04	42.13
0.30	500	18.42	55.26
0.40	425	23.70	71.11
0.50	400	26.10	78.30
0.60	350	32.40	97.20
0.70	325	36.62	109.85
0.80	300	41.86	125.58
0.90	275	48.51	145.53

where S is the grid spacing in meters, P is the detection probability
function (see sections 4.2, 4.3 & 4.4), and C is the cost function. (See
Table 2.1 of Chapter 2).

In order to obtain a meaningful efficiency function when the spacing S
varies within a realistic range, it is evident that both terms of the
function must be kept within the same domain: while P(S) itself varies in
the interval [0, 1], C(S) could vary from almost 0 to very large values.
This situation requires a balancing of terms by means of appropriate
weighting factors, k_1 and k_2. We could choose, for example:

$$k_1 = 1/0.90,$$

and $k_2 = 1/(90\%\text{ile of } C)$, for a specific type of target,

where the 90% ile cost is defined as the cost at a detection probability
level of 0.90. For example, in the case of the Volcanogenic sulfide deposits
of the North American Shield, Appalachian Belt and Tasman Geosyncline pool
of regions, C = 68.56 for percussion, and 205.67 for diamond drilling
(thousands of $US). We could also choose

          $k_1 = 1/0.95$,

and $k_2 = 1/(95\%\text{ile of } C)$, for another specific type of target,

where the 95% ile cost is defined as the cost at a detection probability
level of 0.95. If we choose the Volcanogenic sulfide deposits of the Kuroko
Belt of Japan and Eastern Mediterranean regions as an example, we have,
C = 9.55 for percussion, and 28.65 for diamond drilling. As a result of the
use of the appropriate weighting factors, we can see from Figure 4.2 that

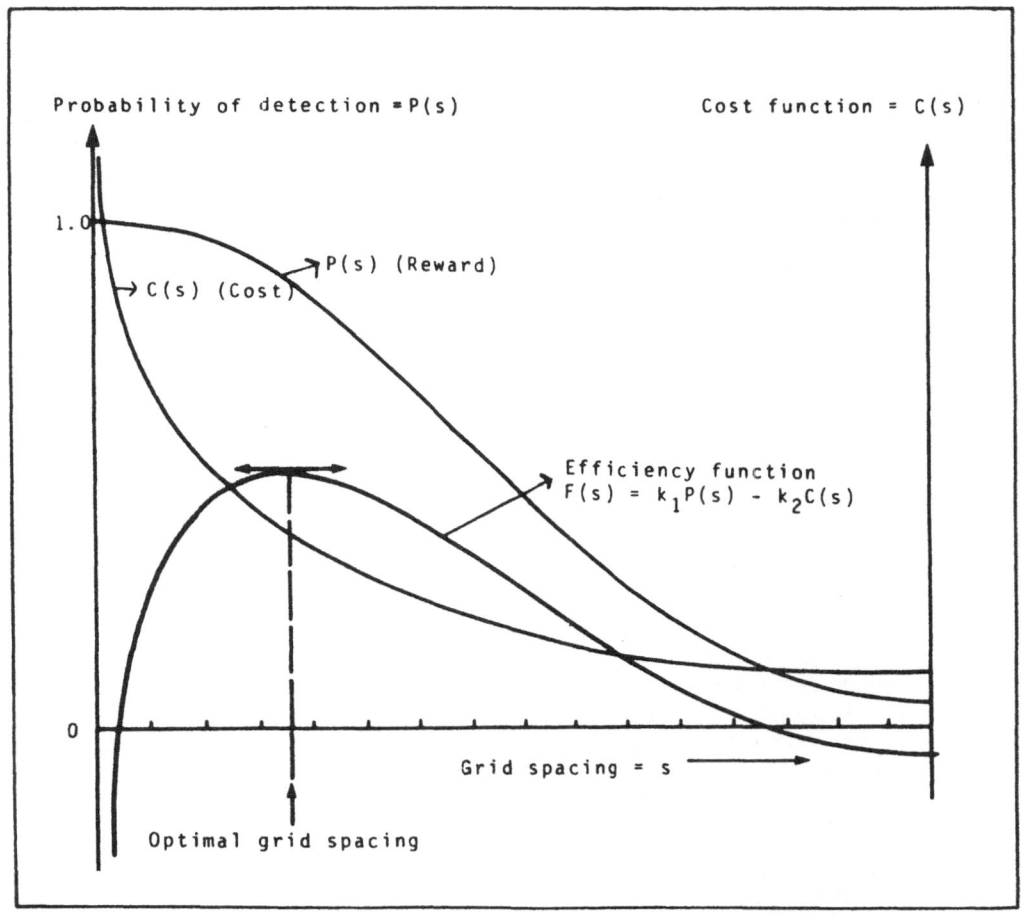

Figure 4.2. Diagram of the optimization of search grid design
based upon the efficiency criterion.

Table 4.16. Optimization of field programs for the detection
of porphyry-Cu-Mo deposits of the North & South
American Cordillera and Australasia.

Optimization based upon efficiency model & Dynamic Programming

Type of deposit: Porphyry Cu-Mo

Pool of Regions: North & South American Cordillera, B.C. Cordillera,

South Pacific Island Arc, Tasman Geosyncline

Confidence intervals: all results are reported as 95% confidence intervals respectively

Expected shape ratio, $R = B/L = 0.37$

Unit cost   = $50 per line km

(i) Airborne geophysical surveys

Grid design: (a) Parallel flight-lines with spacing S meters

	l.c.l.	geom. mean	u.c.l.
Expected target length in meters	775 m	950 m	1180 m
Optimal grid spacing in meters	730 m	880 m	1100 m
Cost in $ per km square	$130	$120	$108
Probability of detection	0.79	0.79	0.79

Grid design: (b) Square grid with spacings S x S meters

	l.c.l.	geom. mean	u.c.l.
Expected target length in meters	775 m	950 m	1180 m
Optimal grid spacing in meters	1070 m	1210 m	1370 m
Cost in $ per km square	$155	$145	$135
Probability of detection	0.82	0.86	0.91

the efficiency model is kept within bounds and becomes a usable function for
the purpose of grid optimization.

4.5.1.2. Introducing the dynamic programming methodology

The choice of the spacing S, which maximizes the Efficiency Function,
EFF, is found by Dynamic Programming, which is computerized in Appendix 4.
The Dynamic Programming models search every path or set of paths through a
decision tree, and give the outcome for each choice. In this particular
problem, the different paths were the different values for S, the grid
spacing, and the optimum choice is the one where EFF is maximized. For
example when S is very small then the cost is very high, and the evaluation
referable to use the square grid design, and program 4.5 is used instead.
of EFF at such points is negative; as S increases through increments of 25
meters say, then EFF reaches 0, and eventually rises to a maximum, and then
declines again as P approaches 0. The value of S which maximizes EFF is the
optimum choice.

Table 4.16. (Continued) Optimization of field programs for the
detection of porphyry-Cu-Mo deposits of the
North & South American Cordillera and Australasia.

Optimization based upon efficiency model & Dynamic Programming

Type of deposit: Porphyry Cu-Mo

Pool of Regions: North & South American Cordillera, B.C. Cordillera,
        South Pacific Island Arc, Tasman Geosyncline

Grid design: Square grid
with spacings S x S meters

Confidence intervals: all results are reported as 95% confidence intervals respectively

Expected shape ratio, R = B/L = 0.37

Unit cost   = $6 per station

(ii) Ground geophysical surveys

	l.c.l.	geom. mean	u.c.l.
Expected target length in meters	775 m	950 m	1180 m
Optimal grid spacing in meters	410 m	520 m	640 m
Cost in $ per km square	$340	$275	$230
Probability of detection	0.82	0.80	0.88

(iii) Vertical drilling programs

  Percussion drilling: unit cost = $2190 per 100 metre hole

  Diamond drilling   : unit cost = $6560 per 100 meter hole

	l.c.l.	geom. mean	u.c.l.
Expected target length in meters	775 m	950 m	640 m
Optimal grid spacing in meters	410 m	520 m	640 m
Percussion drilling Cost in $1,000's per km square	20	14	11
Diamond drilling Cost in $1,000's per km square	62	43	32
Probability of detection	0.82	0.80	0.88

Parallel grids are generally used as control for airborne surveys
covering search areas where the expected strike direction of the target is
known within a given confidence interval. We can maximize the probability of
detection by orienting the grid orthogonally to the mean strike direction.
The grid optimization problem is reduced to that of finding the best grid
spacing S. This is accomplished by use of our computer program 4.4 of
Appendix 4. When the expected strike direction is not known, it may be
preferable to use the square grid design, and program 4.5 is used instead.

In the case of optimization of ground surveys and drilling programs,
we are faced with three considerations: one of them involves the orientation
of the target, i. e. whether the target may be considered to be randomly
oriented or has an expected strike line within a confidence interval;

Table 4.17. Optimization of field programs for the detection of volcanogenic sulfide deposits of the East Mediterranean region and the Kuroko Belt of Japan.

Optimization based upon efficiency model

Type of deposit: Volcanogenic sulfides

Pool of Regions: Kuroko Belt of Japan, Eastern Mediterranean

Confidence intervals: all results are reported as 95% confidence intervals respectively

Expected shape ratio, R = B/L = 0.40

Unit cost   = $50 per line km

(i) Airborne geophysical surveys

Grid design: (a) Parallel flight-lines with spacing S meters

	l.c.l.	geom. mean	u.c.l.
Expected target length in meters	305 m	370 m	450 m
Optimal grid spacing in meters	305 m	375 m	440 m
Cost in $ per km square	$230	$200	$180
Probability of detection	0.77	0.98	0.98

Grid design: (b) Square grid with spacings S x S meters

	l.c.l.	geom. mean	u.c.l.
Expected target length in meters	305 m	370 m	450 m
Optimal grid spacing in meters	470 m	550 m	625 m
Cost in $ per km square	$280	$250	$225
Probability of detection	0.77	0.98	0.98

another one is that of the attitude of the target as expressed by its dip angle; and the last one involves the attitude of the detector, whether vertical (ground geophysical surveys and drilling) or angled (drilling only). The simplest situation is that of the vertical detection of a randomly oriented elliptical target with a mean dip of α degrees. The optimization of the grid is a one-stage computerized procedure based upon the efficiency model and Dynamic Programming to select the optimal grid spacing.

The most complex case involves the detection by drilling to a specified depth at an angle of β degrees in the search for an elliptical target which has a preferred orientation within a confidence interval and an expected dip of α degrees. The procedure is a typical application of the Dynamic Programming approach, which is carried out sequentially as follows:

(1) optimization of the orientation of the grid with respect to the expected target direction,

Table 4.17. (Continued) Optimization of field programs for the detection of volcanogenic sulfide deposits of the East Mediterranean region and the Kuroko Belt of Japan.

Optimization based upon efficiency model

Type of deposit: Volcanogenic sulfides

Pool of Regions: Kuroko Belt of Japan,   Grid design: Square grid

                 Eastern Mediterranean   with spacings $S \times S$ meters

Confidence intervals: all results are reported as 95% confidence intervals respectively

Expected shape ratio, $R = B/L = 0.40$

Unit cost   = $6 per station

(ii) Ground geophysical surveys

	l.c.l.	geom. mean	u.c.l.
Expected target length in meters	305 m	370 m	450 m
Optimal grid spacing in meters	180 m	170 m	200 m
Cost in $ per km square	$700	$770	$630
Probability of detection	0.77	0.98	0.98

(iii) Vertical drilling programs (*)

    Percussion drilling: unit cost = $2190 per 100 metre hole

    Diamond drilling   : unit cost = $6560 per 100 meter hole

	l.c.l.	geom. mean	u.c.l.
Expected target length in meters	305 m	370 m	450 m
Optimal grid spacing in meters	180 m	170 m	200 m
Percussion drilling Cost in $1,000's per km square	85	95	70
Diamond drilling Cost in $1,000's per km square	250	290	220
Probability of detection	0.77	0.98	0.98

N.B. (*) Angled drilling not considered because of sub-horizontal attitude of deposits

(2) optimization of the spacing of the grid, which has previously been optimally oriented, and

(3) optimization of the drilling angle within the acceptable range of 35 to 90 degrees, when the drill-holes are optimally spaced along the grid, which has in turn been optimally oriented in stage (1).

All situations are covered by program 4.9 of Appendix 4. In all cases, the output of the programs 4.4, 4.5 and 4.9 includes:

(a) optimal grid orientation (omitted when dealing with randomly oriented targets),

Table 4.18. Optimization of field programs for the detection of volcanogenic sulfide deposits of North America and Australasia.

Optimization based upon efficiency model & Dynamic Programming

Type of deposit: Volcanogenic sulfides

Pool of Regions: North American Shield, Appalachian Belt, Tasman Geosyncline

Confidence intervals: all results are reported as 95% confidence intervals respectively

Expected shape ratio, $R = B/L = 0.19$

Unit cost $= \$50$ per line km

Optimal grid orientation: 90 degrees to expected strike line

(i) Airborne geophysical surveys

Grid design: (a) Parallel flight-lines with spacing S meters

	l.c.l.	geom. mean	u.c.l.
Expected target length in meters	255 m	300 m	370 m
Optimal grid spacing in meters	210 m	230 m	320 m
Cost in $ per km square	$300	$285	$215
Probability of detection	0.77	0.79	0.79

Grid design: (b) Square grid with spacings S x S meters

	l.c.l.	geom. mean	u.c.l.
Expected target length in meters	255 m	300 m	370 m
Optimal grid spacing in meters	505 m	570 m	605 m
Cost in $ per km square	$265	$240	$230
Probability of detection	0.60	0.62	0.70

(b) optimal grid size, with associated cost of coverage per unit area and associated optimal probability of detection,

(c) optimal drilling angle (when it applies), and

(d) a print-out of the efficiency function values.

4.5.2. Application to the optimization of airborne and ground search for specific types of ore deposits

4.5.2.1. Construction and organization of optimization tables

All optimization tables are constructed on the basis of the concept of a 95% confidence interval for the expected target length (geometric mean), which is translated into 95% fiducial intervals for the parameters of

Table 4.18. (Continued) Optimization of field programs for the detection of volcanogenic sulfide deposits of North America and Australasia.

Optimization based upon efficiency model & Dynamic Programming

Type of deposit: Volcanogenic sulfides

Pool of Regions: North American Shield, Appalachian Belt, Tasman Geosyncline

Grid design: Square grid with spacings S x S meters

Confidence intervals: all results are reported as 95% confidence intervals respectively

Expected shape ratio, R = B/L = 0.19

Unit cost = $6 per station

Optimal grid orientation: 28 degrees to expected strike line

(ii) Ground geophysical surveys

	l.c.l.	geom. mean	u.c.l.
Expected target length in meters	255 m	300 m	370 m
Optimal grid spacing in meters	90 m	110 m	150 m
Cost in $ per km square	$1715	$1380	$875
Probability of detection	0.94	0.95	0.85

(iii) Vertical drilling programs

Percussion drilling: unit cost = $2190 per 100 metre hole

Diamond drilling : unit cost = $6560 per 100 meter hole

	l.c.l.	geom. mean	u.c.l.
Expected target length in meters	255 m	300 m	370 m
Optimal grid spacing in meters	120 m	120 m	150 m
Percussion drilling Cost in $1,000's per km square	160	160	110
Diamond drilling Cost in $1,000's per km square	480	480	320
Probability of detection	0.65	0.83	0.83

optimized search grids. The headings of the table provide also such pertinent information as type of grid and unit cost. The parameters of optimized grids are listed in the following order:

(a) optimal grid orientation and optimal drilling angle, when either or both apply, and

(b) optimal grid spacing followed by associated coverage cost per unit of area and optimal detection probability.

Obviously, because of format restrictions, we could not possibly manage to present the results of optimization of the search for all the six genetic types of deposits in all groups of pooled regions covered by the whole data-base. We have restricted the presentation to include the

Table 4.18. (Continued) Optimization of field programs for the
detection of volcanogenic sulfide deposits of
North America and Australasia.

Optimization based upon efficiency model & Dynamic Programming

Type of deposit: Volcanogenic sulfides

Pool of Regions: North American Shield, Appalachian Belt, Tasman Geosyncline

Grid design: Square grid with spacings 3 x 3 meters

Confidence intervals: all results are reported as 95% confidence intervals respectively

Expected shape ratio, R = B/L = 0.32

Unit cost   = $66 per meter

Optimal grid orientation: 30 degrees to expected strike line

(iv) 55 degree angled drilling programs

	l.c.l.	geom. mean	u.c.l.
Expected target length in meters	255 m	300 m	370 m
Optimal grid spacing in meters	180 m	180 m	210 m
Cost in $1,000's per km square	280	280	210
Probability of detection	0.67	0.82	0.82

(v) Angled drilling programs at optimal angle = 40 degrees

Diamond drilling   : unit cost = $66 per meter

	l.c.l.	geom. mean	u.c.l.
Expected target length in meters	255 m	300 m	370 m
Optimal grid spacing in meters	180 m	210 m	240 m
Cost in $1,000's per km square	360	270	215
Probability of detection	0.85	0.85	0.85

porphyry-Cu-Mo deposits in their single pool of regions, the volcanogenic
sulfide deposits in their three pools, the Ni-Cu ultramafic in their two
pools, and finally the vein gold deposits in a single region. The pyritic
halos of porphyry deposits, the four sub-types of contact metasomatic
deposits, and vein gold deposits in another single region were not included.

For the benefit of readers, the seven optimization tables which are
listed below are organized in the following manner:

A first group including four tables covers the optimization of search

Table 4.19. Optimization of field programs for the detection
of volcanogenic sulfide deposits of Europe,
North America and Australasia.

Optimization based upon efficiency model & Dynamic Programming

Type of deposit: Volcanogenic sulfides

Pool of Regions: W. Australian Shield, Scandinavian Shield, Iberian Peninsula Pyrite Belt,
North American Cordillera, North American Cordillera (exhalative)

Confidence intervals: all results are reported as 95% confidence intervals respectively

Expected shape ratio, $R = B/L = 0.19$

Unit cost   = $50 per line km

Optimal grid orientation: 90 degrees to expected strike line

(i) Airborne geophysical surveys

Grid design: (a) Parallel flight-lines with spacing $ meters

	l.c.l.	geom. mean	u.c.l.
Expected target length in meters	420 m	560 m	650 m
Optimal grid spacing in meters	320 m	400 m	450 m
Cost in $ per km square	$220	$190	$180
Probability of detection	0.79	0.80	0.80

Grid design: (b) Square grid with spacings $ x $ meters

	l.c.l.	geom. mean	u.c.l.
Expected target length in meters	420 m	560 m	650 m
Optimal grid spacing in meters	505 m	640 m	705 m
Cost in $ per km square	$265	$222	$205
Probability of detection	0.86	0.89	0.92

programs for randomly oriented targets with either vertical or sub-
horizontal attitudes, which is the simplest case mentioned above. The
deposit types include porphyry-Cu-Mo deposits in their single pool
(Table 4.16), and the volcanogenic sulfide deposits of the pool of regions
including the Kuroko Belt of Japan and the Eastern Mediterranean
(Table 4.17). The first section of each table covers airborne surveys on
parallel and square grids. The second section includes ground geophysical
surveys followed by vertical drilling programs on square grids.

A second group of fourteen tables covers the more complex case of
oriented and dipping deposits in the context of vertical and angled
detection. The deposit types include (a) volcanogenic sulfide deposits in

Table 4.19. (Continued) Optimization of field programs for the detection of volcanogenic sulfide deposits of Europe, North America and Australasia.

Optimization based upon efficiency model & Dynamic Programming

Type of deposit: Volcanogenic sulfides

Pool of Regions: W. Australian Shield, Scandinavian Shield, Iberian Peninsula Pyrite Belt, North American Cordillera, North American Cordillera (exhalative)

Grid design: Square grid with spacings S x S meters

Confidence intervals: all results are reported as 95% confidence intervals respectively

Expected shape ratio, $R = B/L = 0.19$

Unit cost = $6 per station

Optimal grid orientation: 27 degrees to expected strike line

(ii) Ground geophysical surveys

	l.c.l.	geom. mean	u.c.l.
Expected target length in meters	420 m	560 m	650 m
Optimal grid spacing in meters	140 m	180 m	200 m
Cost in $ per km square	$995	$700	$630
Probability of detection	0.70	0.82	0.86

(iii) Vertical drilling program

Percussion drilling: unit cost = $2190 per 100 metre hole

Diamond drilling : unit cost = $6560 per 100 meter hole

	l.c.l.	geom. mean	u.c.l.
Expected target length in meters	420 m	560 m	650 m
Optimal grid spacing in meters	150 m	170 m	200 m
Percussion drilling Cost in $1,000's per km square	105	90	65
Diamond drilling Cost in $1,000's per km square	320	270	200
Probability of detection	0.70	0.82	0.86

two pools of regions (Tables 4.18. and 4.19), (b) Ni-Cu ultramafic deposits in one pool of regions (Table 4.20) and a single region (Table 4.21), and finally, the vein gold deposits in a single region (Table 4.22). The first section of each table includes airborne surveys on parallel and square grids; the second section covers geophysical ground surveys and vertical drilling programs; and finally, the third section deals with angled drilling, first at the standard 55-degree angle, followed by optimally angled drilling.

A summary of the results given in Table 4.16 for the porphyry-Cu-Mo deposits of the pool including the North & South American Cordillera and

Table 4.19. (Continued) Optimization of field programs for the detection of volcanogenic sulfide deposits of Europe, North America and Australasia.

Optimization based upon efficiency model & Dynamic Programming

Type of deposit: Volcanogenic sulfides

Pool of Regions: W. Australian Shield, Scandinavian Shield, Iberian Peninsula Pyrite Belt, North American Cordillera, North American Cordillera (exhalative)

Grid design: Square grid with spacings S x S meters

Confidence intervals: all results are reported as 95% confidence intervals respectively

Expected shape ratio, R = B/L = 0.32

Unit cost   = $65.60 per meter

Optimal grid orientation: 29 degrees to expected strike line

(iv) 55 degree angled drilling programs

	l.c.l.	geom. mean	u.c.l.
Expected target length in meters	420 m	560 m	650 m
Optimal grid spacing in meters	170 m	180 m	210 m
Cost in $1,000's per km square	335	280	210
Probability of detection	0.95	0.99	0.99

(v) Angled drilling programs at optimal angle = 45 degrees

Diamond drilling   : unit cost = $66 per meter

	l.c.l.	geom. mean	u.c.l.
Expected target length in meters	420 m	560 m	650 m
Optimal grid spacing in meters	180 m	210 m	240 m
Cost in $1,000's per km square	325	250	200
Probability of detection	0.97	0.99	0.99

Australasia, for example, with a pooled expected geometric mean length of 950m, and shape ratio of 0.37, is given below:

Survey type	Optimal spacing	Cost $/km sq.	Detect. Probability
Airborne parallel line	880 m	120	0.79
Airborne square grid	1210 m	145	0.86
Ground Geophysical	520 m	275	0.80
Ground Vertical percus.drilling	520 m	14,000	0.80
Ground Vertical diam. drilling	520 m	43,000	0.80

169

Table 4.20. Optimization of field programs for the detection
of Ni-Cu ultramafic deposits of North America
and Australasia.

Optimization based upon efficiency model & Dynamic Programming

Type of deposit: Ni-Cu Ultramafic

Pool of Regions: North American Shield, Western Australian Shield

Confidence intervals: all results are reported as 95% confidence intervals respectively

Expected shape ratio, $R = B/L = 0.12$

Unit cost   = $50 per line km

Optimal grid orientation: 90 degrees to expected strike line

(i) Airborne geophysical surveys

    Grid design: (a) Parallel flight-lines with spacing S meters

	l.c.l.	geom. mean	u.c.l.
Expected target length in meters	320 m	440 m	605 m
Optimal grid spacing in meters	240 m	365 m	425 m
Cost in $ per km square	$270	$200	$185
Probability of detection	0.78	0.78	0.79

    Grid design: (b) Square grid with spacings S x S meters

	l.c.l.	geom. mean	u.c.l.
Expected target length in meters	320 m	440 m	605 m
Optimal grid spacing in meters	535 m	640 m	725 m
Cost in $ per km square	$250	$220	$205
Probability of detection	0.67	0.75	0.86

4.6. SELECTION OF ORE DEPOSIT TYPES & REGIONS OF SEARCH BASED UPON THE DETECTABILITY CRITERION

4.6.1. Introducing the concept of "detectability"

One of the basic tasks of exploration planners is the proper selection
of ore deposit types to be considered as targets for exploration and regions
of search as well as methods of coverage ensuring a maximum of success in
the search. The risk incurred in carrying out the search for specific types
of deposits in specific regions appears to be an appropriate selection
criterion for economic planning purposes. Naturally, if we wish to base the
selection on risk as a yard-stick, risk should be quantified; one method to
achieve this is to consider the expected loss criterion.

The expected loss criterion combining coverage cost per unit area and

Table 4.20. (Continued) Optimization of field programs for
the detection of Ni-Cu ultramafic deposits of
North America and Australasia.

Optimization based upon efficiency model & Dynamic Programming

Type of deposit: Ni-Cu Ultramafic

Pool of Regions: North American Shield, Western Australian Shield

Grid design: Square grid with spacings S x S meters

Confidence intervals: all results are reported as 95% confidence intervals respectively

Expected shape ratio, R = B/L = 0.12

Unit cost   = $6 per station

Optimal grid orientation: 20 degrees to expected strike line

(ii) Ground geophysical surveys

	l.c.l.	geom. mean	u.c.l.
Expected target length in meters	320 m	440 m	605 m
Optimal grid spacing in meters	90 m	150 m	180 m
Cost in $ per km square	$1715	$870	$700
Probability of detection	0.91	0.72	0.86

(iii) Vertical drilling programs

Percussion drilling: unit cost = $2190 per 100 metre hole

Diamond drilling   : unit cost = $6560 per 100 meter hole

	l.c.l.	geom. mean	u.c.l.
Expected target length in meters	320 m	440 m	605 m
Optimal grid spacing in meters	90 m	150 m	180 m
Percussion drilling Cost in $1,000's per km square	300	115	85
Diamond drilling Cost in $1,000's per km square	905	350	250
Probability of detection	0.91	0.72	0.86

the detection probability is defined by

Expected loss = (coverage cost)* (probability of failure),

where the probability of failure = 1 - (probability of detection).
The goal is to minimize the expected loss, or conversely, to maximize the
"detectability" of the deposit types in their regions of occurrence.
If the region-deposit types are ranked within each continent in order of
increasing expected loss values, then this is equivalent to a ranking in
order of decreasing detectability.

Table 4.20. (Continued) Optimization of field programs for
the detection of Ni-Cu ultramafic deposits of
North America and Australasia.

Optimization based upon efficiency model & Dynamic Programming

Type of deposit: Ni-Cu Ultramafic

Pool of Regions: North American Shield, Western Australian Shield

Grid design: Square grid with spacings 3 x 3 meters

Confidence intervals: all results are reported as 95% confidence intervals respectively

Expected shape ratio, R = B/L = 0.26

Unit cost = $66 per meter

Optimal grid orientation: 29 degrees to expected strike line

(iv) 55 degree angled drilling programs

	l.c.l.	geom. mean	u.c.l.
Expected target length in meters	320 m	440 m	605 m
Optimal grid spacing in meters	150 m	210 m	305 m
Cost in $1,000's per km square	355	270	215
Probability of detection	0.84	0.81	0.77

(v) Angled drilling programs at optimal angle = 45 degrees

Diamond drilling : unit cost = $66 per meter

	l.c.l.	geom. mean	u.c.l.
Expected target length in meters	320 m	440 m	605 m
Optimal grid spacing in meters	180 m	245 m	305 m
Cost in $1,000's per km square	355	215	145
Probability of detection	0.88	0.81	0.80

4.6.2. Examples of selection of region-ore types based on expected loss criterion

If a large mineral exploration company wishes to select the most
favourable ore deposit type and region of search in Europe, the planners can
easily collect all the necessary data from the optimization tables shown in
section 4.5, and calculate the expression of expected loss for each type of
detection method. The results are listed in the upper part of Table 4.23.
Among the volcanogenic massive sulfides deposits of Europe, the Iberian
Peninsula and the Scandinavian Shield show the lowest risk for airborne

Table 4.21. Optimization of field programs for the detection
of Ni-Cu ultramafic deposits of Europe.

Optimization based upon efficiency model & Dynamic Programming			

Type of deposit: Ni-Cu Ultramafic

Region: Scandinavian Shield

Confidence intervals: all results are reported as 95% confidence intervals respectively

Expected shape ratio, R = B/L = 0.31

Unit cost   = $50 per line km

Optimal grid orientation: 90 degrees to expected strike line

(i) Airborne geophysical surveys

Grid design: (a) Parallel flight-lines with spacing S meters

	l.c.l.	geom. mean	u.c.l.
Expected target length in meters	125 m	180 m	265 m
Optimal grid spacing in meters	125 m	180 m	215 m
Cost in $ per km square	$470	$345	$300
Probability of detection	0.75	0.75	0.81

Grid design: (b) Square grid with spacings S x S meters

	l.c.l.	geom. mean	u.c.l.
Expected target length in meters	125 m	180 m	265 m
Optimal grid spacing in meters	300 m	300 m	485 m
Cost in $ per km square	$400	$400	$275
Probability of detection	0.61	0.71	0.66

detection, and the Eastern Mediterranean region shows the lowest risk for
ground vertical detection. The Scandinavian Shield Ni-Cu ultramafic deposits
show the highest risk for airborne as well as vertical ground detection. The
switch from parallel to square grid for airborne surveys leads to a loss
risk reduction of 36% for the higher ranking Iberian and Scandinavian, and
to a loss risk increase of 36% for the lower ranking Ni-Cu-Scandinavian.

The same exercise can be carried out for the five ore deposit types
occurring in five regions of Australasia and East Asia. The results are
shown in the lower half of Table 4.23. The porphyry Cu-Mo deposits of the
South Pacific Island Arc and the Tasman Geosyncline offer the lowest risk

Table 4.21. (Continued) Optimization of field programs for the detection of Ni-Cu ultramafic deposits of Europe.

Optimization based upon efficiency model & Dynamic Programming

Type of deposit: Ni-Cu Ultramafic

Region: Scandinavian Shield

Grid design: Square grid with spacings S x S meters

Confidence intervals: all results are reported as 95% confidence intervals respectively

Expected shape ratio, R = B/L = 0.31

Unit cost   = $6 per station

Optimal grid orientation: 29 degrees to expected strike line

(ii) Ground geophysical surveys

	l.c.l.	geom. mean	u.c.l.
Expected target length in meters	125 m	180 m	265 m
Optimal grid spacing in meters	75 m	105 m	150 m
Cost in $ per km square	$2200	$1400	$870
Probability of detection	0.64	0.69	0.72

(iii) Vertical drilling programs

Percussion drilling: unit cost = $2190 per 100 metre hole

Diamond drilling   : unit cost = $6560 per 100 meter hole

	l.c.l.	geom. mean	u.c.l.
Expected target length in meters	125 m	180 m	265 m
Optimal grid spacing in meters	75 m	105 m	150 m
Percussion drilling Cost in $1,000's per km square	425	225	115
Diamond drilling Cost in $1,000's per km square	1280	675	350
Probability of detection	0.64	0.69	0.72

for airborne detection, while the Kuroko volcanogenic sulfide deposits provide the lowest detection risk for ground vertical detection, and the Western Australian Shield volcanogenic comes first for angled ground detection. The highest detection risk is shown by the Tasman Geosyncline volcanogenic sulfide deposits for airborne and ground vertical detection. The switch from airborne parallel to square grid leads to a beneficial relative cut in loss risk of 27% for the higher ranking types (porphyry-Cu-Mo deposits of the South Pacific Island Arc and the Tasman Geosyncline, & the volcanogenic sulfides of the Western Australian Shield). However, a counter-productive increase in loss risk is obtained for the lower ranking types (52% for the Tasman Geosyncline volcanogenic sulfides).

Table 4.21. (Continued) Optimization of field programs for
the detection of Ni-Cu ultramafic deposits of
Europe.

Optimization based upon efficiency model & Dynamic Programming

Type of deposit: Ni-Cu Ultramafic

Region: Scandinavian Shield

Grid design: Square grid with spacings 3 x 3 meters

Confidence intervals: all results are reported as 95% confidence intervals respectively

Expected shape ratio, R = B/L = 0.61

Unit cost   = $66 per meter

Optimal grid orientation: 45 degrees to expected strike line

(iv) 55 degree angled drilling programs

	l.c.l.	geom. mean	u.c.l.
Expected target length in meters	125 m	180 m	265 m
Optimal grid spacing in meters	120 m	150 m	210 m
Cost in $1,000's per km square	645	430	235
Probability of detection	0.67	0.73	0.66

(v) Angled drilling programs at optimal angle = 45 degrees

Diamond drilling   : unit cost = $66 per meter

	l.c.l.	geom. mean	u.c.l.
Expected target length in meters	125 m	180 m	265 m
Optimal grid spacing in meters	120 m	150 m	215 m
Cost in $1,000's per km square	745	495	300
Probability of detection	0.84	0.80	0.77

4.7. OPTIMAL SELECTION OF REGION-ORE DEPOSIT TYPES BASED ON THE DISCOUNTED PAY-OFF CRITERION

4.7.1. Introducing the discounted pay-off as a criterion for optimal planning

In the previous section we sought to select ore deposit types and
regions of search based solely upon the minimization of risk without taking
into account the magnitude of the reward. If we want to optimize the
selection, we should balance the risk against the reward in a probabilistic
context. This is accomplished by the pay-off model, already described in
section 2.2.3 of Chapter 2, and illustrated by Figures 2.5. and 2.6.

Table 4.22. Optimization of field programs for the detection
of vein-gold deposits of Australasia.

Optimization based upon efficiency model & Dynamic Programming			

Type of deposit: Vein Gold

Region: Western Australian Shield

Grid design: Square grid with spacings S x S meters

Confidence intervals: all results are reported as 95% confidence intervals respectively

Expected shape ratio, R = B/L = 0.09

Unit cost = $6 per station

Optimal grid orientation: 18 degrees to expected strike line

(ii) Ground geophysical surveys

	l.c.l.	geom. mean	u.c.l.
Expected target length in meters	430 m	595 m	820 m
Optimal grid spacing in meters	150 m	170 m	230 m
Cost in $ per km square	$875	$770	$540
Probability of detection	0.54	0.76	0.78

(iii) Vertical drilling programs

Percussion drilling: unit cost = $2190 per 100 metre hole

Diamond drilling : unit cost = $6560 per 100 meter hole

	l.c.l.	geom. mean	u.c.l.
Expected target length in meters	430 m	595 m	820 m
Optimal grid spacing in meters	150 m	170 m	230 m
Percussion drilling Cost in $1,000's per km square	120	95	55
Diamond drilling Cost in $1,000's per km square	350	295	170
Probability of detection	0.54	0.76	0.78

The quantitative expression of the pay-off model is

$$\text{Pay-off} = (GV)*P_s - C*(1 - P_s),$$

where (GV) is the gross value of the expected prize, P_s is the probability
of success, $(1 - P_s)$ the probability of failure, and C is the cost function,
as in the description of the make-up of the expected loss criterion of
section 4.6.

In all business situations, and particularly in the mineral
exploration business, the time factor is of great importance, and should be

Table 4.22. (Continued) Optimization of field programs for
the detection of vein-gold deposits of
Australasia.

Optimization based upon efficiency model & Dynamic Programming

Type of deposit: Vein Gold

Region: Western Australian Shield

Grid design: Square grid with spacings S x S meters

Confidence intervals: all results are reported as 95% confidence intervals respectively

Expected shape ratio, R = B/L = 0.19

Unit cost   = $66 per meter

Optimal grid orientation: 27 degrees to expected strike line

(iv) 55 degree angled drilling programs

	l.c.l.	geom. mean	u.c.l.
Expected target length in meters	430 m	595 m	820 m
Optimal grid spacing in meters	215 m	245 m	305 m
Cost in $1,000's per km square	235	185	125
Probability of detection	0.73	0.84	0.95

(v) Angled drilling programs at optimal angle = 45 degrees

Diamond drilling : unit cost = $66 per meter

	l.c.l.	geom. mean	u.c.l.
Expected target length in meters	430 m	595 m	820 m
Optimal grid spacing in meters	215 m	275 m	335 m
Cost in $1,000's per km square	270	175	125
Probability of detection	0.87	0.86	0.86

introduced in the expression of the pay-off model. Even when a mineral
exploration program has a successful outcome, normally several years lapse
before the production stage eventually leads to a cash flow over the life of
the mine. In the meantime, the money has to be borrowed at a cost (interest)
to finance the exploration and pre-production programs. Therefore the
expected monetary value (total gross value of the mine) and the cost
function (capital expenditure for exploration and pre-production programs)
cannot be compared just in terms of dollars: they are not made up of the
same dollars. Dollars of today and tomorrow do not have the same value, even
if we disregard the effect of inflation, because of the interest involved
(cost of capital). Therefore "discounting" must be introduced.

Table 4.23. Ranking of ore deposit types and regions in Europe, and Australasia and East Asia based on the expected loss criterion.

Field exploration approach →		Airborne detection			Ground vertical detection		
Continent & Region ↓	Ore deposit type ↓	Type of survey or program & grid geometry					
		Airborne geophysical		Ranking *	Ground program: square grid		Ranking ***
		Parallel grid	square grid		Geophys. survey	Vert. drilling	
		Expected loss in US $ / km square of coverage					
Europe							
Iberian Peninsula } Scandinavian Shield	Volcanogenic massive sulf.	$33	$21	1	$68	$22,305	2
East Mediterranean	Volcanogenic massive sulf.	$37	$45	2	$7	$2,525	1
Scandinavian Shield	Ni-Cu ultramafic	$73	$100	3	$417	$185,985	3
Australasia & East Asia							
South Pacific Island Arc } Tasman Geosyncline	Porphyry - Cu-Mo	$22	$18	1	$46	$7,876	2
West Australian Shield	Volcanogenic massive sulf.	$33	$21	2	$68	$22,305	3
Kuroko Belt of Japan	Volcanogenic massive sulf.	$37	$45	3	$7	$2,602	1
East Asia	Contact metason. (W-Mo)	$43	$31	4	$56	$23,397	4
West Australian Shield	Ni-Cu ultramafic	$43	$46	4	$237	$86,872	6
Tasman Geosyncline	Volcanogenic massive sulf.	$51	$78	6	$67	$90,000	7
West Australian Shield	Vein - Gold	-	-	-	$180	$62,550	5

N.B.   * Parallel grid   *** Vertical drilling

Discounting means that one must decide what is the present value of the future dollars that one expects to generate, and then what the annual discount rate %, r, should be after making allowance for the uncertainty about the amount of income to be generated in the future n years of the life of the mine. The discounting factor is expressed as

$1/(1 + r)^n$.

A proper design of the discounted pay-off criterion should balance the present value of profit generated by the operation over a period of n_1 years obtained by discounting it at the rate of r_1, against the present value of exploration and pre-production costs discounted at a rate r_2 over a period of n_2 years. A realistic estimate of the expected gross profit can be derived from the total gross value of the ore deposit by scaling it down to 20% of the gross value which is considered as a reasonable expectation for an average mining venture.

The discounted pay-off criterion is then defined by

$$\text{discounted pay-off} = 0.20*(\text{gross value})*P_s/(1 + r_1)^{n_1} -$$

$$(1 - P_s)*(\text{cost})/(1 - r_2)^{n_2}.$$

Generally the mine life, n_1, varies between 10 and 25 years, and the exploration and pre-production period, n_2, varies between 3 to 5 years.

In the present financial environment, we take both the discount rates, r_1 and r_2, to be 12%, and the life of the mine to be 20 years for large deposits such as porphyry Cu-Mo, but only 10 years for smaller deposits such as the contact-metasomatic W-Mo of the North American Cordillera. For the exploration and development time, we take $n_2 = 3$ years.

The discounted pay-off criterion is most useful at the planning stage to determine if a specific venture may be expected to be viable, and secondly to compare the relative merits of several projects and select the ones with the largest discounted pay-offs.

4.7.2. Application of the discounted pay-off criterion to the optimal selection of region-ore deposit types in two continents

As we did in Section 4.6, we will consider two types of ore deposits in three regions of Europe, and five types of ore deposits in five regions of Australasia and East Asia, but, this time, we will rank them on the basis of the discounted pay-off criterion instead of the detectability criterion. The numerical values of the pay-off are obtained by combining the mean gross values of the various deposit types as derived from statistical modelling in Chapter 2 with the data culled from corresponding optimization tables of Section 4.5. The results for Europe are listed in the upper portion of Table 4.24, and those for Australasia and East Asia are shown in the lower portion of the table.

Among the European deposits, the Iberian Peninsula pyritic volcanogenic massive sulfide deposits show by far the largest discounted pay-off. This ordering parallels that obtained previously when using the detectability yardstick in Section 4.6, at least in case of airborne detection. If we consider ground detection, however, the East Mediterranean volcanogenic sulfides occupy the first rank for detectability, but the region is demoted to second position in the pay-off ranking list. Similarly, the Scandinavian Shield volcanogenic sulfide deposits are demoted from second position in the ground detectability list to third in the discounted pay-off list.

Turning now to the Australasian and East Asian deposits, we find that the porphyry-Cu-Mo deposits of the South Pacific Island Arc occupy first rank in the list based upon the discounted pay-off criterion, while the East Asian W-Mo deposits come last. The Kuroko volcanogenic sulfide deposits of Japan and Western Australian Ni-Cu deposits occupy a near median-position (median pay-off value = 9.1). If we compare Table 4.23. and 4.24., we notice some major differences between the ranking based on detectability and on pay-off value, with the exception of the porphyry-Cu-Mo deposits of the South Pacific Island Arc and the Tasman Geosyncline. For instance, the East Asian W-Mo deposits which come last in the pay-off list occupy a median

Table 4.24. Ranking of ore deposit types and regions in Europe, and Australasia and East Asia based on the discounted exploration pay-off criterion.

Continent & Region	Ore deposit type	Discounted Pay-off in 1982 US $ Millions	Ranking
Europe			
Iberian Peninsula	Volcanogenic massive sulf.	36.2	1
East Mediterranean	Volcanogenic massive sulf.	11.4	2
Scandinavian Shield	Volcanogenic massive sulf.	9.3	3
Scandinavian Shield	Ni-Cu ultramafic	3.2	4
Australasia & East Asia			
South Pacific Island Arc	Porphyry - Cu-Mo	19.2	1
Tasman Geosyncline	Porphyry - Cu-Mo	12.3	2
Tasman Geosyncline	Volcanogenic massive sulf.	11.4	3
Kuroko Belt of Japan	Volcanogenic massive sulf.	11.0	4
West Australian Shield	Ni-Cu ultramafic	7.3	5
West Australian Shield	Vein - Gold	4.8	6
West Australian Shield	Volcanogenic massive sulf.	4.1	7
East Asia	Contact metasom. (W-Mo)	1.3	8

position in the detectability list. Conversely, the Ni-Cu deposits of the Western Australian Shield, which lie near the median position in the pay-off list are ranked much lower in the ground detectability list.

4.8. CHAPTER 4: Exercises

E4 Exercises

Based upon the results of the E2 exercises, for the geometric mean of the length, and the expectation for the shape ratio, B_t/L_t, find the following detection probabilities, or other detection measures, using the appropriate progam in the OPTGRID suite, given in Appendix 5.

E4.1. Contact Metasomatic (Cu-Fe-Au) data set 8.1.10 : find the airborne detection probabilities with grid spacings from 300 to 1000 meters in 100 m steps (1) with parallel grid, and (2) with square grid flight patterns.

E4.2. Volcanogenic sulfides (Iberia) data set 8.1.8 : find the airborne parallel grid spacings required to achieve confidence level for the detection probability of from 0.1 to 0.9 in 0.1 steps.

E4.3. Vein Gold (North American Shield) data set 8.1.4 : find the detection probabilities with a vertical drilling program with grid sizes ranging from 100 to 300 meters, step 50 m.

E4.4. Ni-Cu Ultramafic (N. American Shield) data set 8.1.6: determine the grid sizes required in 55°-angled drilling on square grids to achieve a detection probability level of from 0.1 to 0.9, step 0.1.

E4.5. Vein Gold (W. Australian Shield) data set 8.1.4: using the Optimal Search program 5.6. Dynamic Programming, find the optimal grid size and angle for angled drilling to maximize the efficiency model.

E4.6. Ni-Cu Ultramafic (W. Australian Shield) data set 3.1.6: using the Optimal Search programs 5.4 & 5.5 calculate the grid size to maximize the Efficiency function for airborne surveys on (i) parallel and (ii) square grid). for this region-deposit type.

E4.7. Vein Gold (N. American Shield) data set 8.1.3: using the Optimal Search program 5.6. Dynamic Programming, find the optimal grid size and angle for angled drilling to maximize the efficiency model.

BIBLIOGRAPHY FOR CHAPTER 4

1.   AGOCS, W.B., 1955, Line spacing effect and determination of
optimum spacing illustrated by the Marmora, Ontario, magnetic
anomaly; Geophys., B.20, N.4, pp.871-885.

2.   BELLMAN, R.E. and DREYFUS, S.E., 1962, Applied Dynamic
            Programming; Princeton University Press, New Jersey.

3.   CELASUN, M.M., 1964, The allocation of funds to reconnaissance
            drilling prospects; Colorado Sch. of Mines Quarterly,
            V.59, No.4, pp.169-186.

4.   CHUNG, C.F., 1981, Application of the Buffon needle problem
            and its extensions to parallel-line search sampling
            schemes; Jour. Math. Geol., V.13, No.5, pp.371-390.

5.   DE GEOFFROY, J. and WU, S.M., 1970, Design of an optimal
            sampling plan for regional geochemical surveys;
            Econ. Geol., V.65, pp.340-347

6.   DREW, L.J., 1967, Grid-drilling exploration and its
            application to the search for petroleum; Econ. Geol.,
            V.62, pp.698-710.

7.   DREW, L.J., 1979, Pattern drilling exploration: optimum
            pattern types and hole spacing when searching for
            elliptical targets, Jour.Math.Geol., V.11, No.2,
            pp.223-254.

8.   ELLIS, R.M. and BLACKWELL, J.G., 1959, Optimum prospecting
            plans in mineral exploration; Geophys., V.24, No.2,
            pp.344-358.

9.   GRAYSON, C.J. Jr., 1960, Decisions under uncertainty, drilling
            decisions by oil and gas operators; Graduate School
            of Business Administration, Division of Research,
            Harvard University, Boston, Mass.

10.   GRIFFITHS, J.C., 1966, Grid spacing and success ratio in
            exploration for natural resources; Pennsylvania State
            Univ., Min.Ind.Expmt.Stat., Special Publ.No.1.

11.   HARBAUGH, S.W. and BONHAM-CARTER, G.R., 1970, Computer
            simulation in Geology, (Chapter 8); J. Wiley, New
            York.

12.   JACOBS, O.L.R., 1967, An introduction to Dynamic Programming:
            the theory of multistage decision processes; Chapman
            and Hall, London, U.K.

13.   KENDALL, M.G. and MORAN, P.A.P., 1963, Geometrical
            probability; Chas. Griffith, London, U.K.

14.   KOCH, G.S. and LINK, R.F., 1970, Statistical analysis of
            geological data, V.2, (Chpts. 12 & 14); J. Wiley, New
            York, USA.

15.   McCAMMON, R.B., 1977, Target intersection probabilities for
            parallel line and continuous grid types of search;
            Jour. Math. Geol., V.9, No.4, pp.369-382.

16. MICKEY, M.R. and JESPERSEN, H.W., 1954, Some statistical problems of uranium exploration; U.S. Atomic Energy Commission Report, RME-3105.

17. PETERSON, E.L., 1961, Statistical analysis and optimization of systems; John Wiley, New York.

18. SAVINSKII, I.D., 1965, Probability tables for locating elliptical underground masses with a rectangular grid; Consult. Bureau, New York.

19. SHURYGIN, A.M., 1976, Discovery of deposits of given size by boreholes with pre-selected probability; Jour. Nath, Geol.., V.8, No.1, pp.85-88.

20. SHURYGIN, A.M., 1976, The probability of finding deposits and some optimal search grids; Jour.Math.Geol., V.8, No.3, pp.323-330.

21. SINCLAIR, A.J., 1975, Some considerations regarding grid orientation and sample spacing, in "Geochemical Exploration 1974", Elliot, I.L. and Fletcher, W.K., (Eds.), Elsevier, Amsterdam, Holland.

22. SINGER, D.A. and WICKHAM, F.E., 1969, Probability tables for locating elliptical targets with square, rectangular, and hexagonal point-nets; Pennsylvania State Univ.Miner.Sc.Expmt.Stat. Special Publ. No.1-69.

23. SINGER, D.A., 1972, ELLIPGRID, a FORTRAN IV program for calculating the probability of success in locating elliptical targets with square, rectangular, and hexagonal grids; Geocom Programs, No.4, pp.1-16.

24. SINGER, D.A., 1975, Relative efficiencies for square and triangular grids in the search for elliptically-shaped resource targets; U.S. Geol. Survey, Jour. of Research, V.3, No.2, pp.163-167.

25. SINGER, D.A., 1976, RESIN, a FORTRAN IV program for determining the area of influence of samples or drill holes in resource target search; Computers and Geosciences, V.2, No.2, pp.249-260.

26. SIVAZLIAN, B.D. and STANFEL, L.E., 1975, Optimization techniques in Operations Research; Prentice-Hall, Englewood Cliffs, N.J.

27. SLICHTER, L.B., 1955, Geophysics applied to prospecting of ores; Economic Geol., Jubilee Vol., No.50, pp.885-969.

28. SOLOMON, H., 1978, Geometric probability; Soc.Ind.Appl Math. (S.I.A.M.), Pennsylvania, USA.

29. STONE, L.D., 1975, Theory of optimal search; Academic Press, New York, USA.

30. TRUEMAN, R.E., 1974, An introduction to quantitative methods for decision-making, (Chpt.4); Holt-Rinehart-Winston, New York, USA.

31. WIGNALL, T.K. and DE GEOFFROY, J., 1985, OPTGRID, an improved
     program in BASIC for locating elliptically shaped
     resource targets by ground surveys and drilling
     programs on square grids; Jour.Math.Geol. (in press).

32. WILDE, D.J., 1964, Optimum seeking methods; Prentice Hall,
     Englewood Cliffs, N.J.

CHAPTER FIVE

APPLICATION OF THE GENERAL LINEAR MODEL (GLM) AND OPERATIONS RESEARCH (OR)
MODELS TO THE OPTIMIZATION OF FIELD EXPLORATION AND DEVELOPMENT PLANNING

5.1. THEORETICAL BACKGROUND FOR THE GLM MODEL

5.1.1. Introduction

It is well recognized that there are important relationships between
quantitative or quantified geo-variates and the geological, geophysical and
geochemical environment of mineral deposits, some of which are diagnostic of
the occurrence of ore. However, since there is still much uncertainty as to
the precise nature of these relationships, we are introducing the GLM in
order to make the best possible use of all correlations and interdependence
which reflect statistically the very complex nature of the environment of
ore deposits. The GLM, itself an optimized model, can be used to advantage
for optimizing various mineral exploration procedures such as the location
and delineation of exploration targets and the evaluation of their economic
worth.

In order to ensure the validity of the GLM, the following assumptions
are made:

1. The dependent variates are normally distributed. Since many geo-
    variates are lognormal, a logarithmic transformation ensures a normal
    distribution in these cases.
2. The "errors", i.e. the least-squares deviations of the GLM model values
    from the observed values, are independent normally distributed with
    mean 0, and common variance (homoscedascity).

However, Agterberg (5) has shown that heteroscedascity may be largely
eliminated by means of logarithmic transformations applied to the
geovariates.

The GLM is based on the least-squares regression theory and is thus an
optimized model. It minimizes the sum of the squares on the error, e, which
implies that the variance is minimized.

The GLM is defined by the following linear relationship:

$Y = XA + e,$

where Y is an (n x s) matrix of s dependent variates, n being the sample
size,

X is an (n x r) matrix of r independent variates of full rank (no variate
may be a linear combination of the others) representing quantified
geological factors and geophysical and geochemical survey readings.

A is the (r x s) coefficient matrix of unknown parameters which we wish to estimate to minimize the sum of the squares on each of the s components of the errors.

Finally, e is the (n x s) error matrix.

When dealing with the evaluation and selection of exploration targets, the s dependent variates are of an economic nature, such as gross dollar value, tonnage or dollar grade of ore deposits. However, if we are interested in the location and delineation of exploration targets, Y is a matrix of s survey variates, such as magnetic, radiometric and seismic readings, observed at n locations over the search area. X is an (n x r) matrix of independent variates of full rank, such as geologic coding and location and topographic variates, where r < n.

For the purpose of testing the significance of the regression or trend, the matrix Y is assumed to comprise a set of multivariate normally distributed dependent variates, some of which may be a linear combination of the others, for example the R-mode factors. The error matrix, e, comprises s independent normal variates, each with expectation 0. As previously stated, the matrix of design and independent variates, X, must be of full rank, and this implies that no variate shall be a linear combination of the others.

The GLM has many other applications of interest to explorationists. One of them which will be discussed below is the optimal selection of independent variates to be used for the construction of geo-statistical models. The relevance and significance of each of the independent variables are diagnosed so that they can be ranked in order of importance, and discarded if desired. Due to uncertainty, however, it is a good practice to include all information in the model, knowing that some may be discarded later.

5.1.2. Formulation of the General Linear Model

5.1.2.1. General statement

The GLM is a very versatile model which can easily be computerized for application to many geomathematical problems encountered in mineral exploration situations. We will briefly cover the following cases below:

(a) least squares regression analysis,
(b) trend analysis,
(c) factor analysis (R and Q modes), and
(d) trend factor & residual trend factor analyses.

5.1.2.2. Least-squares regression analysis

The general theory of least-squares regression analysis is to minimize

$$Z = e'e = (Y - XA)'(Y - XA)$$

Partially differentiating with respect to each x_j, j = 1, ... , r, and equating the derivative of each y_i, i = 1, ... , s, to zero in order to minimize Z, yields:

$$XA = Y$$

thus

$$X'XA = X'Y,$$

Hence

$$A = (X'X)^{-1} X'Y$$

where ' and $^{-1}$ indicate transposition and inversion, respectively.

For example, the least squares estimators, a, b, for α, β in the regression line model

$$y = \alpha + \beta x + e$$

are given by

$$(a, b) = [(1, X)'(1, X)]^{-1}X'Y$$

where 1 is the (n x 1) design vector of 1's, which is common for all GLM models, and X, Y are the (n x 1) vectors of independent and dependent variates respectively.
Hence

$$a = (\Sigma x^2 \ \Sigma y - \Sigma x \Sigma xy)/(n\Sigma x^2 - \Sigma x \Sigma x),$$
$$b = (n \ \Sigma xy \ - \ \Sigma x \Sigma y)/(n\Sigma x^2 - \Sigma x \Sigma x),$$

where, here and elsewhere in the text, unless otherwise stated

$$\Sigma = \Sigma_1^n \ ,$$

The least squares sum, Z, may now be found by substituting

$$A = (a, b),$$

in the model with

$$XA = a + bx$$

and from the Gauss-Markov theorem, we hold that a, b are the best linear unbiased estimators (BLUE) of α, β, and yield the minimum variance, s_e^2, the error variance, given by

$$s_e^2 = Z/(n - 2).$$

To test the significance of the regression, an ANOVA test is applied, using the following definitions:

Total sum of squares, SST, is given by

$$SST = \Sigma(y - \bar{y})^2,$$

the error sum of squares, SSE, is given by

$$SSE = Z, \text{ (see above)}$$

and the regression sum of squares, SSR, is given by

$$SSR = SST - SSE$$

The ANOVA table is then formulated as in Chapter 3 as follows:

Variation source	Degrees of freedom	SS	MS	$F(r, n-r-1)$
Regression	$r = 1$	SSR	SSR/r	
Error (Residual)	$n - r - 1 = n - 2$	Z	Z/(n-r-1)	MSR/MSE
Total	$n - 1$	SST		

The statistic $F = MSR/MSE$ is F-distributed with $r, n - r - 1$ degrees of freedom.

The coefficient of determination, R^2, is given by

$$R^2 = SSR/SST.$$

R^2 measures the explanatory power of the GLM, giving the proportion of total variability accounted for by the regression model. Furthermore, in the bivariate case, (x, y), R is the correlation coefficient $r_{x,y}$, and is the multiple correlation coefficient in the multiple regression model.

To test the significance of the difference between a and a_0, we use the t-statistic expressed as

$$t = (a - a_0)/s_a,$$

where s_a is given by

$$s_a^2 = s_e^2(1 + \bar{x}^2/s_x^2)/n,$$

where: $\bar{x}$ and s_x^2 are the mean and variance of x respectively.

The statistic t is Student-t distributed with $n - 2$ degrees of freedom.

Similarly, the significance of the difference between b and b_0 is tested with s_b^2 given by

$$s_b^2 = s_e^2/n/s_x^2.$$

The field examples given in Section 5.2 will illustrate the formulation of the GLM and the testing of its quality.

5.1.2.3. Trend analysis

The GLM is also used in trend analysis, with the independent variate, x, being distance or time. In the case of multivariate, X is a matrix of geographical and sometimes also topographical co-ordinates, as well as design geologic variates. The model when formulated is evaluated to estimate Y by substitution of intermediate values of X, that is by interpolation. The Cape York bauxite example given in Sub-section 5.3.3.2 will demonstrate this procedure.

5.1.2.4. R-mode factor analysis

The variance-covariance (s x s) matrix, S, is formulated from the Y dependent (n x s) matrix, by calculating its (i, j)th element, s(i, j), using the definition

$$s(i, j) = (n\Sigma y_i y_j - \Sigma y_i \Sigma y_j)/n/(n-1), \text{ for } i, j = 1, 2, ..., s$$

The standard deviation, s_i, of variate y_i, is defined by

$$s_i = [s(i, i)]^{0.5}, \qquad \text{for } i = 1, 2, ..., s$$

The correlation (s x s) matrix, R, is then formulated from the S matrix, by calculating its (i, j)th element, r(i, j), using the definition

$$r(i, j) = s(i, j)/s_i/s_j, \qquad \text{for } i, j = 1, 2, ..., s$$

The eigen-values and associated eigen-vectors are then calculated from the R matrix, by forming the characteristic polynomial equation of degree s defined by

$$\text{Det}(R - \lambda I) = 0,$$

where Det = Determinant.

The eigenvalues are the roots of this equation, but only the significant ones, i.e. the largest ones, are required.

The Principal Component Factors, C, are the eigen-vectors associated with the significant eigenvalues. They are calculated by substituting each eigenvalue into the matrix $(R - \lambda I)$, and solving the matrix equation

$$(R - \lambda I)C = C.$$

Finally, the R-mode factors are calculated from the Principal factors, using varimax rotation. However, the above procedures are usually carried out using a statistical package, such as the SPSS system program given in the appendix. The factors are then used to evaluate the factor scores at each location, which assists in classifying locations as anomalous or otherwise, as shown in Chapter 6.

5.1.2.5. Q-mode factor analysis

The procedure used is similar to the one given in 5.1.2.4., but the factors are calculated from an (n x n) S matrix of generalized cosine similarity measures, which is formulated from the Y' matrix. An example from petroleum exploration will give a clear exposition of the technique and its power. Suppose r electric log readings are taken for a sample of n wells or locations, with resulting vectors

$$\underline{x}_i = (x_{i1}, x_{i2}, \ldots, x_{ir}), \quad i = 1, 2, \ldots, n,$$

then define the coefficient of similarity, s_{ij} between well i and well j by

$$s_{ij} = \cos(\underline{x}_i, \underline{x}_j) = \underline{x}_i \cdot \underline{x}_j / \text{Mod } \underline{x}_i / \text{Mod } \underline{x}_j$$

$$= \Sigma_{k=1}^{r} x_{ik} x_{jk} / [(\Sigma_{k=1}^{r} x_{ik}^2)(\Sigma_{k=1}^{r} x_{jk}^2)]^{\hat{}} 0.5$$

for   i, j = 1, 2, ..., n,
and the similarity matrix is defined by

$$S = \{ s_{ij} \}, \quad i = 1, 2, \ldots, n; \quad j = 1, 2, \ldots, n.$$

Factor analysis is now applied to the S matrix, with the calculations similar to R-mode, but operating now on an n x n matrix. The resulting factors, starting with F_1, the factor accounting for the highest percentage of variation between well, then F_2, 2nd highest, etc., but usually with only the first three or four factors being significant in field examples.

This time the classification, discussed at length in Chapter 6, is made by observing which coefficient loading, a_i, is the highest for each factor. For example

$$F_1 = a_1 x_1 + a_2 x_2 + a_3 x_3 + \ldots + a_n x_n$$

is the factor associated with the geo-variate accounting for the largest proportion of variation between the n wells, and the well of key interest on that factor is the one with the largest a_i. In fact, the program POPMIX, given in Appendix 7 classifies two groups of wells on each variate factor. This aspect will be covered more thoroughly in Chapter 6.

5.1.2.6. Trend factor analysis and residual trend factor analysis

The GLM is used in the following manner. The Y dependent matrix is augmented with the Principal Factor scores evaluated at each of the n locations; thus it is now an [n x (s + k)] matrix, where k is the number of significant factors, usually 1, 2, or 3.

The least-squares estimate of each Principal Factor is evaluated for each of the n locations. It is then subtracted from the Principal Factor score at the location to yield the Residual Trend Factor. The GLM is now applied to calculate the Residual Trend Function for each of the Principal Factors to detect anomalous locations significantly above or below the trend. A 10% significance level is usually chosen for this purpose, and thus if the absolute value of the residual exceeds $1.645*(S_e^2/n)^{0.5}$, where S_e^2 is the variance of the Residual Trend scores, then the location is defined to be anomalous, and is termed "local anomaly".

5.2. APPLICATION OF THE REGRESSION ANALYSIS TO VARIOUS MINERAL EXPLORATION PROBLEMS

5.2.1. Application of the GLM to development planning

5.2.1.1. Introduction

The capital investment required by the development of new mining ventures is huge: for example, a new bauxite project requires over $500 million if an alumina plant is to be installed in situ; even small open pit gold mines require an investment of several million dollars. Furthermore, estimates of the required amount of capital are subject to uncertainty due to the changing world demand and economic conditions.

Thus an important application of the GLM is the prediction of the capital requirement for a project under consideration based on the actual capital outlay and average production rate data collected from existing mines. Peters (22, section 2), and De Geoffroy & Wignall (25, section 1) have gathered valuable data on both underground and open pit mines, which we are using them to tackle the prediction problem by means of the linear bivariate regression model.

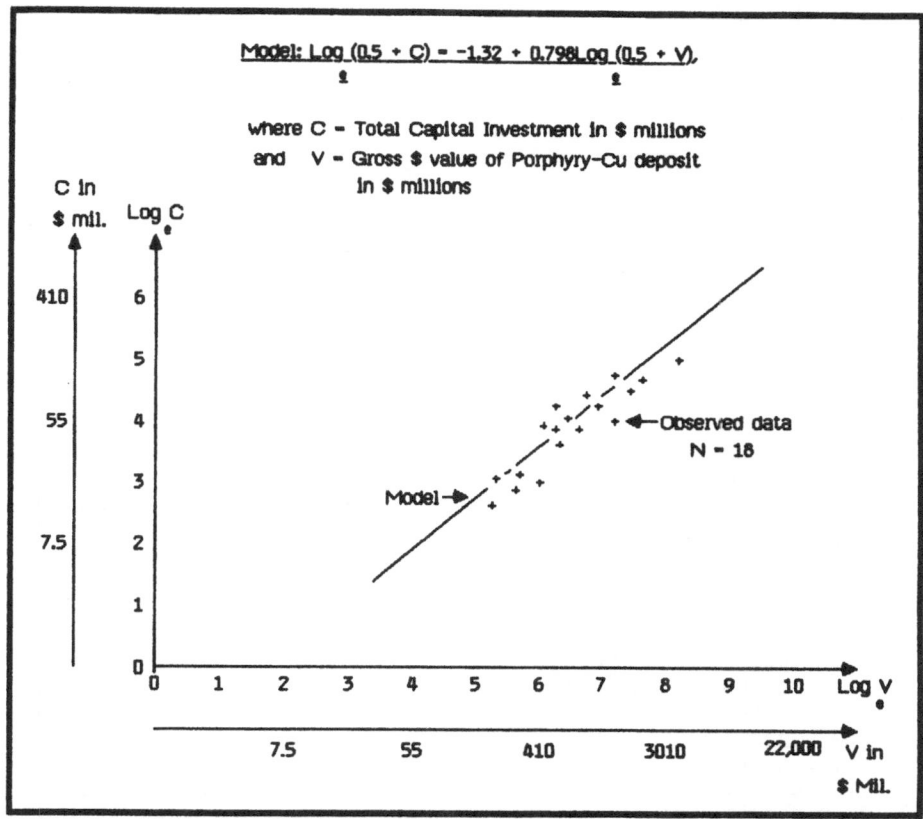

Figure 5.1. Fitting of linear regression model to observed relationship between total capital investment and gross value of porphyry-Cu-Mo deposits of the North and South American Cordillera.

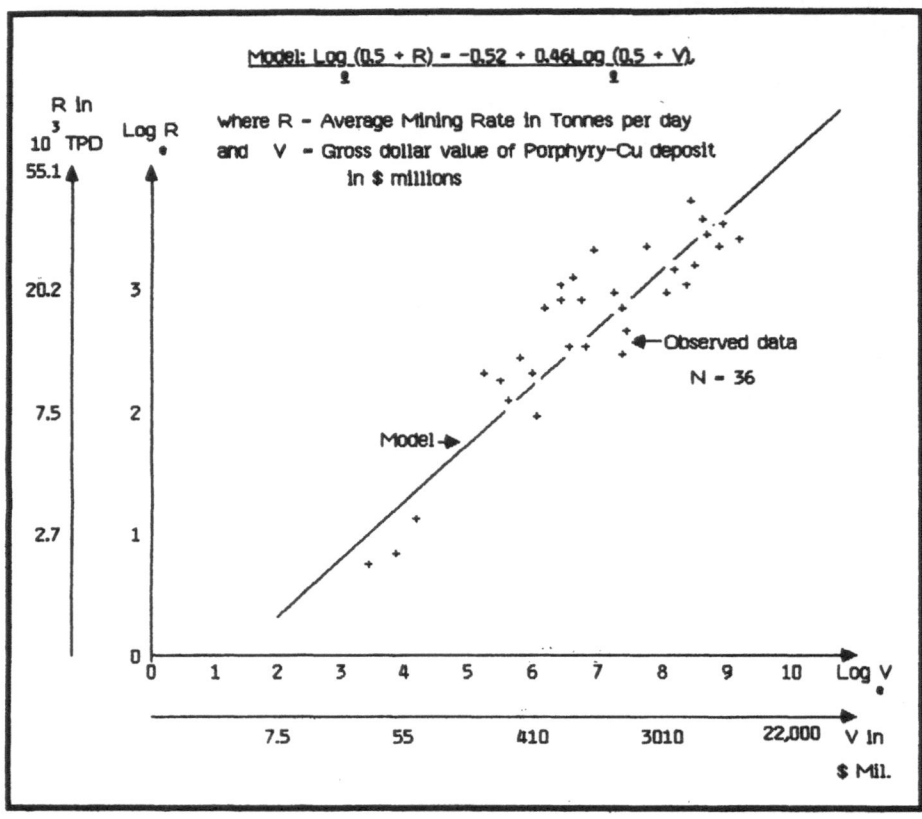

Figure 5.2. Fitting of linear regression model to observed
relationship between the average mining rate and
gross value of porphyry-Cu-Mo deposits of the
North and South American Cordillera.

Using the formulation given in 5.1.2.2., where r = 1, with two
independent variates including the unit vector column, the GLM yields the
BLUE estimates, a, b, where a is the y-axis intercept at x = 0, and b is the
slope of the line, measuring rate of change.

5.2.1.2. Application to development planning of porphyry-Cu-Mo deposits of the North American Cordillera

This example is taken from De Geoffroy & Wignall (25, section 1), and
the results are summarized in Fig. 5.1 and 5.2. The gross values of the 36
porphyry-Cu-Mo deposits of the North American Cordillera in the study were
shown to be Lognormal. Furthermore, in order to achieve significant linear
regression, logarithmic transformations were carried out also on the capital
cost requirement for development, C, and on the mining rate in tonnes per
day, R. As a result, the regression functions of Log(C) on Log(V), and of
Log(R) on Log(V) proved to be useful planning tools. For example, for a
typical porphyry-Cu-Mo orebody in the North American Cordillera Belt with an
average gross value of $2.45 billion, the expected capital outlay is $123
million (1968) for an expected mining rate of 21,000 tonnes per day.

5.2.1.3. Application to development planning of mine projects of various types on a world-wide basis

As mentioned above, the data for this example are provided by Tables 8-7 and 8-8 listed in Peters' book (22, section 2). The results of the application of the GLM to Peters' data are summarized in Fig. 5.3 for the lead time between discovery and production vs. average daily rate, and Fig. 5.4 for capital requirement vs. average daily rate, for both open pit and underground mines. It will be observed that the bivariate regression was formulated as Log(Lead time between discovery and development) on Log(Average mining rate in tonnes per day). The logarithmic transformations were made on both X and Y in order to achieve a significant regression.

Applying the results given in 5.1.2.2, the ANOVA results for the underground mines analysis yielded:

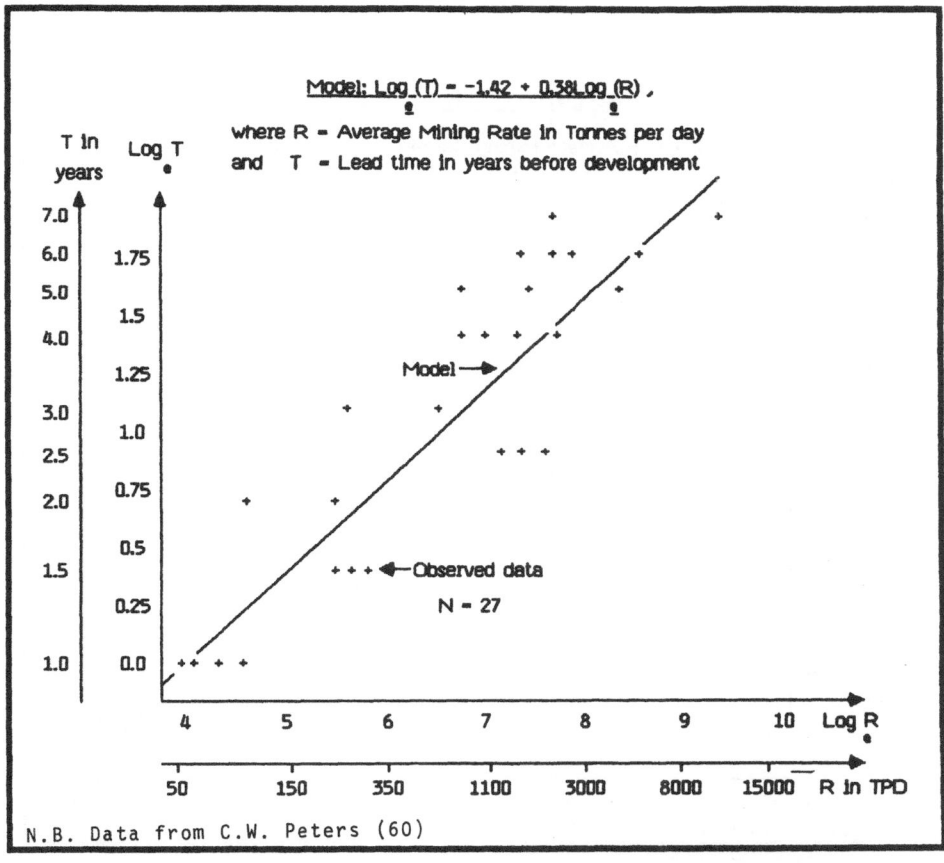

Figure 5.3 (a) Fitting of linear regression model to observed relationship between average production rate and pre-production lead time for underground mines.

193

Variation source	Degrees of freedom	SS	MS	F(1, 25)
Regression	r = 1	7.91	7.91	
Error (Residual)	n - r - 1 = 25	2.68	0.11	73.84
Total	n - 1 = 26	10.59		

The result is thus highly significant at the $\alpha \approx 0.0001$ level, as could be expected when the coefficient of determination R^2 equals 0.75 (= 7.91/10.59 ≈ 0.75).

Similarly for the open pit mines, the results for Log(T) on Log(R) were:

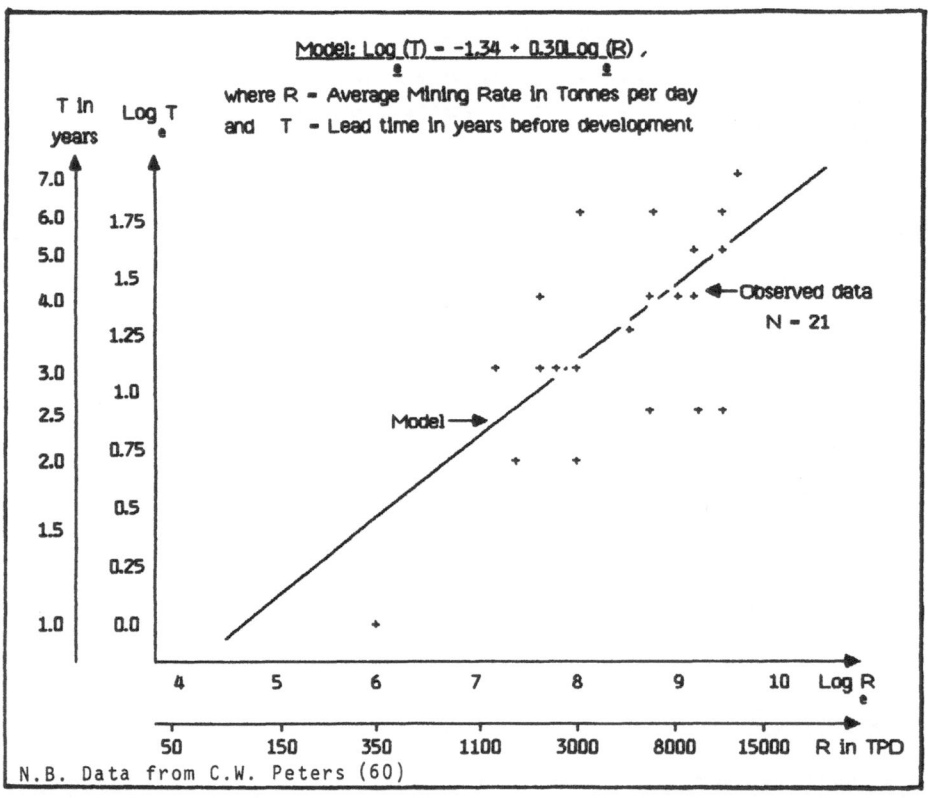

N.B. Data from C.W. Peters (60)

Figure 5.3 (b) Fitting of linear regression model to observed relationship between average production rate and the pre-production lead time for open pit mines.

Variation source	Degrees of freedom	SS	MS	F(1, 19)
Regression	$r = 1$	1.71	1.71	
Error (Residual)	$n - r - 1 = 19$	2.56	0.14	12.74
Total	$n - 1 = 20$	4.27		

which is again highly significant at the $\alpha \approx 0.0001$ level.

From the comparison of Fig. 5.3 (a) and 5.3 (b), it appears that the results for underground mines differ little from the results for open pit mines. It is now appropriate to demonstrate whether the regression line for open pit mines is significantly different from that for underground mines. For this purpose we test the null hypotheses, $\alpha_1 = \alpha_2$ and $\beta_1 = \beta_2$. Let us first test the null hypothesis: $\alpha_1 = \alpha_2$, for the intercepts. We have

$$s_a^2 = s_e^2 (1 + \overline{x}^2/s_x^2)/n,$$

where: $\overline{x}$ and s_x^2 are the mean and variance of x, respectively.

For the underground pits

$$\overline{x}_1 = 6.57, \text{ and } \qquad s_{x_1}^2 = 2.02,$$

hence the variance of α_1 is

$$s_{a_1}^2 = 0.088,$$

Similarly for the open pit mines

$$\overline{x}_2 = 8.43, \text{ and } s_{x_2}^2 = 0.89,$$

hence the variance of a_2 is

$$s_{a_2}^2 = 0.521;$$

Assuming independence of the two parameters, the variance of the difference, $a_1 - a_2$, is the sum of the variances $= 0.609$;

also under the null hypothesis, the statistic is t-distributed with 44 degrees of freedom and is expressed as

$$t = (a_1 - a_2 - 0)/(0.609)^{0.5}$$
$$= (-1.42 + 1.34)/(0.609)^{0.5}$$
$$= -0.10$$

The significance level is ≈ 0.92, inferring that we certainly may not reject the null hypthesis that the two intercepts are equal.

Turning to the testing of the null hypothesis: $\beta_1 = \beta_2$, for the slopes of the two regression lines, we have

$$s_b^2 = s^2/s_x^2/n,$$

where: $\bar{x}$ and s_x^2 are the mean and variance of x respectively.

and: for the underground mines, we have: $s_{x_1}^2 = 2.02$,

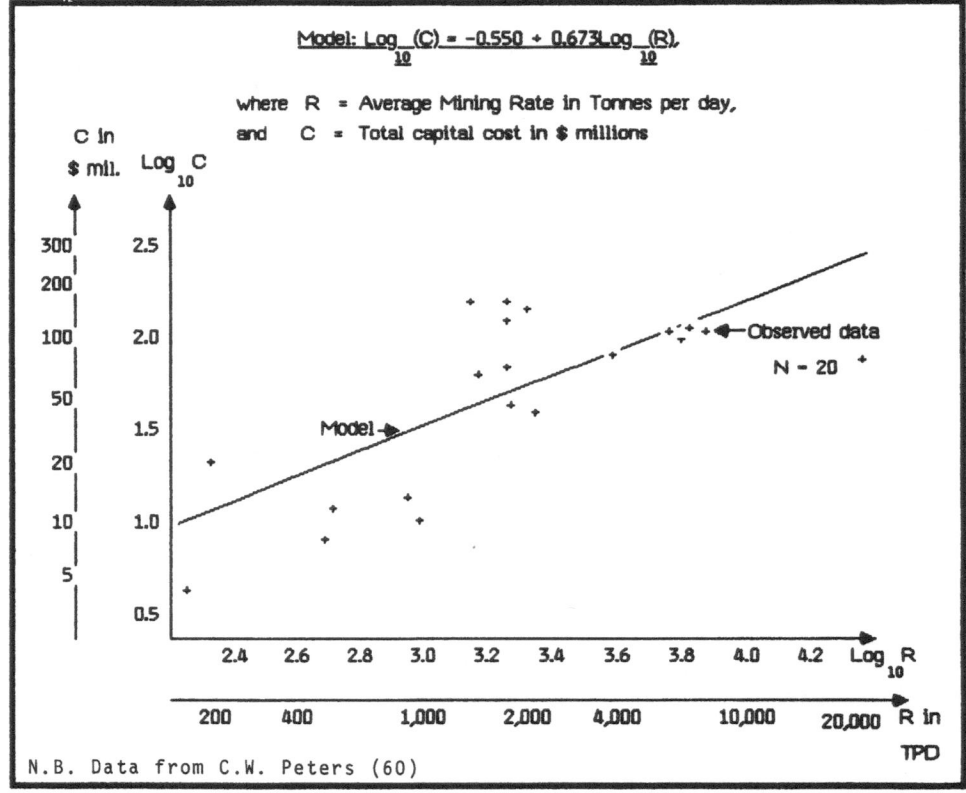

Figure 5.4 (a) Fitting of linear regression model to observed
relationship between average mining rate and
total capital cost for underground mines.

hence the variance of b_1 is $\qquad s_{b_1}^{2} = 0.053$,

Similarly, for the open pit mines $\qquad s_{x_2}^{2} = 0.89$,

and the variance of b_2 is $\qquad s_{b_2}^{2} = 0.152$;

Assuming independence of the two parameters, the variance of the difference, $b_1 - b_2$, is the sum of the variances $= 0.205$;

also under the null hypothesis, the statistic t is t-distributed with 44 degrees of freedom, where

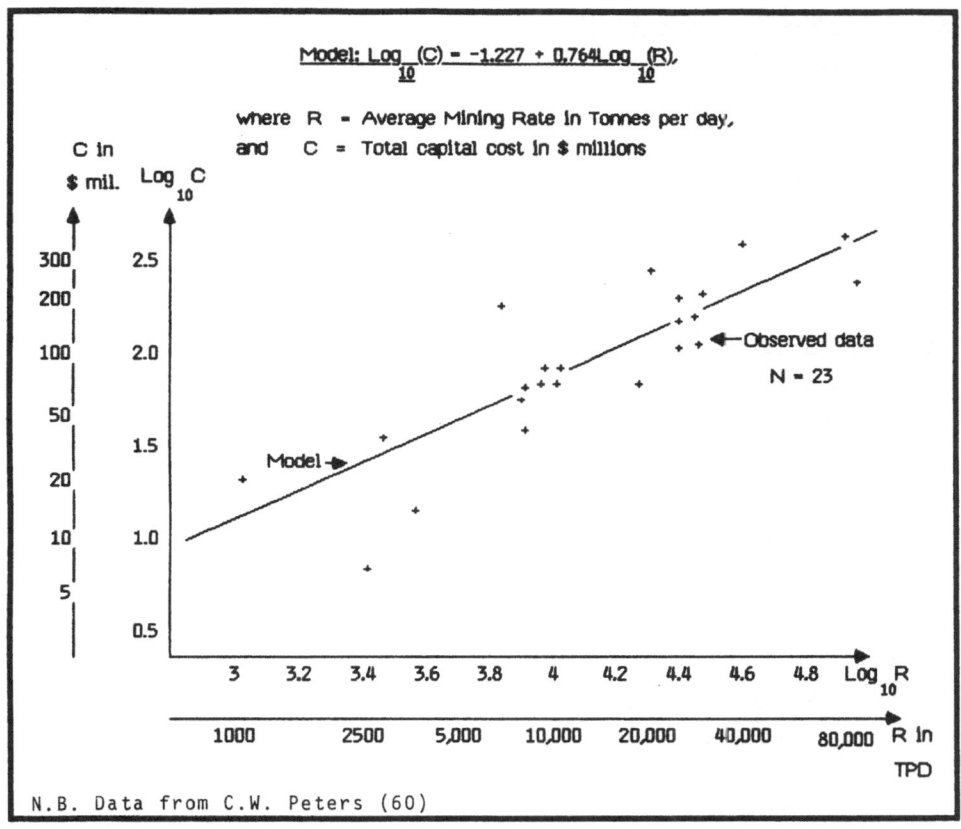

Model: $\mathrm{Log}_{10}(C) = -1.227 + 0.764 \mathrm{Log}_{10}(R)$,

where R = Average Mining Rate in Tonnes per day,
and C = Total capital cost in \$ millions

C in \$ mil. Log$_{10}$ C

Observed data
N = 23

Model →

N.B. Data from C.W. Peters (60)

Figure 5.4 (b) Fitting of linear regression model to observed
relationship between average mining rate and
total capital cost for open pit mines.

$$t = (b_1 - b_2 - 0)/(0.205)^{0.5}$$
$$= (0.38 - 0.30)/(0.205)^{0.5}$$
$$= 0.18$$

The significance level is ≈ 0.85, inferring that we certainly may not reject the null hypthesis that the two slopes are equal.

In conclusion, for planning purposes, it appears that there is no significant difference between the regression lines for underground and open pit mines, if we wish to predict the lead time between discovery and production in terms of the daily mining rate.

We are now turning our attention to the relationship between development capital requirement and daily production rate, and we will determine whether there is a significant difference between the regression functions for underground and open pit mines. (See Figs. 5.4 (a) and 5.4 (b)).

The logarithmic transformations were again essential on both X and Y in order to achieve a significant regression. Applying the results given in 5.1.2.2, the ANOVA result for the underground mines analysis yielded:

Variation source	Degrees of freedom		SS	MS	$F(1, 18)$
Regression	$r = 1$		2.41	2.41	
Error (Residual)	$n - r - 1 = 18$		2.21	0.12	19.62
Total	$n - 1 = 19$		4.62		

The result is thus highly significant at the $\alpha \approx 0.0001$ level, as expected with the coefficient of determination $R^2 = 2.41/4.62 \approx 0.52$

Similarly for the open pit mines, the results for Log(C) on Log(R) were:

Variation source	Degrees of freedom		SS	MS	$F(1, 21)$
Regression	$r = 1$		3.04	3.04	
Error (Residual)	$n - r - 1 = 21$		1.66	0.79	38.52
Total	$n - 1 = 22$		4.70		

The result is again highly significant at the $\alpha \approx 0.0001$ level.

We next test the null hypothesis: $\alpha_1 = \alpha_2$, for the intercepts.

We have
$$s_a{}^2 = s_e{}^2(1 + \bar{x}^2/s_x{}^2)/n,$$

where: $\bar{x}$ and $s_x{}^2$ are the mean and variance of x respectively.

For the underground pits

$$\bar{x}_1 = 3.26, \text{ and } s_{x_1}{}^2 = 0.27,$$

hence the variance of a_1 is

$$s_{a_1}^2 = 0.324,$$

and similarly, for the open pit mines:

$$\bar{x}_2 = 4.13, \text{ and } s_{x_2}^2 = 0.23,$$

hence the variance of a_2 is:

$$s_{a_2}^2 = 0.293;$$

Assuming independence of the parameters, the variance of the difference, $a_1 - a_2$, is the sum of the variances = 0.617;

also under the null hypothesis, the statistic t is t-distributed with 44 degrees of freedom and expressed as:

$$t = (a_1 - a_2 - 0)/(0.617)^{0.5}$$
$$= (-0.55 + 1.23)/(0.617)^{0.5}$$
$$= 0.88$$

The significance level is = 0.49, inferring that we certainly may not reject the null hypthesis that the two intercepts are equal.

Finally, we wish to test the null hypothesis: $\beta_1 = \beta_2$, for the slopes of the lines representing underground and open pit mines.

$$s_{x_1}^2 = 0.27,$$

hence the variance of b_1 is:

$$s_{b_1}^2 = 0.023,$$

and similarly for the open pit mines:

$$s_{x_2}^2 = 0.23,$$

hence the variance of b_2 is:

$$s_{b_2}^2 = 0.015;$$

Again assuming independence of the parameters, the variance of the difference, $b_1 - b_2$, is the sum of the variances = 0.038;

also under the null hypothesis, the statistic t is t-distributed with 44 degrees of freedom.

$$t = (b_1 - b_2 - 0)/(0.038)^{0.5}$$
$$= (0.67 - 0.76)/(0.038)^{0.5}$$
$$= -0.47$$

The significance level is ≈ 0.65, inferring that we certainly may not reject the null hypthesis that the two slopes are equal.

The conclusion of this study is that

(a) the regression functions of the log transforms of lead time to production vs. daily production rate, and of capital requirement for development vs. daily production rate are both highly significant from the statistical standpoint, and thus may be used effectively for interpolation for predictive purposes in development planning, and
(b) there is no significant difference between open pit and underground mine functions in both cases.

5.2.2. Application of GLM to field exploration planning

5.2.2.1. Introduction

The choice of dependent and independent geo-variates is crucial when constructing a GLM for the purpose of screening of exploration targets to be tested by drilling. The dependent variates should reflect some measurements of economic worth, and their nature will be largely dictated by the aims of the project and by constraints such as the availability of information. Expected gross dollar value, total ore tonnage, ore grade, and discounted net profit are the most common measures of economic worth that management may wish to focus upon. However, field explorationists may wish to predict the magnitude of geo-survey variates, such as seismic or geochemical readings, which are of great importance in the delineation and screening of exploration targets.

Independent geo-variates normally consist of qualitative data, such as geologic rock types and formations, as well as quantitative data including geographical and topographic coordinates, and geophysical and geochemical survey data. The qualitative data are best quantified by experimental design. For example, if there are two rock types of importance, and the rest adjudged to be of no interest, then the locations with rock type 1 would be given the coding 1, 0; for rock type 2, the coding would be 0, 1; and all others of no interest would be assigned the symbols -1, -1.

Generally, a large array of quantitative or qualitative geo-variates may be subjectively related to the occurrence of sought-after resources in the geologic environment prevailing in the search area. Objective statistical procedures are then used to screen the geo-variates for relevance and statistical significance. The first step is referred to as "data compression", and the second one as "feature selection", according to Pattern Recognition terminology. Data compression is used to reduce in an objective manner the size of the initial array of geo-variates to facilitate further statistical manipulations.

Feature selection is a critical stage of the procedure leading to the construction of a geostatistical model. Its main objective is to select from the array of independent variates only those which are statistically significant in the GLM for prediction of the dependent variates. This is achieved by step-wise regression analysis, when the GLM is a linear multiple regression function. It is designed to yield the BLUE for the dependent variate, giving the least-squares estimate for the dependent variate based upon the most relevant set of independent variates (5).

The GLM for multiple regression is written

$$Y_{(n \times 1)} = X_{(n \times r)} A_{(r \times 1)} + e_{(n \times 1)}$$

where the subscripts are the dimensions of the matrices. Generally speaking, the economic dependent variate, y, is lognormally distributed, and so it is advisable to transform logarithmically to ensure a normal distribution. Transformations of quantitative independent variates are often useful to achieve significant prediction results. In the instance that the dependent variate is lognormal, then this implies a multiplicative process, so that the independent variates in order to achieve significance will also need the logarithmic transformation before applying the GLM. In step-wise regression analysis programs, the regression function is formulated on the whole set, r, of the m dependent variates initially, and then re-calculated on the (r - 1) remaining ones, when each in turn is discarded, until an ANOVA test, conducted after each step, reveals which is the least significant result. This leads to the variate being dropped. The procedure is then continued on the remaining set of (r - 1) variates, until the best significance is achieved. However, Factor Trend Analysis will achieve the same goal, as will be shown in section 5.3.

5.2.2.2. Application to the screening of geo-variates for the selection of exploration targets

Figure 5.5 illustrates the screening of significant independent geo-variates in the case of the North & South American Cordillera porphyry-Cu-Mo deposits, (23, section 1). An array of 170 independent geo-variates were subjectively selected to describe productive porphyry-Cu-Mo deposits & their immediate environment. The variates were encoded by means of binary notation, (0, 1), and were "compressed" by Characteristic Analysis into 16 geological factors subsequently quantified in a manner relevant to field work. The 16 independent variates were subjected to a step-wise regression analysis in which the dependent variate was the gross dollar value of deposits. As a result, a total of twelve geological variates significantly related to economic worth and directly measurable in the field were screened out to assist in the selection of porphyry-Cu-Mo prospects in the Cordillera of North and South America.

5.2.2.3. Application to the evaluation of economic potential of prospecting areas

The selection of geo-variates closely related to the occurrence of ore deposits was again critical in the selection of prospecting areas in the Grenville region of the Canadian Shield, as described by De Geoffroy & Wignall (14, Chapter 6). A significant regression resulted from the logarithmic transformation of the dependent variate, the aggregate gross value of ore deposits occurring within the unit area (cell), and the logarithmic transformation of nine quantified independent geo-variates.

The significance of each independent variate was determined by a t-test, using the standard error of each coefficient, a_i, in a manner

201

analogous to that given in section 5.1.2.2. However, the expressions are complex and the procedures very tedious, thus requiring the use of computer packages, such as SPSS (see Appendix 5). Insignificant variates other than x_1, the dummy, were discarded, reducing the number of independent variates from 10 to 5 and increasing the ANOVA F-value while maintaining the same significance at 0.001. Trend Factor Analysis would achieve the same result, but the meaning of the factors would be more difficult to interpret.

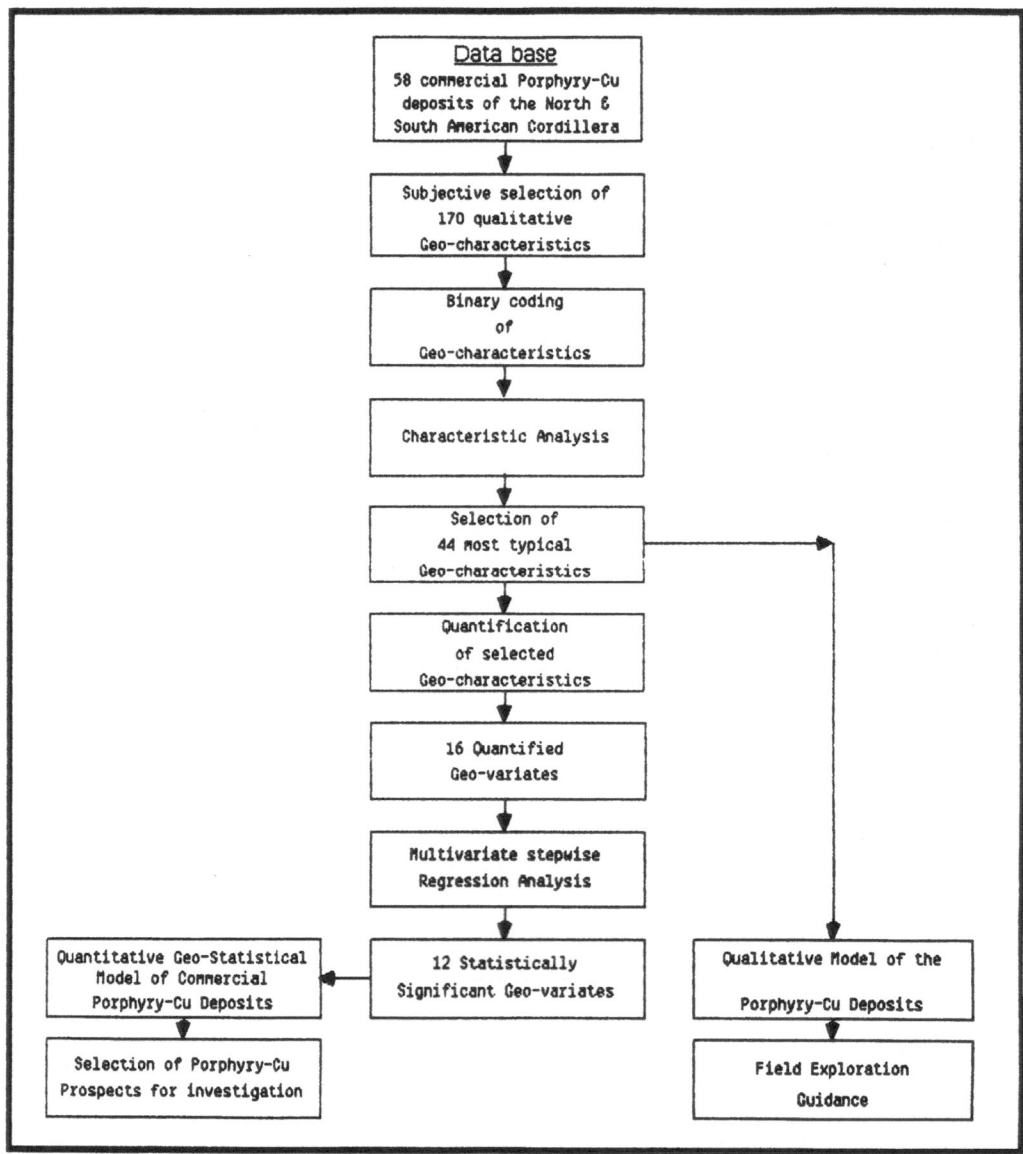

Figure 5.5. Flow diagram of the selection of statistically significant geo-variates for the optimized search for porphyry-Cu-Mo deposits of the Cordilleran Belt of North and South America.

Table 5.1. Selection of statistically significant geo-variates to guide the search for porphyry-Cu-Mo deposits of the British Columbia section of the North American Cordillera.

No.	INITIAL LIST OF QUANTIFIED GEOLOGICAL FACTORS	STATISTICAL SIGNIF.
1	Angle of departure of average structural trend in cell from average Canadian Cordilleran trend (N38°W), in degrees	99%
2	Number of intersecting structural trends within cell (fold axes faults, bedding trend, etc.)	94%
3	Area of Triassic and Lower Jurassic in cell (Takla, Karmutsen, Nicola, Hazelton) (%)	76%
4	Area of Late Cretaceous - Early Tertiary formations (Kingsvale, Skokum, Spence's Bridge, Carmacs) (%)	0
5	Area of volcanics older than mid-Tertiary (%)	99%
6	Area of Early Mesozoic intrusives (200-150 m.yrs) (%)	76%
7	Area of mid-Mesozoic intrusives (150-80 m.yrs) (%)	84%
8	Area of Tertiary intrusives (80-40 m.yrs) (%)	76%
9	Area of acid intrusives (granite) (%)	0
10	Area of intermediate intrusives (granodiorite, monzonite, quartz monzonite) (%)	79%
11	Area of intermediate-basic intrusives (quartz diorite, diorite) (%)	75%
12	Area of syenite intrusives (%)	90%
13	Area of true porphyry (%)	0
14	Area of breccia and diatremes (%)	8%
15	Area of metamorphic aureola (%)	4%
16	Area of hydrothermal alteration (%)	10%
17	Ratio of length of perimeter of intrusive to area of intrusive	91%
18	Number of mineral occurrences in cell	99%

N.B. cell - 60 km x 60 km prospecting area

The regression function was formulated from 43 control cells each 100 square miles in area. An adjoining region with a similar geological environment was gridded into 170 cells of the same size, from each of which the field measurements of the 5 significant geo-variates were collated. The regression function of the control set was then evaluated for each of the 170 cells, and the predicted gross dollar values on the top twenty cells ranged from $84.2 millions to $4.3 millions (1968). The values of the other 150 cells are each below $2 millions, which is adjudged as uneconomic. The total expected value of the ore in the 20 cells amounts to $502.3 millions, with a 95% confidence interval between $96.4 millions and $2603 millions (1968).

Table 5.1. (Continued) Selection of statistically significant geo-
variates to guide the search for porphyry-Cu-Mo deposits
of the British Columbia section of the North American
Cordillera.

No.	LIST OF SIGNIFICANT GEO-VARIATES SELECTED BY MEANS OF MULTIPLE REGRESSION ANALYSIS
1	Angle of departure from average Cordilleran trend
2	Number of intersecting structural features in cell
3	Area of Triassic & Lower Jurasic % cell
4	Area of volcanics older than mid-Tertiary % cell
5	Area of Early Mesozoic intrusive % cell
6	Area of Mid-Mesozoic intrusives % cell
7	Area of Tertiary intrusives % cell
8	Area of intermediate intrusives % cell
9	Area of intermediate-basic intrusives % cell
10	Area of syenitic intrusives % cell
11	Ratio of perimeter over area of intrusives
12	Number of mineral occurrences in cell

N.B. cell - 60 km x 60 km prospecting area

Table 5.1. lists a series of 18 quantified gelogical independent
variates which were chosen subjectively to describe the environment of
productive porphyry-Cu-Mo deposits in the British Columbia section of the
North American Cordillera for the purpose of selecting prospecting areas (60
km x 60 km cells) of greatest merit. Variates 1, 2 and 18 were found to be
approximately normally distributed, while the distribution histograms of the
remaining 15 variates were symmetrized by means of logarithmic
transformations.

The 18 independent variates and the dependent variate, logarithm of the
gross dollar value of the deposits, were subjected to a stepwise regression
procedure which indicated that 8 independent variates were significantly
correlated with the gross dollar value (cofficient at least 80%
significant). Among these, variates 1 and 2 are of a structural nature,
variates 5, 7, 10, 12 and 17 are of a petrologic nature, and variate 18
reflects mineralization features. The 8 variates were used to construct a
geo-statistical model relating the geological make-up of the cells under
investigation to their potential mineral wealth with regard to porphyry-Cu-
Mo deposits.

The purpose of the Superior Province study in the Canadian Shield was
the evaluation of the economic potential and selection of the most
favourable prospecting area units (60 km x 60 km cells) in the search for
Ni-Cu-ultramafic deposits and volcanogenic sulfide deposits. A regression

204

Table 5.2. Selection of statistically significant geo-variates to guide the search for Ni-Cu ultramafic and volcanogenic sulfide deposits of the Superior Province of the Canadian Shield

LIST OF SIGNIFICANT GEOLOGICAL VARIATES SELECTED BY MEANS OF
MULTIPLE REGRESSION ANALYSIS

	Statistical significance		
	N.W. Quebec	N.E. Ontario	N.W. Ont. & S.E. Manit.
x_1 = % area of acid flows in cell	90 %	90 %	81 %
x_2 = % area of basic flows in cell	0 %	70 %	65 %
x_3 = % area of basic intrusive in cell	85 %	90 %	72 %
x_4 = % area of alkalic complex in cell	0 %	79 %	70 %
x_5 = length of contact of acid & basic flows in km in cell	81 %	99 %	99 %
x_6 = length of shear zones & regional breaks in km in cell	0 %	0 %	75 %
x_7 = number of mineral occurrences in cell	72 %	98 %	98 %
x_8 = number of fold axes in cell	0 %	62 %	68 %

N.B. cell = 60 km $\times$ 60 km prospecting area

function was formulated on 74 well-prospected control cells known to include a total of 126 deposits of the types of interest worth a total of $15 billions (1968), in order to estimate the gross ore value expected to occur within each of the 2270 cells under investigation.

A total of 8 independent geological variates were subjectively selected to describe the geological environment of productive cells (see Table 5.2.). As a result of stepwise regression analysis, three geo-variates were found to be significant (coefficients at least 80% significant) in the N. W. Quebec and N. W. Ontario - S. E. Manitoba portions of the Superior Province, while the N. E. Ontario portion yielded 5 significant variates. The regression function of significant variates was used to calculate individual estimates of the gross ore value for the 513 cells, which were found to yield more than a threshold of $20 millions (1968) each. The aggregate estimate for these cells was found to be $43 billions (1968), with a confidence interval of $16 billions to $96 billions (1968).

5.3. APPLICATION OF TREND FACTOR & RESIDUAL TREND FACTOR ANALYSES TO FIELD EXPLORATION PROGRAMS

5.3.1. Introduction

The mineral exploration sequence starts with preliminary planning involving the selection of ore deposit types and regions of search (Chapters 2 and 4), and the prospecting areas within the regions which are to be covered. The next stage is to plan the coverage of prospecting areas using optimal survey designs (Chapter 4). This is followed by the collation of data acquired from field surveys generally complemented by previous surveys which may be available from government or private files.

There are two different types of geo-data: qualitative and quantitative. Most geologic data are qualitative in nature, and so require the use of experimental design to quantify them: for example, if the search is centered on one geologic rock type of interest, and the others are irrelevent or not desirable, then the one of interest would be given the score of 1 on that geovariate, and those of no interest, -1. Often there are two of interest, and these would be designated 1, 0; and 0, 1. The others would be -1, -1. These designs are orthogonal, and ensure that the X matrix of data is well-conditioned (not singular). Geophysical and geochemical surveys data are always quantitative in nature; the former are deterministic, while the latter are mainly stochastic.

The crucial task of interpretation of field results may be assisted by the use of a powerful multivariate method known as Factor Analysis. Factor Analysis makes full use of all the interactions and correlations among the geovariates in the case of R-mode, and amongst the locations in the case of Q-mode. We will return to Q-mode Factor analysis again in Chapter 6. Here, we focus our attention on the results of the R-mode.

The number of survey variates and the strengths of direct and negative inter-correlations will determine the number of significant factors. Usually, there are at least two of them, one a straight linear combination, and the other having one or more linear contrasts. In both cases, the field exploration team will recognize their significance. These factors are evaluated at each sampling location, and are usually expressed as statements in the computer program, as shown in the Trend Factor programs in Appendix 5.1. The two main applications of Factor Analysis and its derivatives, namely Trend Factor and Residual Trend Factor Analyses, are the delineation and economic evaluation of exploration targets, prior to their selection for expensive systematic drill testing programs which are to be covered in Chapter 6.

5.3.2. Delineation of exploration targets

5.3.2.1. Introduction

The delineation of exploration targets is essentially sequential in nature, as illustrated by Figure 1.1 of Chapter 1. The procedure calls for the investigation of areas of decreasing size to be covered by technologies of increasing resolution and increasing coverage cost per unit area, in order to delineate exploration targets of increasing merit. Figure 5.6 illustrates the three main paths generally chosen to achieve this goal. The first and third paths depend mainly on ground coverage, while the second one combines both airborne and ground coverages.

5.3.2.2. Application to the search for porphyry-Cu-Au in the Fiji Islands

The purpose of the exploration project was to locate a Cu-Au-type deposit in a 3 square mile sub-area, which had been selected from a 100 square mile prospecting lease in the Fiji Islands on the basis of reconnaissance geological and geochemical (stream sampling and soil sampling) surveys. The region is underlain by a northwesterly-trending series of volcanic agglomerate and greywacke of late Eocene and early Miocene ages which were intruded by elongated bodies of monzonite of early Pliocene age.

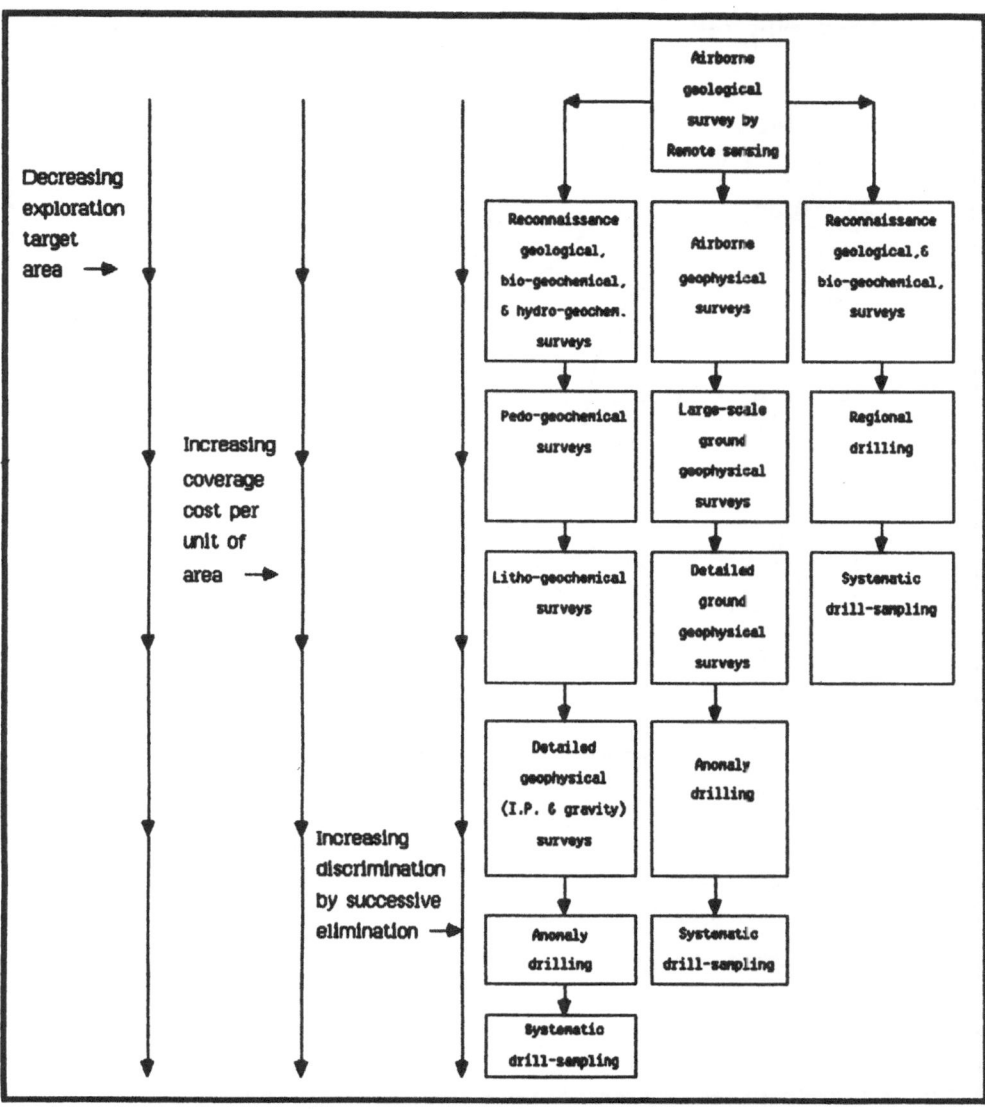

Figure 5.6. Flow diagram of the three main paths to the
sequential delineation of exploration targets.

The input for the Factor, Trend & Residual Trend programs consisted of the following types of field survey data:

(a) Geologic data: the three main types of rock formations known in the area, namely monzonite, agglomerate and greywacke, were coded using experimental design as 1, 0; 0, 1; and -1, -1, respectively. The inferred areal extent of the three rock types is shown on Figure 5.7.

(b) Geochemical data: abundance of Cu, Pb, Zn, expressed parts per million in soil samples taken at 1234 locations, situated at 50 meter intervals along parallel grid lines 100 meters apart.

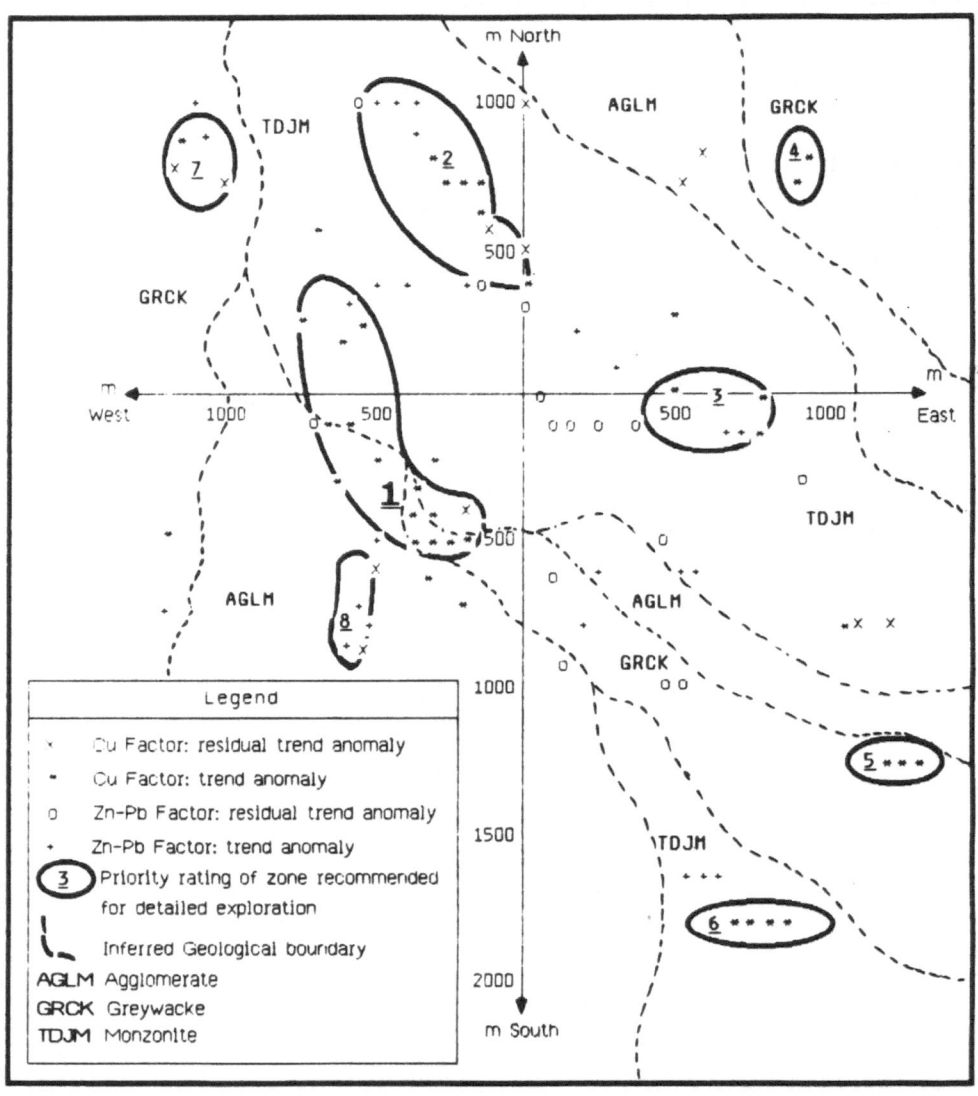

Figure 5.7. Trend factor analysis for porphyry-Cu-Au search in Fiji.

(c) <u>Geographic data</u>: northerly and easterly coordinates, and elevations in meters.

Factor, Factor Trend, Residual Factor Trend programs were used to screen the optimal "anomalous" locations. The latter were grouped to delineate target areas for detailed exploration and drilling. Figure 5.7 shows the location and extent of eight target areas outlined as a result of statistical processing of the field data.

The detailed results of the application of the FACTOR-TREND suite of programs (Appendix 5) are as follows:

(1) <u>Factor analysis</u>: the correlation matrix for the variates (1): Cu, (2): Pb, and (3): Zn abundances yielded the following results:

r_{12} = -0.03:   no significant Cu-Pb correlation

r_{23} = 0.44:   highly significant Pb-Zn

r_{13} = 0.17:   significant Cu-Zn correlation.

A Principal Component Analysis followed by rotation yielded two significant Factors:

Factor 1: a linear combination had high loadings on variates (2), (3): Zn-Pb;

Factor 2: a linear contrast yielded a high loading on variate (1): Cu, and a negative loading on variate (2): Pb.

(2) <u>Trend analysis</u>: The Trend analysis program was formulated with:

Dependent y-variates: y_1 = Log(Cu ppm), y_2 = Log(Pb ppm), y_3 = Log(Cu ppm), y_4 = Log(Factor 1), and y_5 = Log(Factor 2).

Independent Variates: x_1 = dummy = 1 (design variate),

x_2 = northerly coordinate in m,

x_3 = easterly coordinate in m,

x_4, x_5, x_6 = 1, 0; 0, 1; -1, -1, for the three geological types: (experimental design variates).

The trend analysis showed that no significant trends in the northerly or easterly directions were diagnosed on either the factors or the variates.

In the Residual Trend analysis, the residual on each factor at each location is calculated by subtracting the least-squares trend estimate from the factor evaluation based on the observed geochemical values. Only locations significantly above the trend were of interest, so the program was set up to detect and print out those locations which were 1.645 standard errors above the mean on any variate or factor at the α = 0.05 significance level. The program screened out a total of 75 such locations, eight of which were found to be significant on both factors.

5.3.2.3. Search for U-Au deposits, Alligator Rivers area, Northern Territory, Australia

The Alligator Rivers area is the location of some of the World's richest uranium deposits. The best known deposits include the huge Ranger uranium mine, the Jabiluka U-Au mine, the smaller Nabarlek deposit and the Koongara prospect. The geology of the region, shown in Figure 5.8., is inferred from airborne magnetic and radiometric surveys carried out by the

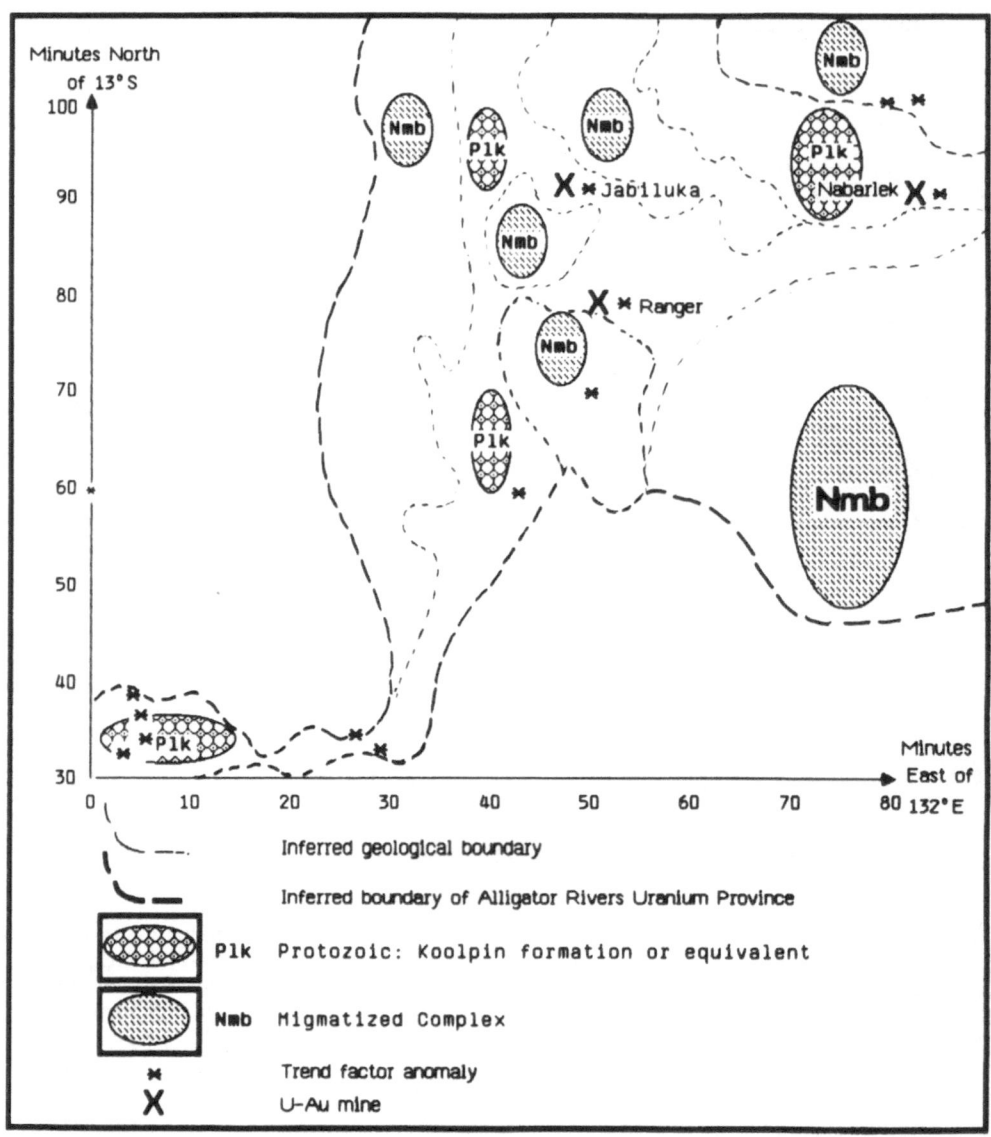

Figure 5.8. Trend factor analysis for U-Au search at Alligator Rivers, Northern Territory, Australia.

Australian Bureau of Mineral Resources in the 1970's. For example, the
Nabarlek deposit recorded 700 counts/ sec on the airborne spectrometer, and
the diagnostic ratio of the Uranium channel 3/ Thorium channel 4 = 20, which
is 20 times larger than the ordinary anomalous value (cut-off = 1). The
anomalies discussed in this study are therefore of great commercial
interest. Jabiluka was not detected by the airborne survey, but was found by
a reconnaissance ground survey crew, using a spectrometer, who found a
single anomalous float boulder. The resulting follow-up drilling revealed an
enormously rich U-Au deposit.

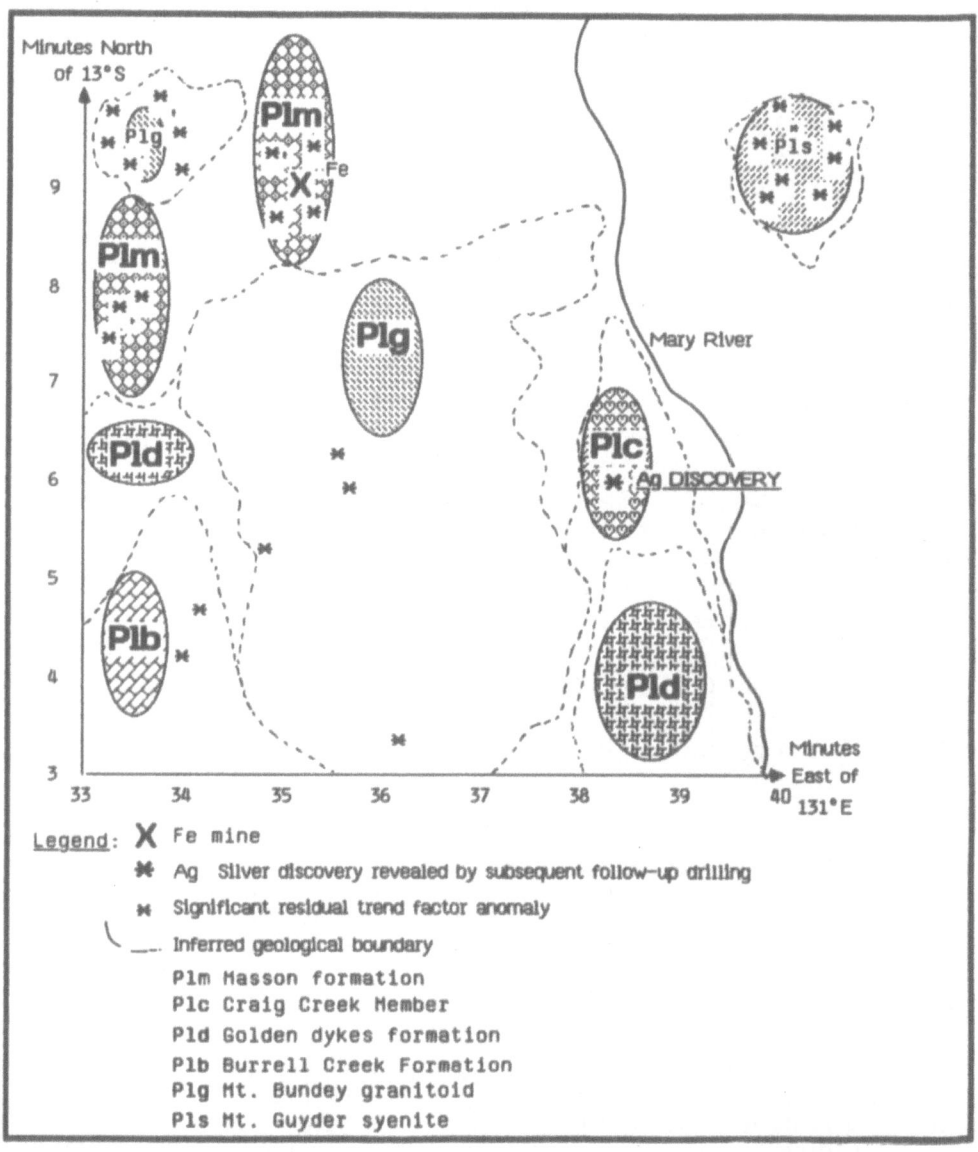

Figure 5.9. Trend factor analysis for U-Fe-Ag contact metasomatic
deposits at Mt. Bundey, Northern Territory, Australia.

Readings from a total of 429 locations within a 5000 square mile region were analysed. The dependent variables were airborne radiometric readings in count/sec, as follows:

y_1 = total count, y_2 = Potassium, K;

y_3 = Uranium; y_4 = Thorium; y_5 = U/Th ratio.

The independent variables were the easterly and northerly coordinates in minutes east of the 132° E meridian, and north of the 13° S parallel of latitude.

The results of the R-mode Factor Analysis program were:

(1) Correlation matrix:

	y_1	y_2	y_3	y_4	y_5
y_1	1	0.95	0.91	0.58	0.31
y_2		1	0.86	0.46	0.33
y_3			1	0.35	**0.54**
y_4				1	**−0.37**
y_5					1

The diagnostic nature of y_5 = U/Th is clearly shown by the highly significant correlation with y_3 = U, and inverse correlation with y_4 = Th.

The factor analysis identified two main factors summarized as the linear combination of the radiometric data. Factor (1) accounts for 76% of the variation, and factor 2, the Uranium factor, is loaded positively on the two Uranium variates, y_3 and y_5, and negatively contrasted with the K & Th variates, y_2, & y_4. They are expressed as

$$F_1 = 0.99\ y_1 + 0.95\ y_2 + 0.94\ y_3 + 0.50\ y_4 + 0.37\ y_5,$$

and $F_2 = -0.10\ y_1 - 0.01\ y_2 + 0.21\ y_3 - 0.72\ y_4 + 0.73\ y_5.$

The main result of the residual trend analysis for diagnostic factor (2) is shown in Figure 5.8. In all, 16 locations were identified as significantly high above the trend, but several were clustered so appear as only 13 on the map. The reader will note that three were recorded in the immediate vicinity of rich U-Au deposits. This highlights the economic potential of the other 10 locations which have not been tested yet as far as is known.

5.3.2.4. Search for Contact metasomatic (U-Fe-Ag) deposits, Mount Bundey area, Northern Territory, Australia

The background for this project was given in Volume 1, Section 12.3.9, De Geoffroy & Wignall; and the results are given in Figure 5.9. The interesting feature of this study is that we do have reports of follow-up drilling leading to the discovery of a commercial silver deposit, with values around 40 oz/ tonne.

5.3.2.5. Search for petroleum, On-shore Gippsland Basin, Victoria, Australia

The Lakes Entrance oil-pools occur in the extreme Eastern boundary of Figure 5.10 (eastern sector). The target stratigraphic zones are in the Latrobe group of Jurassic to Cretaceous age. Structurally speaking, the 1200 sq. km survey area covers an East-West-striking monocline with Southerly dip.

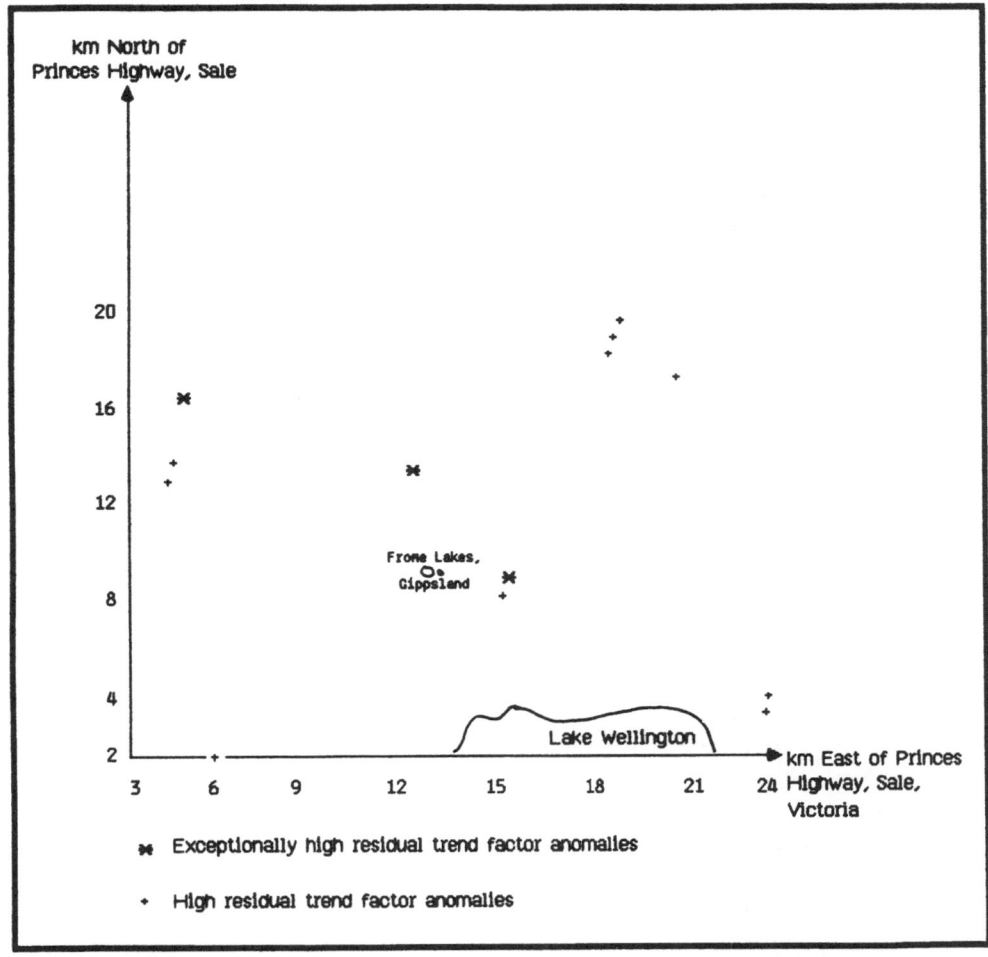

Figure 5.10. Trend factor analysis for hydrocarbon search in the on-shore Gippsland Basin, (Western section), Victoria, Australia.

The data for the study consisted of the geo-chemical gas content of soil-samples measured at 1117 locations from a depth of 1.8m (6 feet) in 1974. The dependent variates are

y_1 = methane, y_2 = ethane; y_3 = propane; and y_4 = butane,

and the independent variates, x_2 and x_3, are the easterly and northerly coordinates in km, respectively.

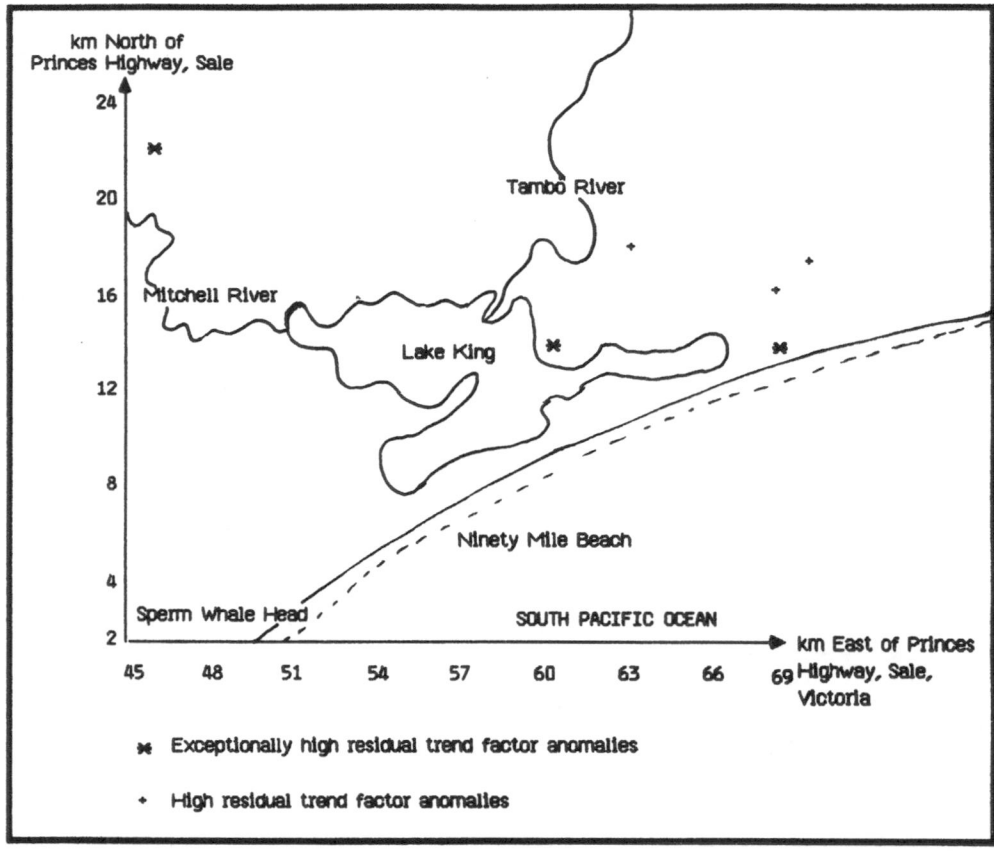

Figure 5.10. (Continued) Trend factor analysis for hydrocarbon search in the on-shore Gippsland Basin, (Eastern section), Victoria, Australia.

(1) Factor analysis:

The Factor Analysis was based upon the following correlation matrix:

(1) Correlation matrix:

	y_1	y_2	y_3	y_4
y_1	1	0.59	0.65	0.59
y_2		1	0.73	0.87
y_3			1	0.76
y_4				1

The results of the Factor analysis were not so satisfactory in this example as a diagnostic ratio was lacking, consequently there was only one factor which accounted for 100% of the variation resulting from the nature of the positive significant correlations in the R-matrix. The factor is expressed as follows:

$$F_1 = 0.89\ y_1 + 0.99\ y_2 + 0.93\ y_3 + 0.70\ y_4.$$

(2) Residual trend factor analysis:

As a result of the Residual Trend Factor Analysis a total of 6 highly significant anomalies, and 12 significant ones were screened out from the 1117 locations. [See Figure 5.10 (a) & (b)].

5.3.2.6. Search for petroleum, Saskatchewan, Canada

The target stratigraphic zones in this project were in the Red Sandstone formations of Mississipian age. A suite of seismic reflection times on three zones of interest were taken and recorded at 1164 locations. The factors determined on the analysis of the field data were

$$F_1 = 0.43\ y_1 + 0.73\ y_2 + 0.53\ y_3,$$

and $$F_2 = 0.90\ y_1 - 0.38\ y_2 - 0.20\ y_3,$$

where, y_i is the seismic reflection time to zone$_i$, i = 1, 2, 3.

The independent variates were the coordinates measured in latitude and longitude.

Nine clusters of interest containing from 2 to 25 locations were screened as low "anomalies", since the search was for minimum reflection times, corresponding to structural highs.

5.3.3. Economic Evaluation of exploration targets

5.3.3.1. Introduction

A second and also very important application of the versatile factor-trend analysis procedure is the evaluation of the economic potential of target areas which have been delineated through the first round of the procedure. However, this time, instead of taking survey readings as dependent variates to be predicted in terms of geographic independent variates, we use the percentage or dollar grade as the dependent variate. The Trend Analysis approach appears to be most successful when dealing with bedded deposits, particularly those yielding industrial minerals such as coal, iron, phosphates and bauxite. A very interesting application of the Trend Analysis approach to the prediction of ash contents of coal beds in the Cape Breton region of Nova Scotia, Canada, was done by Agterberg (13, section 1). Whitten (51, section 1) describes the application of the

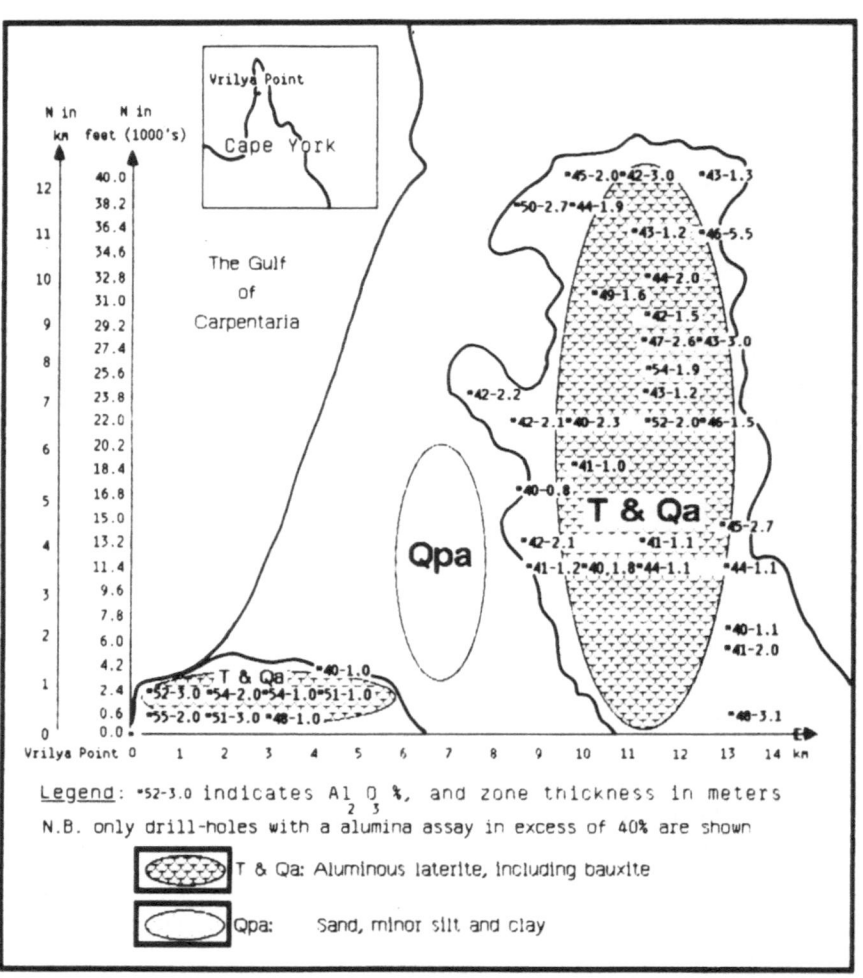

Figure 5.11. Map of the Vrilya Point, Cape York Peninsula, Queensland, Australia, bauxite area (Northern section), showing the results of drilling programs at 1800 foot spacings (550m), with the origin, (0, 0) at Vrilya Point.

GLM to grade prediction in the bedded gold deposits of South Africa. However, a second approach, which is gaining increasing favour for the economic evaluation of ore deposits, is that known as "kriging" which is at the core of the theory of Geostatistics developed by the French School of Fontainebleau, especially by Matheron, in the early 1960's. Davis (19, section 1) gives in his Chapter 6 a very readable account of the statistical theory of "regionalized variables" on which the Kriging procedure is based. Rendu (47, section 1) describes the application of Geostatistics to the economic evaluation of mineral deposits.

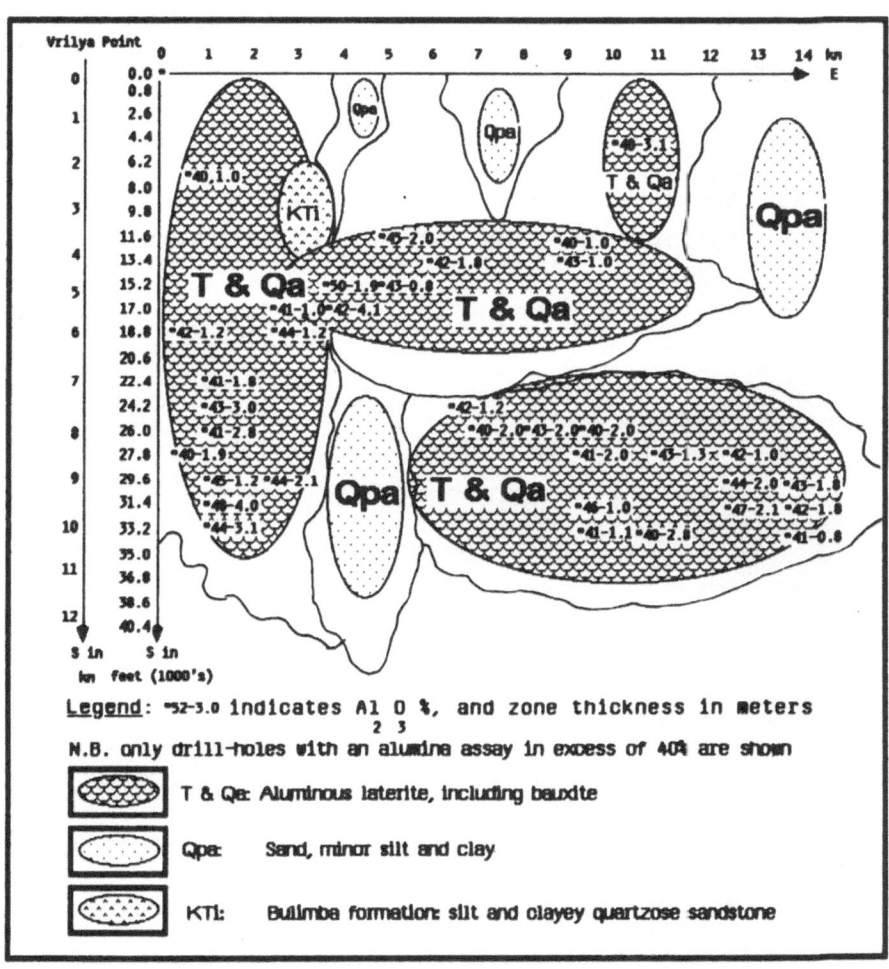

Figure 5.11. (Continued) Map of the Vrilya Point, Cape York Peninsula, Queensland, Australia, bauxite area (Southern section), showing the results of drilling programs at 1800 foot (550m) spacings, with the origin, (0, 0) at Vrilya Point.

5.3.3.2. Example of application: Cape York bauxite, Northern Queensland,
         Australia

       Insearch, Ltd., the research company of the New South Wales Institute
of Technology, was commissioned to do a bulk sampling survey of the Vrilya
Point bauxite deposit, just to the north of the huge bauxite deposit mined
by Comalco at Weipa, Cape York Peninsula, Queensland, Australia. The Comalco
tonnage factor estimate of 25 cu ft per tonne or 1.25 tonnes/cu m. for
screened bauxite pisolites was used in our study. A map of the study area is
shown on
Figure 5.11.

       Dr. R. F. G. MacMillan was appointed to manage the field work at this
stage, and reported:

(1) The overburden varies from 60 to 180 cm (2 to 6 ft.),
(2) The deposit consists of pisolitic bauxite in loose soil, thus enabling
     the work to be done by hand. As a result, a total of 10 tonnes of ore
     were dug from 6 locations covering the 250 square km licence area (see
     Fig.5.12),
(3) The Tertiary lateritic plateau is only 1 to 2 meters higher than the
     swampy regions, where the bauxite pisolites have also been deposited.
(4) X-ray analysis of the samples in the laboratory showed the Al_2O_3

     content of the ore to be in the form of the trihydrate, known as
     Gibbsite.

Figure 5.12. The photograph depicts one of the Vrilya Point locations,
             where Dr. R. F. C. MacMillan and his assistant, a
             Mornington Islander, hand excavated the bauxite bulk
             samples. They reported that the Al_2O_3 pisolites seemed
             to cover the entire region; the swampy areas, as well as
             the plateaux.

(5) Rejection of fine material less than 3 mm yields an average upgrading of 5% in Al_2O_3 content.

(6) Mining the deposit could be achieved by a straightforward stripping of the overburden by bulldozing, to be followed by screening for the upgrading process.

A total of 429 locations were sampled over the 500 square km study area. The maps shown on Figures 5.11. depict only the location of samples which returned more than 40% Al_2O_3. The dependent variables were chosen as overburden in feet, bauxite thickness in feet, and Al_2O_3 grade %:

$$y_1 = \text{overburden}, \quad y_2 = \text{bauxite thickness}, \quad \text{and} \quad y_3 = Al_2O_3\%,$$

and independent variates were the easterly and northerly coordinates in km, x_2 and x_3, respectively.

(1) Factor analysis:

The correlation matrix on which the Factor Analysis study is based was calculated as:

	y_1	y_2	y_3
y_1	1	-0.14	-0.15
y_2		1	0.27
y_3			1

The results of the Factor Analysis were very satisfactory in this example as the main factor accounting for 100% of the variation was a linear contrast of the economic factors and the overburden. The resulting factor, the bauxite factor, was found to be

$$F_1 = -0.16 \, y_1 + 0.38 \, y_2 + 0.41 \, y_3,$$

(2) Trend and trend factor analysis:

The least-squares estimates of the three dependent variates and the bauxite factor for the study area covering the covering the western bauxite plateau at Vrilya Point are obtained from the evaluation of the regression functions formulated on the results of drilling at 40 control locations. The 40 control locations had the following means:

$$\begin{aligned}
Al_2O_3 &= 41.05\% \\
\text{Thickness of economic bauxite zone} &= 4.75 \text{ feet,} \\
\text{Bauxite Factor score} &= 17.9.
\end{aligned}$$

The resulting GLM analysis yielded the following control functions, with the origin, (0, 0), situated at Vrilya Point, as shown on Figure 5.11.

$$\begin{aligned}
Al_2O_3 &= 44.37 - 0.20x_1 + 0.45x_2, \\
\text{Bauxite zone thickness (feet)} &= 3.85 + 0.08x_1 - 0.09x_2, \\
\text{Bauxite Factor score} &= 19.44 - 0.06x_1 + 0.18x_2, \\
\text{Overburden} &= 1.56 + 0.50x_1 - 0.20x_2.
\end{aligned}$$

Evaluating these grades for 1800 feet spacings in that part of the SBML1 area covering the western coastal bauxite plateau yielded the following results:

Location: E, N.	Al_2O_3 Trend	Bauxite thickness (ft.)	Factor score
0 0	44.4	3.8	19.4
0 -1	43.9	3.9	19.3
0 -2	43.5	4.0	19.1
.	.	.	.
.	.	.	.
.	.	.	.
1 0	44.3	3.9	19.2

etc., and the estimates for the final location in the SBML1 area were

6, -1	43.8	4.4	18.9

The grades in the area are thus all predicted as favorable, and the sum of the predicted bauxite thicknesses in the area = 455 feet. Thus the predicted tonnage in the area, using the Weipa measure of 25 cu ft. per tonne of ore (considered conservative) is given by:

Predicted tonnage = 1800 x 1800 x 455/ 25 = 59 million tonnes.

5.4. APPLICATIONS OF OPERATIONS RESEARCH MODELS TO DEVELOPMENT PLANNING

5.4.1. General statement

After presenting the GLM and describing its application to the optimization of mineral exploration (Section 5.1 through 5.3), we wish to introduce in this section several Operations Research models which can be used to advantage to optimize the production side of the mineral industry. Although the GLM and OR models are quite different in nature and structure, we decided to include them in the same chapter because their fields of application overlap. As seen above, the GLM is very versatile, being suitable for application to the optimization of field exploration (5.2.2 and 5.3.2) as well as development planning (5.2.1 and 5.3.3). We will find below, however, that the principal field of application of OR models is restricted to the optimization of development planning.

The term "Operations Research" (OR) was coined by British scientists who were involved in solving problems of optimal allocation of scarce material and man-power resources during the World War II period. During the past three decades, the field of application of OR has been considerably broadened to encompass the consideration of problems of allocation of many types of resources, be it in money, or materials, or time, or skills, in any type of business or engineering situations. Operations Research is essentially a problem-solving discipline seeking to optimize complex man-machine systems by means of mathematical models (8, Section 2).

Operations Research models are formulated to optimize a mathematically-defined objective function which is to be maximized or minimized subject to constraints such as limits of available resources of the types mentioned above. This goal can be achieved through the use of two different methodologies which were designed to tackle two distinct types of situation, namely Dynamic Programming and Linear Programming. Dynamic Programming is designed to handle multistage sequential systems with relatively few variates and constraints, as is the case in Chapter 4 when we seek to design optimal search programs for the detection of ore deposits. The second type of methodology, Linear Programming, is algorithmic in nature, therefore quite different from the first one, and is designed to optimize single stage systems involving a large number of variates and constraints.

We will first introduce below three types of relatively simple OR models. They include two models which are designed to optimize systems involving distances, namely the transportation model (5.4.2) and the Networking model (5.4.3), and a third one used to optimize problems involving completion time, namely the Project Evaluation & Research Technique-Critical Path Method, known as the PERT-CPM model (5.4.4). Finally, we will present a more complex but easily computerized model, known as Linear Programming, which is designed among other things to minimize cost subject to time, budget and man-power constraints (5.4.5).

5.4.2. The Transportation model

5.4.2.1. Introduction

The transportation model is devised to tackle the problem of shipping goods from source i, i = 1, 2, ..., m; to destination j, j = 1, 2, ..., n; so as to minimize the total cost for a specified period of time, say daily; and subject to a cost matrix:

$$C = \{c_{ij}\}, \quad i = 1, 2, \ldots, m; \quad j = 1, 2, \ldots, n;$$

where c_{ij} is the unit cost of shipping from source i to destination j.

The other restrictions are the limited resources available at source i:

A_i, representing for example the daily production rate there, for $i = 1, 2, \ldots, m$; and the capacity:

B_j, at destination j, for example to process the amount received, $j = 1, 2, \ldots, n$.

This problem is known as the Primal problem, and the variables introduced to solve it are the Primal variables. The Dual problem is that of maximizing the use of the resources expended in solving the Primal problem, and the variables introduced are the Dual variables. Whenever the Primal problem is solved with the minimum value of the primal objective equal to C, then simultaneously the Dual problem is also solved with its maximum value also equal to C. Some theoretical examples from the mining industry will illustrate the application of a set of algorithms to solve this problem.

5.4.2.2. Example of a balanced transportation model

Suppose ore is being produced at 3 goldmines, g_1, g_2, and g_3 (m = 3); with respective daily production rates, $A_1 = 200$, $A_2 = 300$, and $A_3 = 180$ tonnes of ore per day, to be shipped to batteries, b_1, b_2, b_3 and b_4 (n = 4); with respective daily processing rates for gold extraction, $B_1 = 150$, $B_2 = 240$, $B_3 = 190$ and $B_4 = 100$ tonnes of ore treated. The model is referred to as "balanced" bacause $\Sigma A_i = \Sigma B_i = 680$ tpd. The transportation cost/tonne matrix, C, is given in Table 1.

Table 1

Source (Mine)	b_1	Destination (battery) b_2	b_3	b_4	Capacity (tonnes/day)
g_1	$7	$6	$9	$6	200
g_2	$8	$10	$7	$11	300
g_3	$15	$12	$6	$7	180
Capacity t/d.	150	240	190	100	Total = 680

Solution

Let $X_{(3 \times 4)} = \{x_{ij}\}$, $i = 1, 2, 3$; $j = 1, 2, 3, 4$

be the matrix of daily shipments in tonnes, then the objective function, z, in this example, is:

$$\text{Minimize } z = \Sigma \Sigma c_{ij} x_{ij} = 7x_{11} + 6x_{12} + \ldots + 7x_{34},$$

subject to the constraints

$$\Sigma x_{1j} = A_1 = 200,$$

and for $i = 2, 3$: $\Sigma x_{ij} = A_i = 300$, for $i = 2$, and 180, for $i = 3$,

Similarly, the total production rate at each battery should = the amount of ore received from the mines:

thus: for $j = 1$, $\Sigma x_{i1} = B_1 = 150$,

and for $j = 2, 3, 4$:

$$\Sigma x_{ij} = B_j.$$

Vogel's approximate method (VAM) is now used to derive an initial (starting) solution, for which the algorithmic steps are:

Step 1: Subtract the lowest cost from the next lowest in each row and column to derive a penalty cost for each row and column.
Step 2: Find a row or column with the highest penalty cost, which is column 2 in the example with a penalty cost of 4*. Load the square with the lowest cost in this column with the maximum possible tonnage subject to constraints.
N.B. * indicates a highest.
In this case, it is square$_{12}$ with c_{12} = \$6, and we load it with 200 tonnes as A_1 = 200, and B_2 = 240. Now subtract 200 from A_1, and B_2, and since A_1 = 0 now, we ignore row 1.

Step 3: Check whether all capacities are reduced to zero. If so, then go to step 5. If not, go to step 4.
Step 4: Go to step 1, ignoring any rows or columns that have been deleted

The resulting table is the initial Iterative Tableau 1:

Iterative Tableau 1

Source (Mine)		Destination (battery)			Capacity (t/d.)	Penalty cost			
	b_1	b_2	b_3	b_4					
g_1	\$7	\$6	\$9	\$6	200 0	–	–	–	
		200			–				
g_2	\$8	\$10	\$7	\$11	300 \$1	\$1	\$3	\$3	
	150		150		150				
g_3	\$15	\$12	\$6	\$7	180 \$1	\$1	\$1	\$6	
		40	40	100	80				
Capacity t/d.	150	240	190	100	Total = 680				
		40	40						
Penalty cost	\$1	\$4*	\$2	\$1					
	\$7*	\$2	\$1	\$4					
	–	\$2	\$1	\$4*					
	–	\$2	\$1	–					

Step 5: Check whether you have m + n - 1 = 6 entries (basic variables), if not set one of the non-basic entries (or more if necessary) = 0, then calculate the dual variables:

u_i, i = 1, 2, 3, and v_j, j = 1, 2, 3, 4, associated with these basic variables, using the theorem that if x_{ij} is non-zero (basic), then

$$c_{ij} = u_i + v_j.$$

In order to use this calculating theorem, we need to set one u_i, or v_j = 0, and to simplify the calculations, choose the row or column with the most entries. In this case, u_3 = 0, since row 3 has the most basic variables, and therefore

$$v_2 = 12 - 0 = 12,$$
similarly

$$v_3 = 6, \text{ and } v_4 = 7.$$
Hence

$$u_1 = 6 - 12 = -6, \text{ and } u_2 = 7 - 6 = 1$$
Finally

$$v_1 = 8 - 1 = 7.$$

Go to step 6.

The results of step 5 are given in Iterative Tableau 2.

Iterative Tableau 2

Source (Mine)	b_1	b_2	b_3	b_4	Capacity (t/d.)	Dual u_i
g_1	$7	$6 200	$9	$6	200	-$6
g_2	$8 150	$10 +	$7 -150	$11	300	$1
g_3	$15	$12 -40	$6 +40	$7 100	180	$0
Capacity t/d.	150	240	190	100	Total = 680	
Dual v_j	$7	$12	$6	$7		

Step 6: Utilizing the dual of the minimization model, check whether the present solution is optimal, since by the L.P. duality theorem

$$u_i + v_j >= c_{ij}, \quad i = 1, 2, 3; \ j = 1, 2, 3, 4.$$

Hence, for the non-basic variable squares, search for the largest negative in the calculation:

$$c_{ij} - u_i - v_j,$$

if there is a negative then go to step 7, otherwise the present solution is optimal.

In this case, in the square$_{22}$,

$$10 - 1 - 12 = -3,$$

and this turns out to be the only negative, so we proceed to step 7.

Step 7: Insert a + sign in square$_{22}$, to indicate x_{22} is the variable entering the basis, and complete a loop-path of basic variables forming a vertices of rectangle(s), including square$_{22}$, inserting -, +, -, ..., alternately at the vertices. In Iterative Tableau 2, the entries are: +, -150, +40, -40.

We choose the smallest negative, and add its absolute value to the two + entries, and subtract it from the + value of the two - entries in Iterative Tableau 2, so that $x_{22} = 40$, and $x_{32} = 0$ leaves the basis. The result of this change is given in Iterative Tableau 3. Go to step 5.

Iterative Tableau 3

Source (Mine)	b_1	Destination (battery) b_2	b_3	b_4	Capacity (t/d.)	Dual u_i
g_1	$7	$6	$9	$6	200	-$4
		200				
g_2	$8	$10	$7	$11	300	$0
	150	40	110			
g_3	$15	$12	$6	$7	180	-$1
			80	100		
Capacity t/d.	150	240	190	100	Total = 680	
Dual v_j	$8	$10	$7	$8		

This time, the most basic variables occur in row 2, so $u_2 = 0$, hence

$$v_1 = 8, \quad v_2 = 10, \text{ and } v_3 = 7.$$

Next

$$u_1 = 6 - 10 = -4.$$

Calculating the $c_{ij} - u_i - v_j$ values for the non-basic squares, we find there are no negative values, so that the present solution is optimal.

Thus:

the minimum cost = z_{min} = 200*$6 + 150*$8 + 40*$10 + 110*$7 + 80*$6 + 100*$7

= $4750 is the solution to the primal problem.

The solution to the dual problem of maximizing the use of the resources at the minimum cost is given by:

maximize

y = -$4*200 + $0*300 + -$1*180 + $8*150 + $10*240 + $7*190 + $8*100
 = $4750,

where y is the dual objective function.

The tonnages to achieve the minimum cost are also read from the final table:

From Gold Mine Number	To Battery Number	Tonnage
1	2	200
2	1	150
2	2	40
2	3	110
3	3	80
3	4	100

5.4.2.3. Unbalanced Transportation model

Suppose ore is being produced at 2 goldmines, g_1 and g_2 (m = 2); with respective daily production rates, A_1 = 250 and A_2 = 450 tonnes of ore per day, to be shipped to batteries, b_1, b_2, b_3 and b_4 (n = 4); with respective daily processing rates for gold extraction, B_1 = 200, B_2 = 250, B_3 = 150 and B_4 = 250 tonnes of ore treated.

We readily see that the model is unbalanced since ΣA_i = 700, while ΣB_j = 850.

To balance the model, introduce a dummy mine with capacity A_3 = 150, so that now m = 3. Set all the transportation costs from it equal to 0. The transportation cost/tonne matrix, C, is given in Table 1.

Table 1:

| Source (Mine) | Destination (battery) | | | | Capacity (tonnes/day) |
	b_1	b_2	b_3	b_4	
g_1	$5	$4	$7	$5	250
g_2	$7	$4	$8	$3	450
dummy	$0	$0	$0	$0	150
Capacity t/d.	200	250	150	250	**Total = 850**

Solution:

Let $X_{3 \times 4} = \{ x_{ij} \}$, $i = 1, 2, 3$; $j = 1, 2, 3, 4$

be the matrix of daily shipments in tonnes, then the objective function, z, in this example, is

$$\text{Minimize } z = \Sigma \Sigma c_{ij} x_{ij} = 5x_{11} + 4x_{12} + \ldots + 0x_{34},$$

subject to the constraints

$$\Sigma x_{1j} = A_1 = 250,$$
and for $i = 2, 3$: $\Sigma x_{ij} = A_i = 450$, for $i = 2$, and 150, for $i = 3$,

Similarly, the total production rate at each battery should = the amount of ore received from the mines:

thus: for $j = 1$, $\Sigma x_{i1} = B_1 = 200$,
and for $j = 2, 3, 4$:

$$\Sigma x_{ij} = B_j.$$

Applying Vogel's Approximate method (VAM) to derive an initial (starting) solution, for which the algorithmic steps are:

Step 1: Subtract the lowest cost from the next lowest in each row and column to derive a penalty cost for each row and column.

Step 2: Find a row or column with the highest penalty cost, which is column 3 in the example with a penalty cost of 7*. Load the square with the lowest cost in this column with the maximum possible tonnage subject to constraints. In this case, it is square$_{33}$ with $c_{33} = \$0$, and we load it with 150 tonnes as $A_3 = 150$, and $B_3 = 150$. Now subtract 150 from A_3, and B_3, and since $A_3 = 0$ now, we ignore row 3.

However, $B_3 = 0$ also, so in order to have the requisite number of basic variables, load the next lowest in column 3 with 0, namely square$_{13}$, so that $x_{13} = 0$ is basic.

Step 3: Check whether all capacities are reduce to zero. If so, then go to step 5. If not, go to step 4.

Step 4: Go to step 1, ignoring any rows or columns that have been deleted The resulting table is the initial Iterative Tableau 1:

Iterative Tableau 1

Source (Mine)	b_1	Destination (battery) b_2	b_3	b_4	Capacity (t/d.)	Penalty cost			
g_1	$5	$4	$7	$5	250	$1	$1	$1	–
	200	50	0						
g_2	$7	$4	$8	$3	450	$1	$1	$1	–
		200		250					
dummy	$0	$0	$0	$0	150	$0	–	–	–
			150						
Capacity t/d.	200	250	150	250	Total = 850				
Penalty cost	$5	$4	$7*	$5					
	$2*	$0	–	$2					
	–	$0	–	$2*					
	–	$0	–	–					

Step 5: Check whether you have

$$m + n - 1 = 6 \text{ entries (basic variables)},$$

if not set one of the non-basic entries (or more if necessary) = 0, then calculate the dual variables, u_i, i = 1, 2, 3, and v_j, j = 1, 2, 3, 4, associated with these basic variables, using the theorem that if x_{ij} is non-zero (basic), then

$$c_{ij} = u_i + v_j.$$

In order to use this calculating theorem, we need to set one u_i, or v_j = 0, and to simplify the calculations, choose the row or column with the most entries. In this case, u_1 = 0, since row 1 has the most basic variables, and

therefore v_1 = 5 - 0 = 5,
similarly v_2 = 4, and v_3 = 7.
Hence: u_2 = 4 - 4 = 0, and u_3 = 0 - 7 = -7
Finally v_4 = 3 - 0 = 3. Go to step 6.

The results of step 5 are given in Iterative Tableau 2.

Iterative Tableau 2

Source (Mine)	b_1	Destination (battery) b_2	b_3	b_4	Capacity (t/d.)	Dual u_i
g_1	$5	$4	$7	$5	250	$0
	200	50	0			
g_2	$7	$4	$8	$3	450	$0
		200		250		
dummy	$0	$0	$0	$0	150	-$7
		-40	+40	100		
Capacity t/d.	200	250	150	250	Total = 850	
Dual v_j	$5	$4	$7	$3		

Step 6: Utilizing the dual of the minimization model, check whether the present solution is optimal, since by the L.P. duality theorem

$$u_i + v_j >= c_{ij}, \quad i = 1, 2, 3; \quad j = 1, 2, 3, 4.$$

Hence, for the non-basic variable squares, search for the largest negative in the calculation:

$c_{ij} - u_i - v_j$, if there is a negative then go to step 7, otherwise the present solution is optimal.

In this case, all are non-negative, so that Iterative Tableau 2 is optimal.

Thus the minimum cost = z_{min} = 200*$5 + 50*$4 + 0*$7 + 200*$4 +250*$3
+ 150*$0
= $2750 is the solution to the primal

problem.

The solution to the dual problem of maximizing the use of the resources is given by:

maximize
y = $0*250 + $0*450 + -$7*150 + $5*200 + $4*250 +
$7*150 + $3*250
= $2750.

The tonnages to achieve the minimum cost are also read from the final table:

From Gold Mine Number	To Battery Number	Tonnage
1	1	200
1	2	50
1	3	0
2	2	200
3	4	250
Dummy	3	150

From the solution table, it is therefore Battery 3 which is under-supplied by 150 tonnes to minimize the transportation costs.

5.4.3. Network models

5.4.3.1. Economic Background

Peters summarized the whole concept of the economics of ore production in Chapter 8 of his book (22, Section 2). He states that: "A market exists at some time and at some place for nearly every reasonably pure mineral". Industrial minerals and a few metallic ores like those of iron and aluminum have a **place value**; they are highly affected by transportation costs, being of large bulk and low values, and can be sold only in areas comparatively close to the place of production, unless cheap transportation by sea or rail is readily available. Metalliferous ores, by contrast, are usally sent semi-processed from the mine-site to yield high value, low-bulk products. Their sale is not restricted to local demand only, but is affected by world market conditions. If a mineral has a high **unit value** rather than a high **place value**, it can be sold at prices that reflect an international balance between supply and demand. Most metallic ores, particularly those of precious and rare metals, fit into this category, provided they are concentrated to economic grades at or near the place of production. An excellent example of study of the impact of the structure of a transportation network on the optimal development of mineral wealth of a region is provided by Harris (12, Section 2).

5.4.3.2. Theoretical Background

The purpose of networking is to find the shortest distance between 0 and n in a directed network, and the minimum span of the network. We wish to find the shortest distance to $node_j$, u_j, for $j = 1, 2, ..., n$, using the following algorithm:

Start at any of the locations, say 0, and select u_j, such that

$$u_j = \text{Minimum } \{u_0 + d_{0j}\},$$

whenever $distance_{ij}$ belongs to the network, and where d_{ij} = distance between nodes, i, j. That is simply select the nearest location. This link is part of the minimum network, and the two locations form a linked set. From either location (node), choose a node at the shortest distance away from either node in the linked set, and join it to the appropriate node, to form a linked network spanning the set of three.

Now choose the nearest node to any in the linked set, and repeat the algorithm until all n nodes are linked. The result is the minimum network spanning them.

5.4.3.3. Example of Cape York bauxite project

An example follows taken from the most economic locations for the development of the Vrilya Point, Cape York, Australia, bauxite project. In Figure 5.13, the locations of interest are given their northerly and easterly coordinates, measured in km. Network diagrams are not drawn to scale, nor are the paths joined orientated in the true direction. Instead of distances, the squares of the distances are used, as the minimum of the squares of distance will also be the minimum of the distance. Starting from any node (0, 0 is a convenient one), 3, 1 is the nearest. From either end, select the nearest, which is 4, -5, and join it to 3, 1. Proceeding from any of the three forming the connected set, 2, -7 is the nearest, and is joined to 4, -5. The minimum spanning network is shown in Figure 5.13, and will assist in planning the development when the project becomes feasible.

5.4.4. PERT - CPM Models

5.4.4.1. Theoretical Background

Project Evaluation and Research Techniques and Critical Path Method, PERT - CPM, are designed to minimize the total time for development projects. The algorithm is structured into six steps, as follows:

Step 1: Draw the flow diagram, the directed network for the project, and number the nodes from left to right beginning with node 0, and   ending with n for a project with n tasks. For any $task_{ij}$, i = 0, 1,   ..., n-1; j = 1, 2, ..., n, draw the directed $path_{ij}$ to represent it   (not to scale).

Step 2: For each task, an optimistic (shortest) time, a, is required; also a   pessimistic (largest) time, b, and a modal time, m. Assuming that the times for the task are β - distributed, the expected time and variance for it are given by

Expected time = (a + 4m + b)/6, and

Variance   = $(b - a)^2/36$.

Calculate these parameters for each task in the project, and mark them on the directed network.

Step 3: Find the earliest starting time at $node_j$, for j = 0, 1, ..., n, ES_j, using the definition

ES_0 = 0, and

ES_j = Maximum $\{ES_i + d_{ij}\}$, for i = 0, 1, ..., n - 1,

whenever $task_{ij}$ belongs to the project, and where d_{ij} = the duration

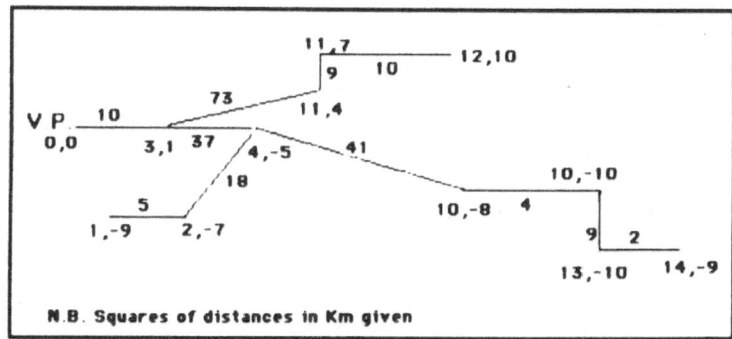

Figure 5.13. Minimum network span covering the nodes of economic interest for economic planning at Vrilya Point, Cape York Peninsula, Queensland, Australia.

of $task_{ij}$.

Simultaneously, find the variance at $node_j$.

Assuming the expected times for each task are independent, we have:

$$Variance_j = Variance_i + Variance_{ij},$$

where the $task_{ij}$ is the selected one from ES_j, where $Variance_0$ is defined as 0.

That is to say, calculate ES_j and $Variance_j$ simultaneously. If there is more than one $task_{ij}$ from which ES_j was calculated, then $Variance_j$ = Maximum of this set. Indicate the earliest starting times and variances at each node as they are calculated.

Step 4: The latest completion time at $node_i$, for $i = n, n-1, \ldots, 0$, LC_i, is defined by

$$LC_n = ES_n,$$
$$LC_i = \underset{j}{\text{Minimum}} \{LC_j - d_{ij}\}, \text{ for } j = n, n - 1, \ldots, 1.$$

Step 5: Find the critical tasks, and hence the critical path(s), where a task is defined as critical if it satisfies the three conditions

(1) $ES_i = LC_i$,

(2) $ES_j = LC_j$,

(3) $d_{ij} = ES_j - ES_i$;

Indicate the critical paths on the project directed diagram.

Step 6: The PERT analysis model is based upon the following criteria:

(1) The expected time to each $node_i$, $\mu_i = ES_i$, $i = 1, 2, \ldots, n$,

(2) The variance time to each $node_i$, $\sigma_i^2 = Var_i$, $i = 1, 2, \ldots, n$,

(3) The times to reach $node_i$ are normal(μ_i, σ_i^2), $i = 1, 2, \ldots, n$, so that the probability that the whole project or a portion of it up to any stage is completed on schedule is calculated, using the standard normal table, as

$$Probability = Pr\{Z < (St_i - \mu_i)/\sigma_i\},$$

where St_i is the scheduled time, which for intermediate nodes is often taken to be LC_i, $i = 1, 2, \ldots, n$.

Confidence intervals may also be calculated, for example a 95% confidence interval for the time to complete the whole project is

$$95\% \text{ confidence interval} = \mu_n \pm 1.96\sigma_n.$$

5.4.4.2. Example: Cape York Peninsula Bauxite project

The geologic and economic background for the Cape York bauxite project were already given in section 5.3.3.2 and Figure 5.11. Suppose the times in weeks for the exploration phase of the bauxite project at Vrilya Point, Cape York Peninsula are given in the following table:

Node number i,j		Task i,j	Times: a,	m,	b weeks		
0,	1	Geological survey	2	3	5		
0,	2	Topographic survey	3	4	6		
0,	3	Magnetometer survey	1	2	3		
1,	2	Dummy					
1,	3	Bulk sampling	1	2	4		
2,	3	Exploratory drilling	3	4	5		
3,	4	Systematic drilling	6	9	10		
4,	5	Feasibility study	3	5	7		

Find

(i)   the critical path(s),
(ii) the 95% confidence interval for the project completion time, and
(iii) the probability of its completion in a scheduled time of 25 weeks.

Solution:

Step 1: Calculate expected time and variance for each task where
Expected time $= (a + 4m + b)/6$, and
Variance $= (b - a)^2/36$.

Node number i,j		Task i,j	Times (wks): a, m, b			Expected	Variance
0,	1	Geological survey	2	3	5	19/6	1/4
0,	2	Topographic survey	3	4	6	25/6	1/4
0,	3	Magnetometer survey	1	2	3	2	1/9
1,	2	Dummy					
1,	3	Bulk sampling	1	2	4	13/6	1/4
2,	3	Exploratory drilling	3	4	5	4	1/9
3,	4	Systematic drilling	6	9	10	26/3	4/9
4,	5	Feasibility study	3	5	7	5	4/9

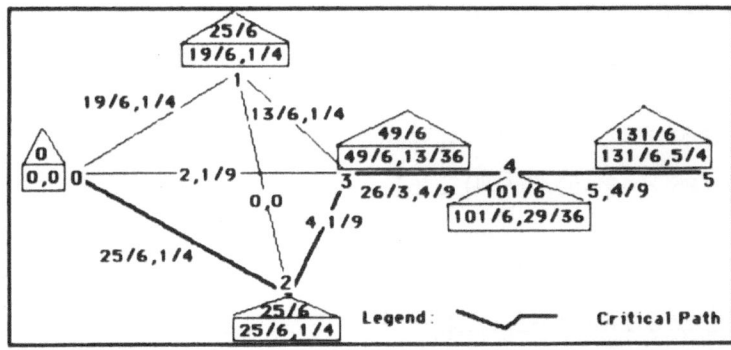

Figure 5.14. PERT-CPM network diagram for the Vrilya Point bauxite project.

Step 2: Draw the flow diagram (see Figure 5.14) and insert expected times for each path_{ij}, variances; and ES_i, Var_i, for each node $i = 0, 1, \ldots, 5$.

Using the definitions

$$ES_0 = 0,$$
and $\quad ES_j = \text{Maximum } \{ES_i + d_{ij}\}$, for $i = 0, 1, \ldots, n - 1$,
$$\text{Variance}_j = \text{Variance}_i + \text{Variance}_{ij},$$

where the task_{ij} is the selected one from ES_j, where Variance_0 is defined as 0.

That is to say calculate ES_j and Variance_j simultaneously. For example:

$$ES_3 = \text{Maximum } \{19/6 + 13/6, \ 0 + 2, \ 25/6 + 4\} = 49/6$$
and $\quad Var_3 = Var_2 + 1/9 = 1/4 + 1/9 = 13/36,$

Step 3: By using the definitions
$$LC_n = ES_n,$$
$$LC_i = \text{Minimum } \{LC_j - d_{ij}\}, \text{ for } j = n, n - 1, \ldots, 1.$$
$$LC_5 = 131/6, \quad LC_4 = 101/6,$$
$$LC_3 = 49/6, \quad LC_2 = 25/6, \text{ and } LC_1 = 25/6.$$

The network is shown in Figure 5.14, with the earliest starting times and variances, which were entered in the boxes as they were calculated, at each node; and the latest completion times entered in the triangles; from the diagram and criteria, it is easily seen which tasks are critical, and as soon as identified, the critical path was drawn.

The critical path is
$$(0, 2, 3, 4, 5) = \text{Ans (i)}$$

Step 4: $ES_5 = 131/6$ & $Var_5 = 5/4$, therefore the 95% confidence interval for the completion time for the project is

(ii) 95% Confidence Interval $= 131/6 \pm 1.96*[(5/4)^{0.5}]$
$$\approx (19.64, \ 24)$$

The probability of completion in 25 weeks is given by calculating

$$Z = (25 - 131/6)/[(5/4)^{0.5}]$$
$$\approx 2.83,$$

(iii) Hence probability ≈ 0.99.

5.4.4.3. Copper exploration project at Ingladahl, India

The data for this example was taken from Sarma (Chapter 2, Section 2 (19)). Sarma used PERT to estimate the total exploration survey costs in the

234

Ingladahl Cu project, India. The results of the survey costs are summarized in Figure 5.15, where for j = 1, 2, 3, the tasks$_{ij}$ are:

Node number$_{i,j}$	Task$_{i,j}$	Costs: $1000: a, m, b	Expected	Variance
0, 1	Aeromag. survey	153 153 153	153.3	0
1, 2	Topographic survey	110 121 170	382/3	100
2, 3	Geophysical survey	5 13/2 7	19/3	1/9
	TOTAL		287	901/9,

which apart from a printing error in the published paper are the results given by Sarma.

Since the three surveys form a sequence, the flow diagram is simply the line shown in Figure 5.15. The expected cost for the whole project, and its variance, is just the sums of these column entries, and for example the probability the total cost does not exceed $300,000 is calculated using the Z standard Normal transformation:

$$z = (300 - 287)/[(901/9)^{0.5}] \approx 1.30,$$

which implies a probability > 0.90.

However, since economic measures are usually lognormal, it is more appropriate to enter the log-transforms of the data; thus, the table becomes:

Node number$_{i,j}$	Log Costs: $1000's: a, m, b			Mean	St. dev.	Geom. Mean
0, 1	2.18	2.18	2.18	2.18	–	153
1, 2	2.04	2.08	2.23	2.098	0.033	125.3
2, 3	0.70	0.81	0.85	0.798	0.025	6.3
	TOTAL			2.455	0.041	284.6

Analyzing the same problem now yields

$$z = (2.478 - 2.455)/0.041 \approx 0.56,$$

and the probability that the total cost does not exceed $300,000 is now 0.70, (a much more conservative estimate) taking into account the lognormal nature of the data.

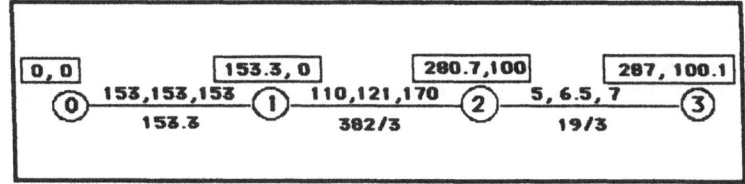

Figure 5.15. PERT-CPM network diagram for the Ingladahl Cu project.

From the solution table, it is therefore Battery 3 which is under-supplied by 150 tonnes to minimize the transportation costs.

5.4.5. Application of Linear Programming

5.4.5.1. The Dual Simplex method

The Linear Program, LP, is designed to optimize a linear objective function, z, subject to a set of constraints. An important constraint is that all the primal decision variables are non-negative. The optimization of the objective function is called the Primal problem, and the variables introduced to solve it are the primal variables. To every Primal problem of minimization, their is a Dual problem of maximization of resources consumed in minimizing the Primal objective function. The problem may be written in matrix form:

$$\text{Maximize or Minimize: } z = C_{(1 \times n)} X_{(n \times 1)},$$

where $x_i >= 0$, $i = 1, \ldots, n$;
subject to m further constraints

$$\{A_{(m \times n)}, I_{(m \times m)}\} X_{(m+n \times 1)} = B_{(m \times 1)},$$

where I is the identity matrix, augmenting the given constraint coefficient matrix, A; and where B is the limiting capacity matrix of resources available. Harbaugh (31, Section 1) gives an excellent account of the use of the SIMPLEX algorithm illustrated by an example pertaining to the mining industry. The listings for the computerized Linear Programming model are given in Appendix 6, using the Dual simplex method. The Dual simplex method requires that all the constraints are less than or equal to inequalities.

5.4.5.2. Example

AAA Resources Corporation operates three oil & gas fields in Southern California, and has contracts to deliver 2000 barrels of oil and 50 million cubic feet of gas daily. It also has a labor contract guaranteeing the employees at oil & gas field number 1 at least 7 hours work per day. The hourly production capacities of oil in barrels per hour and gas in millions of cubic feet per hour are given in Table 1.

Table 1.

Oilfield	Oil: barrels/hr.	Gas: mil. cu. ft./hr.	Operating cost: $1000's/hr.
1	40	3	10
2	120	4	15
3	60	3	8

The problem the AAA management has to face is as follows: how many hours per day must the AAA Resources Corporation operate the 3 oil & gas fields to fill their contracts, and minimize the total operating costs, which must not exceed $300,000?

Formulation of problem:

the objective function in this example is given by

Minimize $Z = 10x_1 + 15x_2 + 8x_3$, where x_1, x_2, $x_3 >= 0$, subject to the constraints

$$40x_1 + 120x_2 + 60x_3 >= 2000 \quad \text{(oil)},$$
$$3x_1 + 4x_2 + 3x_3 >= 50 \quad \text{(gas)},$$
$$x_1 \qquad\qquad\qquad >= 7 \quad \text{(labour contract)},$$
$$10x_1 + 15x_2 + 8x_3 <= 300 \quad \text{(Operating costs)},$$

Solution:

Step 1: Convert all the constraints to <= and add slack variable, x_{n+i}, to the constraint i, for i = 1, 2, ..., m, to form m equations in m + n unkowns, with constraint constant b_i on the right-hand side of equation i.

The slack variables are introduced to take up the slack on the left-hand side of the inequalities, which are thus converted to equations, which can then be solved algebraically.

Write the objective function as

$$z - \Sigma c_i x_i = 0.$$

Go to step 2.

Step 2: feasibility condition: to determine the variable leaving the basis, select the b_i with the largest negative value, if there are no negatives then go to step 4. In example 5.3, -2000 is selected, thus x_4 leaves the basis. Go to step 3.

Step 3: optimality condition: in order to select the variable entering the basis in place of x_4, using only the negative coefficients in the row (if there are none, then the solution is infeasible, and no feasible solution exists), calculate the decision quotients, by dividing the coefficient in the objective row by the coincident negative coefficient chosen in step 2.

If the problem is minimization, then choose a smallest quotient to find the pivot. In the example, -15/-120 is smaller than -10/-40 and -8/-60, thus x_2 enters and -120 is the pivot. Go to step 5.

The resulting Iterative Tableau 1 for step 1,2 & 3 for example 5.3 is:

Iterative Tableau 1

Basis	x_1	x_2	x_3	x_4	x_5	x_6	x_7	Solution	Quotient
z	-10	-15	-8	0	0	0	0	0	
x_4*	-40	-120*	-60	1	0	0	0	-2000*	1/4, 1/8, 2/15
x_5	-3	-4	-3	0	1	0	0	-50	
x_6	-1	0	0	0	0	1	0	-7	
x_7	10	15	8	0	0	0	1	300	

Step 4: If there are no negatives amongst the b_i coefficients, then the present solution is feasible. To determine whether it is also optimal, select an entering variable for minimization problems if there are positive numbers in the objective row (other than the solution), select a largest; if there are none then the present solution is a minimum. For maximization problems select the largest negative. To determine the pivot, calculate the non-negative quotients by dividing the entries in the solution column by the coefficients in the entering variable column, and choose the smallest non-negative quotient. This determines the leaving basic variate and the pivot. Go to the beginning of step 4 again.

Step 5: By multiplying the row which the variate has left by 1/pivot, this coefficient is then 1, by adding or subtracting relevant multiples of the pivot row to each of the other rows in turn, make all the other entries in the pivot column 0. Thus the basic variate column is part of the identity matrix, as is required for the basic variable set.

In the example, division by -120, then multiplication of the resulting row in Iterative Tableau 2, by 15, 4, 0, and -15, and addition respectively to the z row, x_5, x_6, and x_7 rows yields Iterative Tableau 2. Go to step 2.

Iterative Tableau 2

Basis	x_1	x_2	x_3	x_4	x_5	x_6	x_7	Solution	Quotient
z	-5	0	-1/2	-1/8	0	0	0	250	
x_2	1/3	1	1/2	-1/120	0	0	0	50/3	
x_5	-5/3	0	-1	-1/30	1	0	0	50/3	
x_6*	-1*	0	0	0	0	1	0	-7*	
x_7	5	0	1/2	1/8	0	0	1	50	

Applying steps 2 to 5 again yields Iterative Tableau 3.

Iterative Tableau 3

Basis	x_1	x_2	x_3	x_4	x_5	x_6	x_7	Solution	Quotient
z	0	0	-1/2	-1/8	0	-5	0	285	
x_2	0	1	1/2	-1/120	0	1/3	0	43/3	
x_5	0	0	-1	-1/30	1	-5/3	0	85/3	
x_1	1	0	0	0	0	-1	0	7	
x_7	0	0	1/2	1/8	0	5	1	15	

There are no negatives on the right-hand side, so the solution is feasible. From step 4, since it is a minimization problem, we examine the objective row for a positive number other then the solution, and since there is none, the solution is optimal, and Iterative Tableau 3 yields the final solution.

Therefore the minimum operating cost for the AAA Resources Corporation subject to the constraints is:

$285,000 by operating Oil & gas field number 1 for 7 hours per day, and number 2, 14hrs.20mins per day.

The tableau also gives the solution to the dual problem:

Maximize: $Y = 2000y_1 + 50y_2 + 7y_3 + 300y_4$ subject to the following constraints:

$$40y_1 + 3y_2 + y_3 + 10y_4 <= 10$$
$$120y_1 + 4y_2 + 0y_3 \quad 15y_4 <= 15$$
$$60y_1 + 3y_2 + 0y_3 + 8y_4 <= 8$$

where, in the final tableau

$$y_1 = -x_4 = 1/8$$
$$y_2 = -x_5 = 0$$
$$y_3 = -x_6 = 5,$$

and $\quad y_4 = x_7 = 0.$

and $\quad Y_{max} = 2000(1/8) + 0 + 7(5) + 0 = 285$

It will be observed that the resources used for 2 and 4 (gas & expenditure), both have surpluses with slack variables, $-x_5 = 1/8$, and $-x_7 = 5$, whereas the resources 1 and 3, in the basis in the dual function are scarce, and therefore binding. These considerations are used in sensitivity analysis, which often leads to a further improvement in the model. The results given above have been checked on our Linear Programming computer program given in Appendix 6, which prints out all the tableaux, and is therefore extremely valuable for carrying out sensitivity analyses, as well as for the solution of Linear Programming problems in general.

5.4.5.3. Integer Programming model

The Integer LP has the additional constraints that all the variables must be integers. An example will illustrate the Branch & Bound algorithm for solving the problem. Suppose that, in example 5.4.3.2., due to labor contracts part hours must be paid for as full hours; then the additional constraints are x_1, x_2, x_3 must be integers.

Solution

Step 1: From the solution to Example 5.4.3.2., since there are additional constraints which may be binding, the minimal solution ≥ 285; thus 285 is a lower bound for the integer solution. The starting solution is therefore Iterative Tableau 3, with the new constraint that x_2 is an integer, in this case ≥ 15 or ≤ 14. The first branch we consider is with $x_2 \geq 15$. Go to step 2.

Step 2: By making the new constraint $\leq$, and adding slack variable, x_8:

$$-x_2 + x_8 = 15,$$

however this equation cannot join those in Initial Tableau 3, as x_2 is in the basis, so we elimate x_2 by simply adding the x_2 row to it, which yields Iterative Tableau 4.

Iterative Tableau 4

Basis	x_1	x_2	x_3	x_4	x_5	x_6	x_7	x_8	Solution	Quotient
z	0	0	-1/2	-1/8	0	-5	0	0	285	
x_2	0	1	1/2	-1/120	0	1/3	0	0	43/3	
x_5	0	0	-1	-1/30	1	-5/3	0	0	85/3	
x_1	1	0	0	0	0	-1	0	0	7	
x_7	0	0	1/2	1/8	0	5	1	0	15	
x_8^*	0	0	1/2	-1/120*	0	1/3	0	1	-2/3*	

Applying steps 2 to 5 from 5.2.3.1.2 yields:

Iterative Tableau 5

Basis	x_1	x_2	x_3	x_4	x_5	x_6	x_7	x_8	Solution	Quotient
z	0	0	-8	0	0	-10	0	-15	295	
x_2	0	1	0	0	0	0	0	-1	15	
x_5	0	0	-3	0	1	-3	0	-4	31	
x_1	1	0	0	0	0	-1	0	0	7	
x_7	0	0	8	0	0	10	1	15	5	
x_4	0	0	-60	1	0	-40	0	-120	80	

Hence, $z = 295$ at $(7, 15, 0)$ is an upper bound for the solution to the integer program.

Step 3: Branch to $x_2 \leq 14$ from Initial Tableau 3, so that the new constraint to be added to it is:

$$x_2 + x_8 = 14,$$

and since x_2 is already in the basis, subtacting the x_2 row yields:

Iterative Tableau 6

Basis	x_1	x_2	x_3	x_4	x_5	x_6	x_7	x_8	Solution	Quotient
z	0	0	-1/2	-1/8	0	-5	0	0	285	
x_2	0	1	1/2	-1/120	0	1/3	0	0	43/3	
x_5	0	0	-1	-1/30	1	-5/3	0	0	85/3	
x_1	1	0	0	0	0	-1	0	0	7	
x_7	0	0	1/2	1/8	0	5	1	0	15	
x_8	0	0	-1/2*	1/120	0	-1/3	0	1	-1/3*	

Applying steps 2 to 5 from 5.2.3.1.2 yields:

Iterative Tableau 7

Basis	x_1	x_2	x_3	x_4	x_5	x_6	x_7	x_8	Solution	Quotient
z	0	0	0	-2/15	0	-14/3	0	-1	856/3	
x_2	0	1	0	0	0	0	0	1	14	
x_5	0	0	0	-1/20	1	-1	0	-2	29	
x_1	1	0	0	0	0	-1	0	0	7	
x_7	0	0	0	2/15	0	14/3	1	-1	47/3	
x_3	0	0	1	-1/60	0	-2/3	0	-2	2/3	

Hence $z = 856/3$ is a new lower bound for z_{min}, but $x_3 = 2/3$: non-integral.

Step 4: Branch to $x_3 \leq 0$, hence Iterative Tableau 8 is derived:

Iterative Tableau 8

Basis	x_1	x_2	x_3	x_4	x_5	x_6	x_7	x_8	x_9	Solution	Quotient
z	0	0	0	-2/15	0	-14/3	0	-1	0	856/3	
x_2	0	1	0	0	0	0	0	1	0	14	
x_5	0	0	0	-1/20	1	-1	0	-2	0	29	
x_1	1	0	0	0	0	-1	0	0	0	7	
x_7	0	0	0	0	2/15	0	14/3	1	0	47/3	
x_3	0	0	1	-1/60	0	-2/3	0	-2	0	2/3	
x_9 *	0	0	0	1/60	0	2/3	0	2	1	-2/3*	

which is infeasible, since x_9 must leave, but there are no negative coefficients in the row, so no variable may enter.

Step 5: Branch to $x_3 \geq 1$, hence Iterative Tableau 9 follows.

Iterative Tableau 9

Basis	x_1	x_2	x_3	x_4	x_5	x_6	x_7	x_8	x_9	Solution	Quotient
z	0	0	0	-2/15	0	-16/3	0	-1	0	796/3	
x_2	0	1	0	0	0	2/3	0	1	0	14	
x_5	0	0	0	-1/20	1	-1	0	-2	0	29	
x_1	1	0	0	0	0	-1	0	0	0	7	
x_7	0	0	0	0	2/15	0	16/3	1	0	44/3	
x_3	0	0	1	-1/60	0	-2/3	0	-2	0	2/3	
x_9*	0	0	0	-1/60	0	2/3	0	-2*	1	-1/3*	

Applying steps 2 to 5 from 5.2.3.1.2 yields:

Iterative Tableau 10

Basis	x_1	x_2	x_3	x_4	x_5	x_6	x_7	x_8	x_9	Solution	Quotient
z	0	0	0	0	0	-10	0	15	-8	288	
x_2	0	1	0	0	0	0	0	1	0	14	
x_5	0	0	0	0	1	-3	0	6	-3	30	
x_1	1	0	0	0	0	-1	0	0	0	7	
x_7	0	0	0	0	0	10	1	-17	8	13	
x_3	0	0	1	0	0	0	0	0	-1	1	
x_4	0	0	0	1	0	40	0	120	-60	2	

There are no negatives on the right-hand side, so the solution is feasible. From step 4, since it is a minimization problem, we examine the objective row for a positive number other then the solution, and since there is none, the solution is optimal, and Iterative Tableau 10 yields the final solution. Therefore, the minimum operating cost for the AAA Resources Corporation subject to the additional integer constraints is \$288,000 by operating Oil & gas field number 1 for 7 hours per day, and number 2, 14hrs per day, and number 3, 1 hour per day. It will be observed that the minimum cost for the integer problem increased by \$3000 over the cost when this constraint is relaxed. The final tableau gives the solution to the dual too:

Maximize: $Y = 2000y_1 + 50y_2 + 7y_3 + 300y_4 + 14y_5 + y_6$

where, in the final tableau

$$y_1 = -x_4 = 0, \quad y_2 = -x_5 = 0, \quad y_3 = -x_6 = 10,$$
$$y_4 = x_7 = 0, \quad y_5 = x_8 = 15, \quad y_6 = -x_9 = 8,$$

Check: substitution in the dual objective function yields the dual solution

$$Y_{max} = 7*10 + 14*15 + 1*8 = 288.$$

CHAPTER 5: Exercises

E5 Exercises

Based on computer program FACTOR in Appendix 5, 5.1.,
derive the factor loadings for the field data sets
given in Appendix 8

E5.1. The Alligator Rivers project data, 8.2.1, using
variates 3, 4, 5 & 6.

E5.2. The Mount Bundey project data, 8.2.2., using
variates 3, 4, 5 & 6.

E5.3. Use the factor loadings of exercise E5.1. to modify
programs 5.2. & 5.3. to carry out a trend and residual
trend Factor analysis, using variates 1, 2 as the
independent variates, and the significant Factors
as the dependent variates, and select the local
anomalies which are significantly above the trend

E5.4. Use the factor loadings of exercise E5.2. to modify
programs 5.2. & 5.3. to carry out a trend and residual
trend Factor analysis, using variates 1, 2, 7, 8, ..., 10
independent variates, and the significant Factors
as the dependent variates, and select the local
anomalies which are significantly above the trend

E5.5. AAA Resurces also operates two oil & gas fields in North
California, and must deliver 45 barrels of oil and 30
million cu. feet of gas daily. The hourly production rate
for the first field is 3 oil and 1 gas, with operating
cost 7000$/hour; and for the second field 4 oil, and
3 gas, with operating cost 11,000$/hour. The total wage
bill per day must not exceed $125,000. If the wages
for field no. 1 total $4000/hour, and for no. 2,
$2000/hour; minimize the total operating cost. Check
your answer using the Linear Program in Appendix 6.

BIBLIOGRAPHY FOR CHAPTER 5

Section 1: General Linear Model

1.   AGTERBERG, F.P., 1964, Methods of Trend Surface Analysis;
         Colo. School of Mines Quart., V.59, pp.11-130.

2.   AGTERBERG, F.P., 1970, Multivariate prediction equations in
         geology; Jour. Math. Geol., V.2, pp.319-324.

3.   AGTERBERG. F.P., 1971, A probability index for detecting
         favourable geological environments; Can. Inst. M.&
         M., Spec.Vol.12., pp.82-91.

4.   AGTERBERG, F.P., 1974, Automatic contouring of geological maps
         to detect target areas for mineral exploration; Jour.
         Math. Geol., V.6, pp.375-395.

5.   AGTERBERG, F.P., 1974, Geomathematics, (Chpts. 5 & 15),
         Elsevier Publ., Amsterdam, Holland.

6.   AGTERBERG, F.P., 1978, Quantification and statistical analysis
         of geological variables for mineral resource
         evaluation; Bur. Rech. Geol. Min. Memoire No.91,
         pp.399-406.

7.   AGTERBERG, F.P., 1981, Applications of image analysis and
         multivariate analysis to mineral resource appraisal;
         Econ. Geol., V.76, pp.1016-1031.

8.   AGTERBERG, F.P. and CABILIO, P., 1969, Two-stage least-squares
         model for the relationship between mappable
         geological variables; Jour. Int. Assoc. Math. Geol.,
         V.1, pp.37-153.

9.   AGTERBERG, F.P. and FABBRI, A.G., 1978, Statistical treatment
         of tectonic and mineral deposit data; Global
         Tectonics and Metallogeny, V.1, No.1, pp.16-29.

10. BATES, C., 1959, An application of statistical analysis to
         exploration for uranium in the Colorado Plateau;
         Econ. Geol., V.54, pp.449-466.

11. BOSTICK, N.H., 1970, Electronic data processing applied to
         uranium resource prediction and exploration; Trans.
         Soc. Min. Eng., A.I.M.E., V.247, pp.4-10.

12. BOTBOL, J.M., SINDING LARSEN, R., McCAMMON, R.B. and GOTT,
         G.B., 1977, Weighted Characteristic Analysis of
         spatially-dependent mineral deposit data; Jour. Math.
         Geol., V.9, No.3, pp.309-311.

13. CANADIAN INSTITUTE OF MINING AND METALLURGY, 1968, Ore reserve
         estimation and grade control; C.I.M.M. Special Volume
         No.9, Montreal, Canada.

14. CHAPMAN, R.P., 1976, Limitations of Correlation and Regression
         Analysis n geochemical exploration; Trans. Inst.
         Min. Metall. (London), V.85, pp.B279-B283.

15. CHAPMAN, R.P., Geochemical interpretation of a statistical analysis of multi-element stream sediment data collected from New Brunswick; Can. Inst. M.& M. Bull., V.70, No.788, pp.62-69.

16. CHAPMAN, R.P., 1978, Evaluation of some statistical methods of interpreting multi-element geochemical drainage data from New Brunswick; Jour. Math. Geol., V.10, No.2, pp.195-224.

17. CHORLEY, R.J. and HAGGETT, P., 1965, Trend surface mapping in geological research; Trans. Inst. British Geogr. Publ. No.37.

18. CHUNG, C.F., and AGTERBERG, F.P., 1980, Regression models for estimating mineral resources from geological map data; Jour. Math. Geol., V.12, pp.458-473.

19. DAVIS, J.C., 1974, Statistics and data analysis in Geology (Chpts. 5 & 6); J. Wiley, New York.

20. DE GEOFFROY, J.G. and WIGNALL, T.K., 1970, Application of statistical decision techniques to the selection of prospecting areas and drilling targets in regional exploration; Can. Inst. M.& M. Bull., August, pp.893-899.

21. DE GEOFFROY, J.G. and WIGNALL, T.K., 1970, Statistical decision in regional exploration: application of Regression and Bayesian classification analyses in the South-West Wisconsin Zinc Area; Economic Geol., V.65, pp.769-777.

22. DE GEOFFROY, J.G. and WIGNALL, T.K., 1971, A probabilistic appraisal of mineral resources in a portion of the Grenville Province of the Canadian Shield; Economic Geol., V.66, pp.466-479.

23. DE GEOFFROY, J. and WIGNALL, T.K., 1972, A statistical study of the geological characteristics of the porphyry-copper-molybdenum deposits of North and South America; Economic Geol., V.67, pp.656-668.

24. DE GEOFFROY, J. and WIGNALL, T.K., 1973, Design of a statistical data processing system to assist in regional exploration planning; Can. Mining Jour., V.94, No.11, pp.30-35 & No.12, pp.35-36.

25. DE GEOFFROY, J.G. and WIGNALL, T.K., 1973, Statistical models for porphyry-copper-molybdenum deposits of the Cordilleran Belt of North and South America; Can. Inst. M.& M. Bull., May, pp.84-90.

26. DRAPER, N.D. and SMITH, H., 1966, Applied Regression Analysis; J. Wiley, New York.

27. GRAYBILL, F.A., 1961, Introduction to linear statistical models; McGraw-Hill, New York.

28. HARBAUGH, J.W., 1964, Use of Factor Analysis in recognizing
       facies boundaries, Bull. Am. Assoc. Petrol.
       Geologists, V.48, pp.529. (Abstract.)

29. HARBAUGH, J.W., 1964, A computer method for four-variable
       trend analysis illustrated by a study of oil-gravity
       variations in Southeastern Kansas; Kansas Geol. Surv.
       Bull., 171.

30. HARBAUGH, J.W. and MERRIAM, D.F., 1968, Computer applications
       in statigraphic analysis; J. Wiley, New York.

31. HARBAUGH, J.W. and BONHAM-CARTER, G.R., 1970, Computer
       simulation in Geology; J. Wiley, New York.

32. HARMAN, H.H., 1967, Modern Factor Analysis; Univ. of Chicago
       Press.

33. HENLEY, S., 1972, Non-linear mapping and a related R-mode
       technique for compression of multivariate data;
       Bureau of Mineral Resources, Geology and Geophysics,
       Bull. No.124, Canberra, Australia.

34. HENLEY, S., 1972, Plane representation of multivariate data
       structures; Bureau of Mineral Resources, Geology and
       Geophysics, Bull. No.123, Canberra, Australia.

35. KAISER, H.F., 1958, The VARIMAX criterion for analytic
       rotation in Factor Analysis, Psychometrika, V.23,
       pp.187-200.

36. KLOVAN, J.E., 1975, R- and Q-mode Factor Analysis, pp.21-69,
       in Concept in Geostatistics, R.S. McCammon, edit.;
       Springer-Verlag, New York.

37. KOCH, G.S. Jr. and LINK, R.F., 1971, Statistical analysis of
       geological data (Chpt.9); J. Wiley, New York.

38. KRUMBEIN, W.C., 1959, Trend Surface Analysis of contour-type
       maps with irregular control point spacing; Jour.
       Geophys. Res. V.64, pp.823-834.

39. KRUMBEIN, W.C. and GRAYBILL, F.A., 1965, An introduction to
       statistical models in Geology, (Chpts. 10, 12, 13);
       McGraw-Hill, New York.

40. LAWLEY, D.N. and MAXWELL, A.E., 1963, Factor Analysis as a
       statistical method; Butterworth, London.

41. MERRIAM, D.F. and COCKE, N.C., (edit.), 1967, Computer
       applications in the Earth Sciences: colloquium on
       Trend Analysis; Kansas Geol. Survey Comput. Contrib.
       No.12.

42. MILLER, R.L. and KAHN, J.S., 1962, Statistical analysis in the
       geological sciences, (Chpts 8, 14 & 15); J. Wiley,
       New York.

43. MORRISON, D.F., 1967, Multivariate statistical methods;
       McGraw-Hill, New York.

44. OVERALL, J.E. and KLETT, C., Applied Multivariate Analysis;
     McGraw-Hill, New York.

45. PIRKLE, F., CAMPBELL, K. and WECKSUNG, G.W., 1980, Principal
     Component Analysis as a tool for interpreting NURE
     aerial radiometric survey data; Jour. Geol., V.88,
     pp.57-67.

46. RAO, C.R., 1955, Estimations and tests of significance in
     Factor Analysis; Psychometrika, V.20., pp. 93-111.

47. RENDU, J.M., 1978, An introduction to geostatistical methods
     of mineral evaluation; Monograph of the S. Afr. Inst.
     M. & M.

48. SAAGER, R. and SINCLAIR, A.J., 1974, Factor Analysis of stream
     sediment geochemical data from the Mt Nansen area,
     Yukon Territory, Canada; Mineral. Deposita, V.9,
     pp.243-252.

49. SAUNDERS D.F., 1979, Characterization of uraniferous
     geochemical provinces by aerial Gamma-Ray
     spectrometry; Min. Eng., December, pp.1715-1722.

50. SINCLAIR, A.J. and WOODSWORTH, G.L., 1970, Multiple Regression
     as a method of estimating exploration potential in an
     area near Terrace, B.C.; Econ. Geol., V.65, pp.998-
     1003.

51. SINCLAIR, A.J. and GODWIN, C.I., 1979, Application of Multiple
     Regression Analysis to drill target selection, Casino
     porphyry-copper-molybdenum deposit, Yukon Territory,
     Canada; Trans. Inst. M.& M.,(London) V.88, pp.B93-
     B106.

52. WATSON, G.S., 1971, Trend Surface Analysis; Jour. Math. Geol.,
     V.3, No.3, pp.215-226.

53. WHITTEN, E.H.T., 1966, The general linear equation in
     prediction of gold content in Witwatersrand rocks,
     South Africa; Proc. Sympos. Math. Stat. and Comput.
     Appl. Johannesburg, pp.124-141.

BIBLIOGRAPHY FOR CHAPTER 5

Section 2: Operations Research Models

1. BATTERSBY, A., 1964, Network Analysis for planning and scheduling; Macmillan & Co. Ltd., London.

2. BERGE, C., 1973, Graphs and Hypergraphs; Elsevier, Amsterdam.

3. COBB, H., 1960, Operations Research: a tool in oil exploration; Geophysics, 25, 1009-1022.

4. COHN, D.L., 1954, Optimal systems: 1. The vascular system; Bulletin of Mathematical Biophysics, V.16, pp.59-74.

5. COX, J.L., 1961, Cutting costs through Operations Research; Min. Congr. Jour., 47, pp.45-46.

6. COYLE, R.G., 1969, Review of the literature on Operations Research in the mining industry; Trans. Inst. Min. Metall., London, 78, A1-A19.

7. CREIGHTON, R.L., HOCH, I. and SCHNEIDER, M., 1959, The optimum spacing of arterials and expressways; Traffic Quarterly, V.13, pp.447-494.

8. DUCKWORTH, E., 1962, A guide to Operations Research; Methuen, London.

9. FARBEY, B.A., LAND, A. and MURCHLAND, J.D., 1965, The "cascade" algorithm for finding minimum distances on a graph; London School of Economics, LSE-TNT-19.

10. GARRISON, W.L. and MARBLE, D.F., 1962, The structure of transportation networks; U.S. Army Transportation Command, Technical Report, pp.62-11

11. HARRIS, D., 1984, Computer Graphics and Applications; Chapman & Hall, London, and New York.

12. HARRIS, D.P. and EURISTY, D.A., 1973, The impact of the transportation network upon the potential supply of base and precious metals; Proc. 10th Internat. Symposium A.P.C.O.M., Johannesburg, S. Africa, S. Afr. Inst. Min. & Metall., pp.99-108.

13. HAZEN, S.W., Jr., 1968, Operations Research: a growing force in the mineral industries; Min. Eng., 20, pp 88-90.

14. KANSKY, K.J., 1963, Structure of transport networks: relationships between network geometry and regional characteristics; University of Chicago, Department of Geography, Research Papers, 84.

15. KLEINROCK, L., 1964, Communication nets: stochastic message flow and delay, J. Wiley, New York.

16. KU, Y.U. and BEDROSIAN, S.D., 1965, On topological approaches to network theory; Journal of the Franklin Institute, V.279, pp.11-21.

17. LITTLE, J.D.C., MURTY, K.G., SWEENEY, D.W. and KAREL, C., 1963, An algorithm for the travelling salesman problem; Operations Research, V.11, pp.972-989.

18. LOSCH, A., 1954, The economics of location; Yale University Press, New Haven.

19. MIEHLE, W., 1958A, Link-length minimization in networks; Operations Research, V.6, pp.232-243.

20. MOORE, E.F., 1959, The shortest path through a maze, Annals of the Computation Laboratory of Harvard University.

21. PERT Guide for Management, 1963; U.S. Government Print Office, Washington.

22. PETERS, W.C., 1978, Exploration and Mining Geology, pp. 24-36; J. Wiley, New York.

23. POLLACK, M. and WIEBENSON, W., 1960, Solutions of the shortest-route problem: a review; Operations Research, V.8, pp.224-230.

24. PRUSS, D.E, and FREEMAN, G.W., 1961, Mining exploration: An Operations Research and simulation approach; Min.World, 14, pp.42-43.

25. RAPAPORT, H. and ABRAMSON, P., 1959, An analogue computer for finding an optimum route through a communications network; Institute of Radio Engineers, Transactions on Communications systems, CS-7, 37-42.

26. ROUBENS, M. (Editor), 1977, Advances in Operations Research; Elsevier, Amersterdam.

CHAPTER SIX

MULTIVARIATE BAYESIAN CLASSIFICATION MODELS: APPLICATION TO THE
OPTIMAL SELECTION OF PROSPECTING AREAS AND EXPLORATION TARGETS

6.1. GENERAL STATEMENT

The problem of decision-making under uncertainty assumes a critical
importance when costly drilling decisions need to be made, particularly in
petroleum exploration. We find ourselves in two types of situations:

(i) Control data are available in well-prospected areas with similar
geological make-up, or (ii) no control locations are available. The second
situation is increasingly common, as explorationists have to deal with more
remote and little-known regions.

(i) When control data are available, Wignall (40) and (41) proposed
Bayesian classification analysis to assign locations to K classes, generally
three, based on geological, geophysical, geochemical and economic data
obtained in a well-prospected control area. Essentially, the problem of
maximizing the probability of success can be solved by conditional
probability laws. Each location under investigation has an information
vector of geo-variates which can be compared with that of selected locations
in the control area. As a result, each location can be assigned an index
which is based on the probability that the location belongs to each of the K
populations identified in the control area. The procedure is computerized by
means of the CLASSIFICATION program whose listings are given in Appendix 7.

(ii) In situations without control data, then a pre-classification
analysis is needed. Essentially, the first step is an R-mode Factor Analysis
on the whole set of available geovariates, to discern factors which can be
identified by the exploration team as diagnostic of the occurrence of ore,
to be followed by the evaluation of each factor at every location. These
factor scores can then be divided optimally into two populations on each
factor, using the program POPMIX given in Appendix 7. The second step is to
use a Q-mode Factor Analysis, which will give a weighting to each location
on each factor. These weightings can then be processed using POPMIX, and a
preliminary assignment can be made for each location using this second
channel. At this stage, anomalies of various types become apparent; a
control set can be made up by selecting the most typical locations for each
population, and the probabilistic assignments can then be made using the set
of classification functions discussed in (i). Often in situations where
control is available, the procedure given in (ii) can be used to advantage
for testing the quality of the model. This is achieved by observing whether
the control sets defined by procedure (ii) include the locations which
should be used for control in procedure (i).

Table 6.1. Multivariate Normal Bayesian model for the classification of exploration targets.

MODEL FOR THE CLASSIFICATION OF EXPLORATION TARGETS

Probability density function

$$f_1(X) = \{(2\pi)^p |S|\}^{-0.5} \exp[-0.5(X - \bar{X})S_1^{-1}(X - \bar{X})_1^\bullet], \quad i = 1, 2, 3$$

p = number of significant geologic factors

S = variance-covariance matrix, $|S|$ is its determinant, & S^{-1} its inverse

$(X - \bar{X})_1^\bullet$ = transpose

i = 1 for population of ore-bearing targets, class 1

i = 2 for population of ore-bearing targets, class 2

i = 3 for population of barren targets

Bayesian probability of occurrence of commercial ore, class i

$$P_i = \frac{q_1 f_1(X)}{q_1 f_1(X) + q_2 f_2(X) + q_3 f_3(X)} \qquad i = 1, 2$$

$P_3 = 1 - P_1 - P_2$ = the Bayesian probability of failure

q_i = historical record of success for region for class i type $\qquad i = 1, 2$

   = number of class i ore bodies / total number of drilled prospects $\qquad i = 1, 2$

$q_3 = 1 - q_1 - q_2$ is a priori probability of failure

$f_i(X)$ = probability density function for population i, $\qquad i = 1, 2, 3$

6.2. OPTIMAL SELECTION OF PROSPECTING AREAS AND EXPLORATION TARGETS BASED ON CONTROL DATA

6.2.1. Formulation of multivariate Bayesian classification models

6.2.1.1. Theoretical Background

Let $\underline{A}_j$ be the vector of geovariates collated for location $_j$, with j = 1, 2, ..., n; where n is the number of prospective locations or areas (or cells).

Table 6.2. Multivariate Bayesian models for the classification
of exploration targets: Poisson and Exponential models.

MODEL FOR THE CLASSIFICATION OF EXPLORATION TARGETS

Probability function for (1) Poisson distribution on each variate

$$f_i(X) = \prod_{j=1}^{p} \frac{(\bar{x}_j)^{x_j}}{x_j!} \exp[-(\bar{x}_j)], \qquad\qquad i = 1, 2, 3$$

Probability density function for (2) exponential distribution on each variate

$$f_i(X) = \prod_{j=1}^{p} (\bar{x}_j) \exp[-(\bar{x}_j)x_j], \qquad\qquad i = 1, 2, 3$$

p = number of significant geologic factors

i = 1 for population of ore-bearing targets, class 1

i = 2 for population of ore-bearing targets, class 2

i = 3 for population of barren targets

Bayesian probability of occurrence of commercial ore, class i

$$P_i = \frac{q_1 f_1(X)}{q_1 f_1(X) + q_2 f_2(X) + q_3 f_3(X)} \qquad\qquad i = 1, 2$$

$P_3 = 1 - P_1 - P_2 =$ the Bayesian probability of failure

$q_i =$ historical record of success for region for class i type $\qquad\qquad i = 1, 2$

$\qquad =$ number of class i ore bodies / total number of drilled prospects $\qquad i = 1, 2$

$q_3 = 1 - q_1 - q_2$ is a priori probability of failure

$f_i(X) =$ probability or probability density function for population i, $\qquad i = 1, 2, 3$

Let B_i be the probability that $\underline{A}_j \in$ (belongs to) population$_i$,

for i = 1, 2, ..., p; where p is the number of populations.

Then, by the conditional probability law:

Probability$\{B_i | \underline{A}_j\} =$ Probability$\{\underline{A}_j | B_i\}$*Probability$\{B_i\}$/Probability$\{\underline{A}_j\}$

Let f_k be the probability function or probability density function for
population k, for k = 1, 2, ..., p, and let $q_i =$ Probability$\{B_i\}$,
with i = 1, ..., p.

We now have

$$\text{Probability}\{B_i|\ \underline{A}_j\} = q_i f_i(\underline{A}_j)/\Sigma_{k\ =\ 1}^{\ p}\ q_k f_k(\underline{A}_j),\ \text{for } i = 1,\ 2,\ \ldots,\ p;$$

and $j = 1,\ 2,\ \ldots,\ n,$

This expression, directly derived from Bayes' theorem, is the linchpin of our Bayesian classification scheme. Readers should note that the result reduces to Fisher (21), when $p = 2$, and $f_k \propto$ Normal $N(\mu_k,\ \sigma_k^2)$.

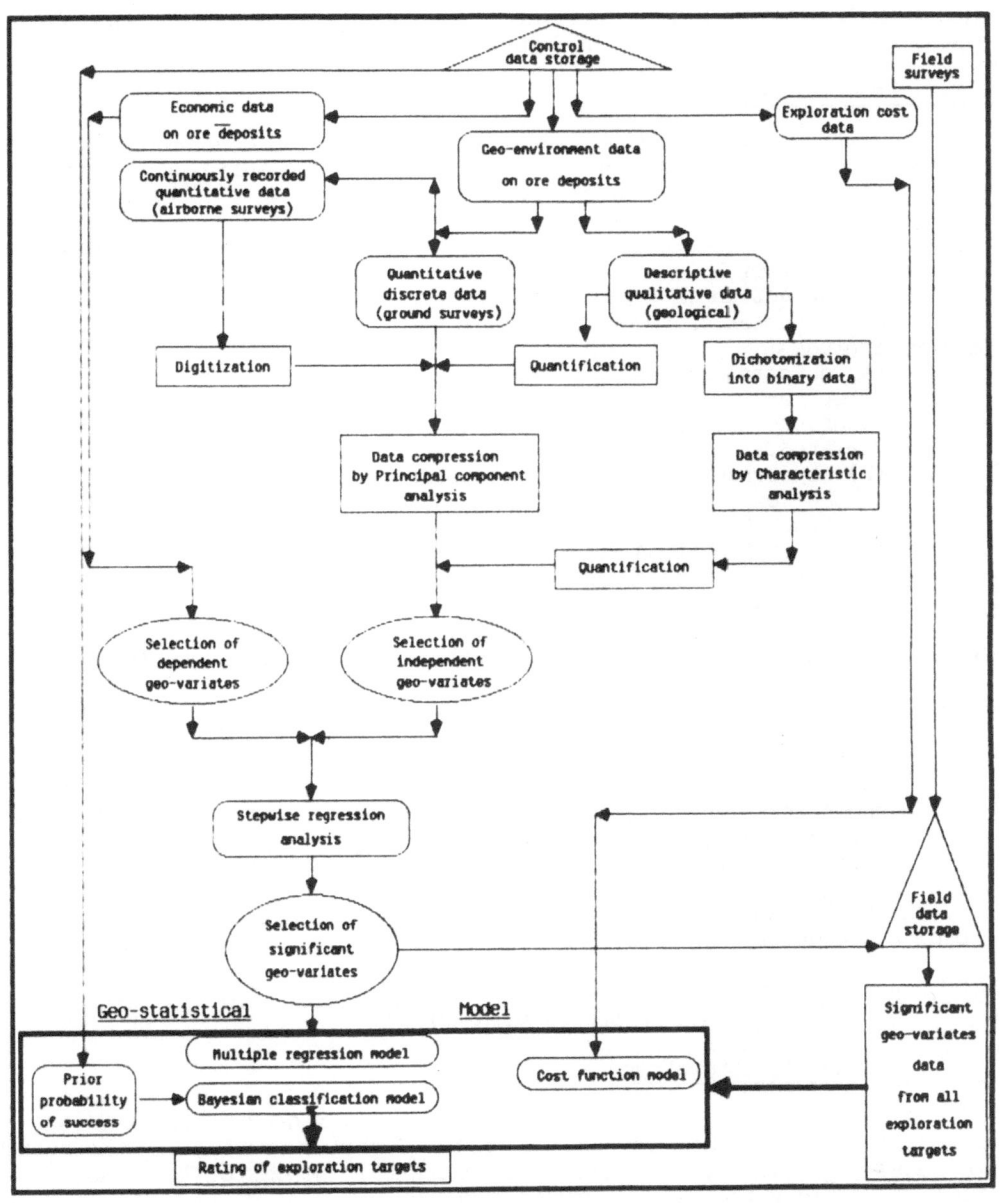

Figure 6.1. Flow-diagram of the optimal selection of exploration targets for drill testing based on control data

6.2.1.2. Application to some well-known distributions

Assuming p multivariate normal distributions, the result is summarized in Table 6.1, for p = 3. Examples are given below and in Wignall (40) and (41). If f_k is lognormal, a logarithmic transformation on the vector $\underline{A}_j$ will bring us back to the normal distribution, however readers are cautioned against ad hoc use of the logarithmic transformation of data, as it may affect the efficiency of the classification scheme.

If we have to deal with p multivariate Poisson distributions, the result is given in Table 6.2, for p = 3, which also describes the result when we have to deal with p multivariate exponential distributions. In both cases, examples can be found in Wignall (40).

6.2.2. Practical methodology

The methodology of target classification based on control sets follows the "pattern recognition" approach which proceeds sequentially in four main steps as follows (see flow diagram on Figure 6.1):

Step 1: (a) data acquisition and processing for all targetss under study and all control locations,
(b) selection of dependent and independent variates required for the construction of the geomathematical model

Step 2: includes:
(a) a "data compression" procedure, followed by a "feature selection" procedure based on a stepwise multiple regression technique, leading to the screening of the significant independent variates,
(b) construction of a geostatistical model with control sets based on ore-bearing and barren locations to formulate classification functions for each class of multivariate populations.
(c) These functions are evaluated for each location and the Bayesian classification model is then used to assign probabilities that each study target belongs to the ore or non-ore class, based on the a-priori and a-posteriori probabilities.

Step 3: Estimates of the expected economic worth of locations classified in the "ore" category are calculated.

Step 4: A pay-off criterion for the optimal selection of exploration targets under investigation is formulated. The pay-off model uses the probabilities of success (ore) and failure (no ore), regression estimates of the economic worth, and an appropriate cost function as shown in Figure 6.1. The pay-offs associated with all study targets are calculated and used to rank the targets in decreasing order of merit for selection subject to budgetary constraints.

The field examples which follow are based on actual exploration projects, in which the theory has been applied.

6.2.3. Application to the optimal selection of prospecting areas in the Superior Province of the Canadian Shield

6.2.3.1. General methodology

A Bayesian classification analysis scheme was used in the Archean greenstone belts of the Western half of the Superior Province in the search for massive sulfides (Cu-Zn) and Cu-Ni ultramafic deposits. The study area was first divided into 3 sub-provinces on the basis of geological

254

Table 6.3. Multivariate Normal Bayesian model for the classification
of prospecting areas in the search for Cu-Zn volcanogenic
sulfide and Ni-Cu ultramafic deposits in the Superior
Province of the Canadian Shield.

MODEL FOR THE CLASSIFICATION OF AREAS TO BE
PROSPECTED FOR Cu-Zn & Cu-Ni DEPOSITS

Probability density function

$$f_i(X) = \{(2\pi)^p |S|\}^{-0.5} \exp[-0.5(X - \bar{X}_i) S^{-1} (X - \bar{X}_i)'], \quad i = 1, 2$$

p = number of significant geologic factors   (p = 4 for N.W. Quebec)

(p = 7 for N.E. Ontario)

(p = 8 for N.W. Ont. & S.W. Man.)

S - variance-covariance matrix, $|S|$ is its determinant, & S^{-1} its inverse

$(X - \bar{X}_i)'$ = transpose

i = 1 for population of areas including commercial deposits

i = 2 for population of areas without commercial deposits

Bayesian probability of occurrence of commercial ore

$$P = \frac{q_1 f_1(X)}{q_1 f_1(X) + q_2 f_2(X)}$$

q_1 = historical record of success for the Superior Province

q_1 = number of ore bodies / total number of drilled prospects = 0.173

$q_2 = 1 - q_1 = 0.827$

$f_i(X)$ = probability density function for population i,   i = 1, 2

considerations and aeromagnetic patterns. Each sub-province was gridded into
36-square mile cells including control cells with known deposits, and little
prospected cells to be rated. The method was to use the multivariate normal
classification model to generate the probability of locating commercial ore
bodies in the study cells, based on the a-priori probability q, which was
measured by the proportion of successes in the Superior Province, and on the
geo-variate values of the individual cells. This complex and tedious
procedure was fully computerized by the first writer; the listings of the
computer program written for that purpose are to be found in section 7.2 of
Appendix 7.

The probabilities and ore value estimates were combined to calculate the expected pay-off, which was used to establish an ordinal rating of the geo-economic potential of the cells. The statistical index was multiplied by two factors reflecting the amount of previous exploration and the land occupancy status of the cells to obtain an objective final rating of the merit of the cells for exploration purposes.

6.2.3.2. Formulation of Classification models

Among the 2344 cells underlain by Greenstone Belts in the study area, there are 74 control cells which include 126 known base metal deposits worth a total of $15.4 billions (1968), and 2270 cells to be rated. The Bayesian Classification model used in this project is summarized in Table 6.3, and the Classification program is given in Appendix 7.2, but with p = 2. The control set for population (1) comprised the best 52 control cells with the highest aggregate ore values ; whilst the control for population (2) consisted of the other 22, with aggregate ore values less than $20 millions. The two functions were then evaluated at each of the study cells. Taking into account the a-priori probabilities measured by the historical success ratio for drilled base metal prospects in the Superior Province (q_1 = 0.173), the Bayesian classification model gave the probability of occurrence of commercial ore in each cell. Hotelling's T^2- test was then used to test the significance of the classification for each of the 3 sub-provinces.

6.2.3.3. Calculation of the pay-off rating of the study cells

The estimated ore values in study cells were obtained through the use of the multiple regression model (GLM), as shown in Chapter 5. For each study cell, the estimated ore values and the probability of occurrence of commercial ore were combined with geographic factors to yield an expected pay-off (EDP), the numerical values of the EDP were used as a rating index for the study cells.

The EDP is defined by

$$EDP = m_1 * P_o / 4 + m_2 * (1 - P_o),$$

where m_1 is the gross value regression estimate for the cell, 1/4 represents the expected return on the gross value, m_2 is a loss function, always negative, which is the average cost of reconnaissance coverage of the cell weighted by the distance factor, which reflects the distance from the center of the cell to the nearest transportation facility (rail, road, etc.). (See volume 1, Chapter 3).

The expected pay-offs in $millions for the 2270 study cells were found to range in value from -5 to +169.5. There are 1757 study cells (77.2% of the total) with negative pay-off values indicating they should not be considered for exploration. This leaves 513 Greenstone cells to be rated for prospecting merit. Among these cells, only 252 show a pay-off in excess of $5 million, which was taken as a cut-off point for the selection of cells for further consideration. These cells are concentrated into 26 clusters coinciding with centers of volcanogenic activity and mafic intrusions which are unequally distributed within the three sub-provinces. These cells were subjected to an additional selection procedure in two stages. The first stage is based on the availability of land for exploration, and the second one on the amount of previous exploration carried out in the region. As a result of the screening, a total of 50 study cells were selected and recommended for airborne geophysical coverage and ground follow-up work.

6.2.3.4. Calculation of the discounted overall pay-off for the Superior Province project

We use an exponential model for the probability of discovery, which is

$$P_d = 1 - Exp(-kx),$$

a commonly accepted heuristic model (see Chapter 2), where x is the exploration expenditure in $millions (reconnaissance + semi-detailed exploration), and k is an empirical coefficient taken as equal to 1.70 for the Canadian Shield. Note that for an average exploration expenditure for a 36 sq. mile cell of $0.168 (1968),

$$P_d = 0.333.$$

The probability of occurrence in the 50 best cells selected for airborne coverage from the 252 cells screened above ranges from 0.75 to 0.95, with a mean = 0.82; so that the average probability of success in any of the 50 cells is 0.82*0.33 = 0.27. The total gross value of the commercial base metal resources expected to occur in the selected cells is $18.8 billion, yielding an expected maximum gross profit of $18.8 * 0.20 = $3.76 billions. The present value of this return is $628 millions, calculated on the basis of a 12% discount rate over an average expected mine life of 16 years. The present value of the total exploration expenditure based on the same discount rate over 3 years is $6.1 millions.

Thus the discounted dollar pay-off is given by

EDP = 628*0.27 - 6.1*0.73
    = 164.8 $millions (1968).

6.2.4. Application to the optimal selection of prospecting areas in the Grenville Province of the Canadian Shield

6.2.4.1. General background

A similar exercise in the Grenville Province of the Canadian Shield by the authors (14) gave a probabilistic appraisal of the mineral resources in a study area shown in Figure 6.2.

Control area

The control area is situated in the S.E. part of Ontario and includes 21,000 sq miles (see Figure 6.2), of which 12,000 sq miles are underlain by the Grenville Group. The basement complex covers the remaining 9000 sq miles. The geology of the area is well known as a result of mapping done by the Canadian Geological Survey and the Ontario Department of Mines over the past 80 years. The area has yielded a great variety of metallic ores and industrial minerals. The deposits appear to be distributed in well defined patterns related to zones of regional metamorphism and intrusions. The mineralization associated with the metasomatic alteration of Grenville carbonate meta-sediments is the most important economically. About 400 small mineral deposits have been detected to date, yielding uranium, iron, molybdenum, feldspar, mica and apatite, but only sporadic commercial production has been recorded to date, with the exception of the Marmora iron deposit.

Study Area

The study area selected for this investigation is situated in Quebec and includes 17,000 sq miles, which was divided into 170 cells. The area is serviced by several highways and two rail-roads. Systematic mapping has been carried out by the Quebec Department of Mines since the mid 1930's, and by 1964 covered 2/3 of the study area. The Grenville Project was initiated by the Geological Survey of Canada in 1964 to update the geological interpretation of the area.

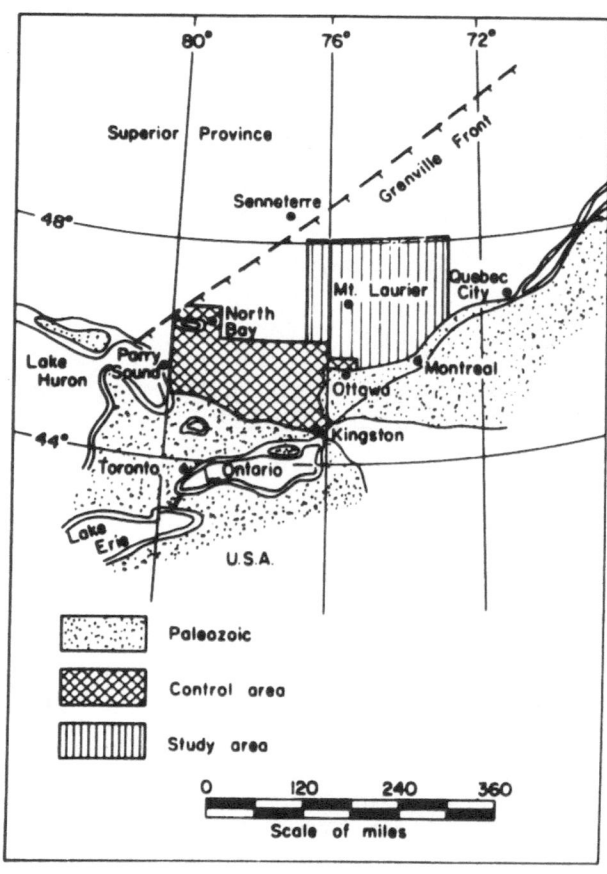

Figure 6.2. Map of a portion of the Grenville Province of the Canadian Shield showing the control and study areas.

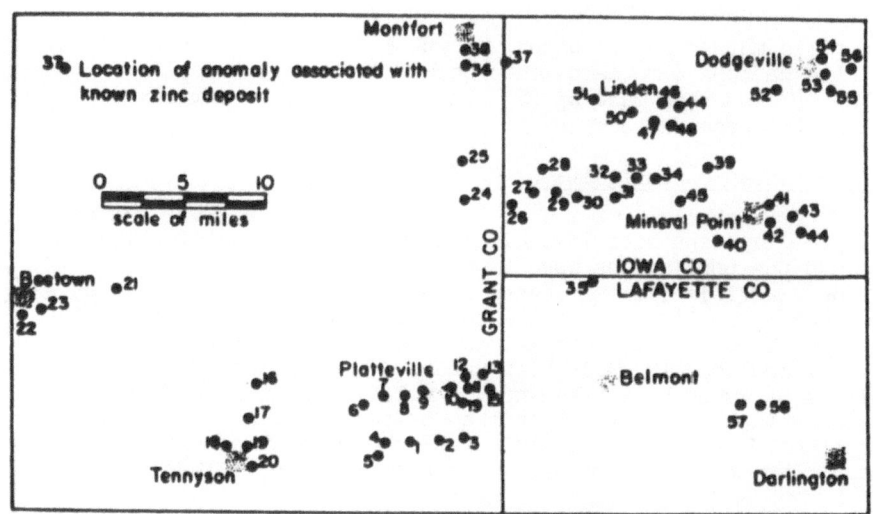

Figure 6.3. Map of a portion of the Upper Mississipi Valley Pb-Zn district showing the control locations.

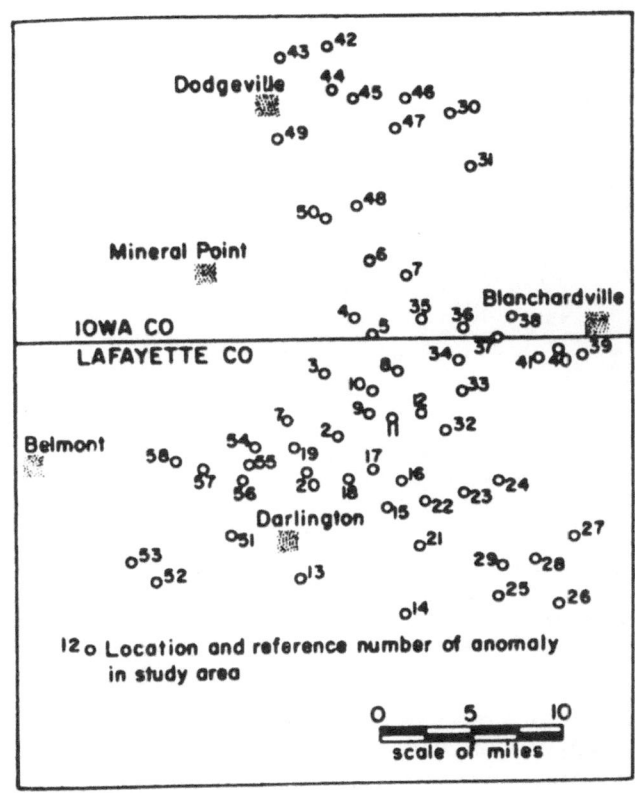

Figure 6.3 (Continued) Map of a portion of the Upper Mississipi Valley Pb-Zn district showing the study locations.

The Grenville metasediments underly about 5000 sq miles of the study area. Most of the carbonate metasediments occur in the S.W. portion of the study area. The remainder is made up of a large ultra-basic complex (1000 sq miles) and the basement complex (11,000 sq miles).

6.2.4.2. Bayesian classification

In the Bayesian classification analysis, the Hotelling T^2-test was applied and the populations were found to differ significantly at the 0.001 level. The classification was based upon 5 geovariates selected by a stepwise regression procedure. The control group for population (1) consisted of 30 cells with ore values > $2 millions; for population (2), the control set included 13 cells with values < 2 millions. The a-priori probability of success is taken as the relative freqency of exploration success over the last 80 years in the control area. From the Ontario Department of Mines files, it was established that the testing of 794 prospects resulted in 109 productive mines, hence q_1 = 109/794 = 0.138.

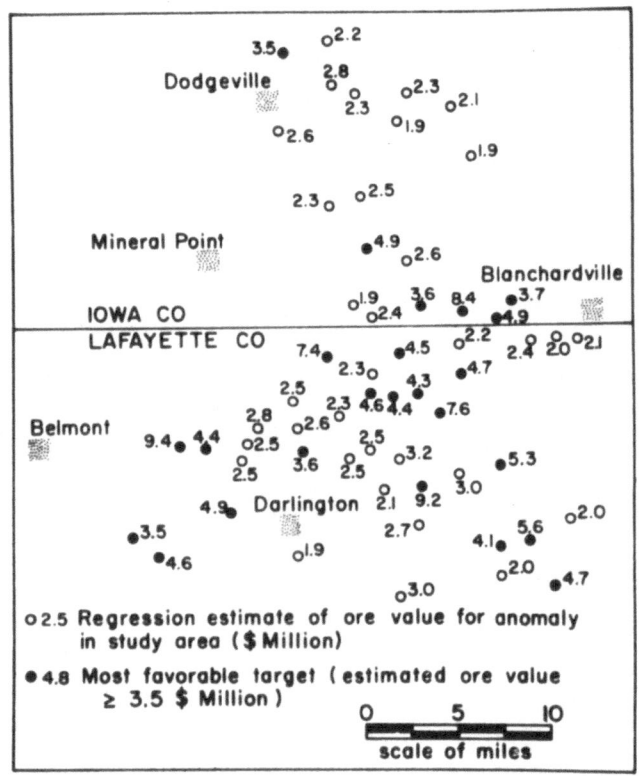

Figure 6.3 (Continued) Map of a portion of the Upper Mississipi Valley Pb-Zn district showing the results of the classification of study cells.

The resulting probabilistic classification for the 170 cells in the study area is as follows:

12 were assigned success probabilities > 0.80, and therefore selected for investigation,
2 were assigned success probabilities between 0.40 and 0.80,
156 were assigned success probabilities < 0.40, and therefore rejected.

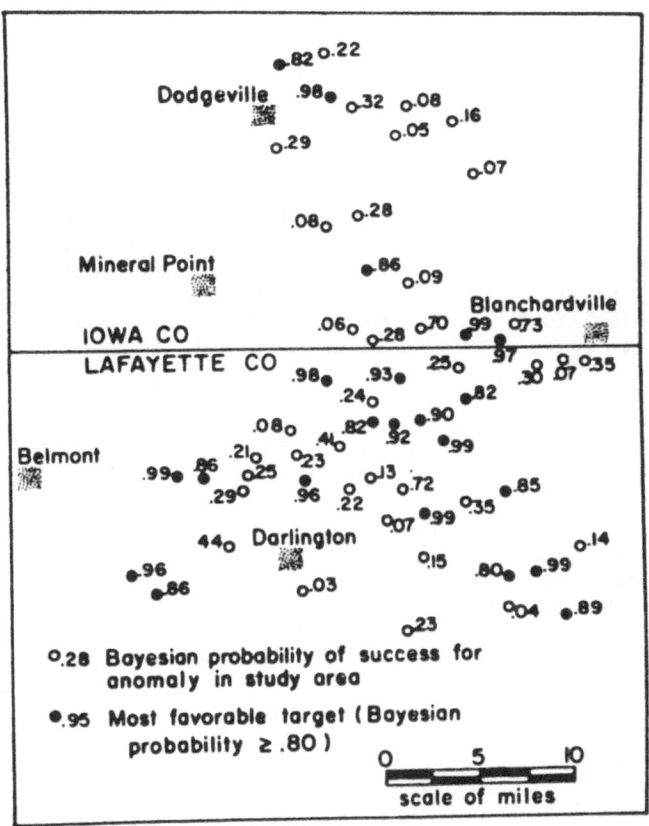

Figure 6.4. Map of selected targets in the study area of the Upper Mississipi Valley Pb-Zn district based on the combined results of Bayesian Classification Analysis and Regression Analysis.

We can see that the classification established a high degree of discrimination, with only two cells in the uncertain area, subject to mis-classification.

6.2.5. Application to the optimal selection of exploration targets for drill testing: Example in the Upper Mississipi Valley Southwest Wisconsin Zinc area

The Bayesian Classification model was applied by the authors (12) to a 225 square mile area of the Southwest Wisconsin Zinc Area shown in Figures 6.3 and 6.4. A total of 58 quarter of a square mile sub-cells, each including a commercial deposit, make up the control area adjacent to the study area. After processing the geochemical data, (heavy metal ppm count in spring water), a total of 58 anomalous sub-cells were indicated. Among these, 22 were assigned to population 1, with probabilities > 0.80, and 33 to population 2, with success probabilities < 0.35. Only 3 locations were assigned probabilities in between the two thresholds, which puts them in the doubtful category. Based on Dr. Heyl's information (30, Section (3), Chapter 1, we learned that lead mineralization had been found on the surface in several of the 22 sub-cells, while zinc mineralization had been intersected at shallow depth in one of these sub-cells in the 1960's.

6.3. OPTIMAL SELECTION OF EXPLORATION TARGETS WITHOUT CONTROL DATA

6.3.1. Introduction

Figure 6.5 is the flow-chart of a sequential procedure which is used to classify every location, whether in a study area or a control area, if any. A location may be anomalous on one of several geophysical or geochemical variates, i.e. showing significantly higher readings than the normal "background". We are interested only in anomalies reflecting the occurrence of mineralized material of ore grade. If there is just one geo-variate, the classification is carried out without Factor Analysis. The Trend Analysis is reduced to a least-squares regression.

For the multivariate model, the first step is to apply Factor Analysis in order to make the best use of the naturally occurring interactions and correlations among the geo-variates. For this purpose the SPSS computer program whose listings are included in Appendix 5 (the first two statements refer to the Honeywell version) is well recommended. If there are not too many locations, it is wise to use Q-mode as well as R-mode, as the two analyses will lead to a better discrimination into anomalous sets and background. The readers should note that, since we have no control ore deposits, we cannot use the same terminology as in the first section of this chapter. Following the Factor Analysis, there are two methods for detecting anomalies. The Trend & Trend Factor analyses have already been dealt with in Chapter 5; so let us focus on the POPMIX model.

6.3.2. Formulation of multivariate Bayesian classification model without control data

As illustrated by Figure 6.5., we can break the procedure into three steps, as follows:

Step 1:

Factor Analysis is the first step in our classification model without control. Factor Analysis has often been used for classification purposes, as shown by Klovan (11), simply because it brings out the maximum variations of the data in the manner discussed in Chapter 5. As a result, when the factors are evaluated for each member of a mixture of populations, separate classes become apparent, depending upon the degree and nature of correlations among the variates. Discrimination is usually much improved when there are some negative interactions. Researchers interested in discrimination problems in varied fields of Science have found that the use of ratios of groups of variates as well as individual variates is highly beneficial, because of the introduction of a built-in negative correlation between the ratio and the variates.

Step 2:

Step 2 covers the pre-classification procedure which resolves the mixed population into component populations. The underlying assumption is that the

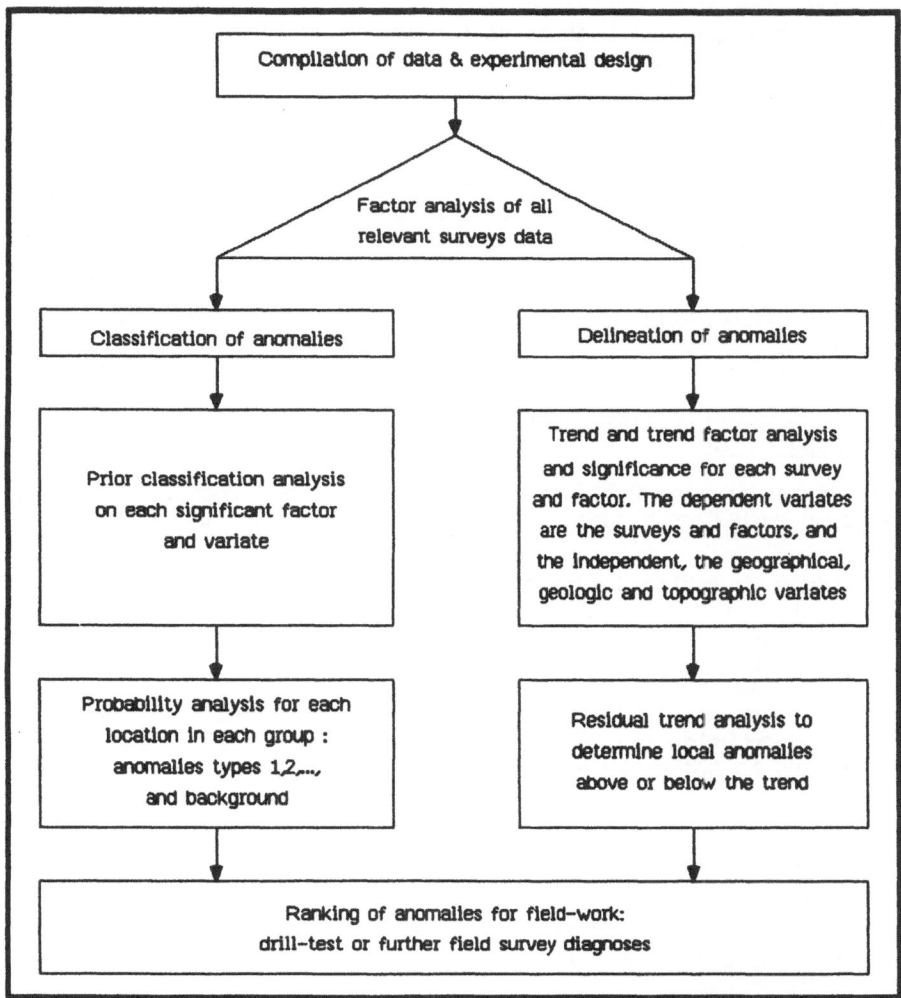

Figure 6.5. Flow diagram of the optimal selection of exploration targets for drill testing without control data

set of locations comprises samples from at least two populations, so that 4 parameters need to be estimated:

(i)   The mean, μ_1, of population (1),

(ii)   The mean, μ_2, of population (2),

(iii)   The common standard deviation, σ,

(iv)   The proportion, p, of locations belonging to population (1).

In order to estimate these parameters, appropriate distribution models must be assumed, and most workers such as Bridges & McCammon (1980), Bhattacharya (1967), and Rao (1952) have concentrated on normal populations. The latter used the method of statistical moments, and this leads to a good resolution; so our approach is based on this method. The program is given in Appendix 7: (POPMIX). It is used on each geovariate and on each significant factor, to assist in selecting control sets by simultaneously printing out factor scores for each location on each factor. Usually three sets become apparent at this stage: anomalies on factor 1, anomalies on factor 2, and background. Samples from these three sets are then used as control for the Classification program.

Step 3:

Using the control sets of locations obtained from POPMIX, but not the factors or factor scores, multivariate normal functions are calculated for each population by means of the Classification Program given in Appendix 7. Every location is then evaluated based on the same geovariates from which the functions were formulated, and the Bayesian classification is then effected (see Table 6.1) using a-priori probabilities if available.

6.3.3. General methodology

6.3.3.1. Resolution of mixed multinormal populations into k component populations

As stated in step 2 of section 6.3.2, there are four parameters to be estimated on each factor and/or geo-variate, namely μ_1, μ_2, σ, and p.

This may be achieved by a simple graphical method which requires the plotting of the histogram of each factor and geo-variate distribution of the mixed populations. The histogram of the factors generally turns out to be bi-modal, and the modes may be used as estimates of the means, and p may also be estimated as the sub-area surrounding the mode, divided by the total area. However, this is too subjective, so we now present the formulation of POPMIX procedure which is computerized in Appendix 7. The POPMIX procedure is carried out in three steps, as follows:

Step 1:

First join the source POPMIX file to the file of geo-data, stating the number of variates at the beginning of the file or interactively. In the example, three factors of variates were defined, F_1, F_2, F_3, and these factors were the output of the factor loadings from the factor program.

Step 2:

Calculate the k-statistics, k_1, k_2, k_3, k_4, the latter 3 being moments about the mean, k_1; thus, for each geovariate and factor, x,

$$k_1 = \Sigma x/n$$

$$c_1 = \Sigma x^2/n$$

$$c_2 = \Sigma x^3/n$$

$$c_3 = \Sigma x^4/n$$

$$k_2 = c_1 - k_1^2$$

$$k_3 = c_2 - 3*k_1*c_1 - 2*k_1^3$$

$$k_4 = \Sigma x^4/n - 4*k_1*c_2 + 6*k_1^2*c_1 - 3*k_1^4.$$

Step 3:

Having found the k-statistics, the estimates for the 4 parameters are made from equations

$$k_1 = pm_1 + qm_2,$$

$$k_2 = n\{s^2 + p(m_1 - k_1)^2 + q(m_2 - k_1)^2\}/(n - 1),$$

$$k_3 = n^2\{p(m_1 - k_1)^3 + q(m_2 - k_1)^3\}/(n - 1)/(n - 2),$$

and $\quad k_4 = n^3\{p(m_1 - k_1)^4 + q(m_2 - k_1)^4\}/(n - 1)/(n - 2)/(n - 3)$

$$- 3n^2\{[p(m_1 - k_1)^2 + q(m_2 - k_1)^2]^2\}/(n - 1)/(n - 1),$$

where $p + q = 1$.

Solution to the 4 equations in 4 unknowns leads to a cubic equation in z, where

$z = (m_1 - k_1)(m_2 - k_1)$, which is necessarily negative,

since: $m_1 < k_1 < m_2$.

Hence let z_1 be the negative root of

$$f(z) = z^3 + 0.5k_4 z + 0.5k_3^2 = 0.$$

This equation is solved using Taylor's theorem with initial solution,

$$z_1 = -k_2,$$

The differential, h, to be added to $-k_2$ is given by

265

$f(z_1 + h) = 0 = f(z_1) + h(3z_1^2 + 0.5k_4)$, neglecting the terms in h^2,

and h^3, where $z_1 = -k_2$ initially,

hence

$$h = -f(z_1)/(3z_1^2 + 0.5k_4),$$

then set $\quad z_1 = z_1 + h$, and repeat the step until $|h| < 0.00001$;

For most field examples the limit set for h in the program is sufficiently small to yield an accurate answer, but the limit may easily be changed if 0.00001 is unsatisfactory.

Statement 70 of the listings in 7.1. in Appendix 7 ensures that the Taylor step is repeated until the error is smaller than this limit.

Having found z_1, substitution into the k_3 equation yields the quadratic

$$w^2 + k_3 w/z_1 + z_1 = 0.$$

The negative solution is

$$w_1 = m_1 - k_1$$

so that

$$m_1 = k_1 + w_1,$$

the other root to the quadratic is

$$w_2 = k_3/z_1 - w_1$$

thus

$$m_2 = k_1 + w_2.$$

We finally obtain

$$p = w_2/(w_2 - w_1),$$

and $\quad s^2 = k_2 + z_1.$

6.3.3.2. Prior classification analysis to select the most anomalous locations on each variate and each factor

The program POPMIX needs some slight modification for each project. For instance when there is a need to supply the number of variates being read in. We have to go to Statement 8 of the listings in Appendix 7 dealing with factors whose weightings were obtained from the FACTOR program in

Appendix 5. Statements 27, 28, 29 have to be modified to take into account the new factor functions. Statement 37 ensures that the factor scores on each factor are given in the output at each location.

As a result of the application of POPMIX, we now have at hand the four statistics listed in step 2 of section 6.3.2, which are required for the resolution of the mixed multinormal populations. The most important statistic for our purpose is the proportion, p, of locations belonging to the anomalous population (1). A multiplication of the proportion p by N, which is the total number of locations in the study area yields the number of locations which are anomalous on that factor or variate. The control for population (1) of anomalous locations can easily be chosen from the top of the list obtained by ranking them in order of decreasing factor scores. Similarly, the proportion for the second Factor or variate yields the control for population (2). A random sample from the remainder, or the whole remainder, will be the control set for the background population of non-anomalous locations.

6.3.3.3. Bayesian classification analysis using the selected locations as control sets for the three populations

As a result of the application of the POPMIX program, we are now in a position to use the Bayesian procedure outlined in Table 6.1. However, we are now dealing with three control sets which are generated from the field data, instead of using the control sets derived from the well prospected control area, as described in Section 6.2 and illustrated by three field examples.

6.3.3.4. Conclusions and remarks

It will be apparent in the application of the Classification programs that POPMIX is really crucial to diagnose the control locations even in field project where control is believed to be established, for POPMIX determines the control sets objectively. In fact, even when control sets are available, the classification procedure should be the sequential set of programs: FACTOR, POPMIX, and CLASSIFICATION. Note that for the CLASSIFICATION program, factors must not be used as their inclusion would make the matrix singular; we use the factor scores only for the purpose of selecting the control sets.

6.3.4. Application to mineral exploration projects

6.3.4.1. Introduction

This section is devoted to the practical application of the method devised by the first writer to select exploration targets in little known regions, without the benefit of the availability of control locations. The geo-economic variety of the examples underscores the versatility and flexibility of the method. The four examples described below are located in Australasia. They include

(a) a porphyry-Cu-Au project in the Fiji Islands,
(b) a uranium-gold exploration project,
(c) a uranium-silver project, both the situated in the Northern Territory of mainland Australia,
(d) a petroleum exploration project in the on-shore area of the Gippsland Basin-Bass Strait field of Victoria, Australia.

Since the projects have already been mentioned as examples of the application of the FACTOR & FACTOR-TREND programs in Chapter 5, the particulars of their geological setting will not be repeated here.

6.3.4.2. Fiji porphyry-Cu-Mo-Au project

The field data used as input for the classification study are the geochemical abundance in Cu, Pb, Zn in soils, and the geographical co-ordinates as well as elevations.

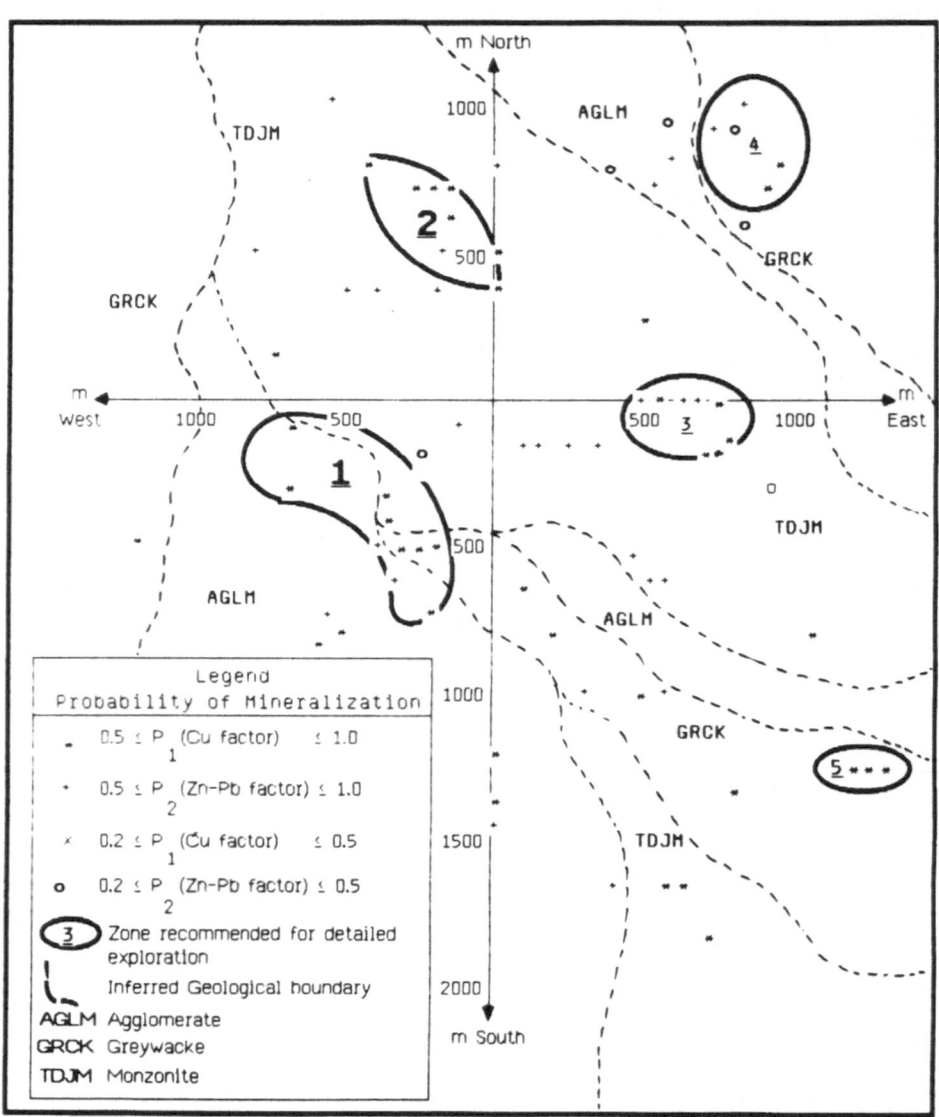

Figure 6.6. POPMIX & Classification Analysis for the porphyry-Cu-Mo-Au search in the Fiji Islands.

Factor Output:

from the Factor Analysis program, which was discussed in Chapter 5, two factors were found. Factor (1) essentially accounted for most of the variation in the Pb-Zn abundances; and factor (2), a shape factor with a contrast of Cu and Pb-Zn, accounted for the Cu abundance variations.

POPMIX Output:

the program diagnosed p_1 = 0.0251 of the 1234 locations belonged to population (1) (1234 x 0.0251 = 31 anomalies on factor (1)); and p_2 = 0.009 belonged to population (2) (1234 x 0.009 = 11 anomalies on factor 2); this left 0.9659 as the proportion for the background.

Hence, the 31 locations with the highest factor (1) scores were selected as control for population (1); and the top 11 on factor (2) became the control set for population (2). The remaining 1192 locations were used as control for the background population.

Classification Output:

The program is designed to select only those locations belonging to the anomalous populations for the printing of coordinates. In the case of projects where control locations are available, the threshold is that the sum of the probabilities of belonging to the anomalous populations is at least 0.5, but the probabilities in projects with no control nearly always approach 1. The Fiji porphyry-Cu-Au output reads in full as follows:

	pop.(1)	pop.(2)	pop.(3)
population size:	31	11	1192
Mean on Variate 1 (Cu):	270	92	73, Std. dev: 47
Mean on Variate 2 (Pb):	20	44	21, Std. dev: 13
Mean on Variate 3 (Zn):	71	535	93, Std. dev: 63
Mean on Variate 4 (Elev):	412	371	416, Std. dev: 97

The interactions amongst the 4 variates were given as follows:
Correlation matrix:

	(1)	(2)	(3)	(4)
(1)	1	0.047	0.067	-0.107
(2)		1	0.249	0.169
(3)			1	-0.014
(4)				1

The correlation matrix indicates that Cu is negatively correlated with elevation, whereas Pb is positively correlated with both elevation and Zn, and Zn is negatively correlated with elevation. The study clearly underlines the downward mobility of Cu and Zn, as opposed to the relative immobility of Pb in the conditions of rugged topography and heavy monsoonal rainfall prevailing in the Fiji Islands.

The inverse of the variance-covariance matrix was printed, and was well-conditioned.

Comparison of population mean differences and significance levels:

Significance level	Probability
Variate 1: F = 28.7	0.00000, pops 1, 2
Variate 2: F = 6.9	0.00008, pops 1, 2
Variate 3: F =109.9	0.00000, pops 1, 2
Variate 4: F = 0.4	0.82736, pops 1, 2
Variate 1: F = 0.4	0.78274, pops 2, 3
Variate 2: F = 8.5	0.00002, pops 2, 3
Variate 3: F =133.4	0.00000, pops 2, 3
Variate 4: F = 0.6	0.66128, pops 2, 3
Variate 1: F =131.0	0.00000, pops 1, 3
Variate 2: F = 0.1	0.99305, pops 1, 3
Variate 3: F = 1.0	0.58642, pops 1, 3
Variate 4: F = 0.1	0.99846, pops 1, 3
Variate 1: F=4676.5	0.00000, pops 1,2,3
Variate 2: F= 157.8	0.00000, pops 1,2,3
Variate 3: F= 151.8	0.00000, pops 1,2,3

The set of F-tests on the differences between the populations taken both pairwise and then all three together show highly significant differences on the Cu, Pb, and Zn variates, but no significant difference on elevations.

The resulting clear classifications were therefore to be expected.

Classification of Locations into 3 populations

The results shown below were obtained from 1234 locations. Because of format constraints, we selected a random sample of only 12 for display in the following table. The reader will note that coordinates are given only for locations assigned to anomalous populations (1) and (2) for the sake of simplification of the presentation of the results.

Loc. Ident.	Coordinates	Probabilities: Pop. 1	Pop. 2	Pop. 3
a 9147		0.110	0.000	0.890
a 9148		0.054	0.000	0.946
a 9633	-300, -650	1.000	0.000	0.000
a 8919		0.000	0.000	1.000
a 8922	-1400, 0	0.000	1.000	0.000
a 9719	-500, -450	0.000	1.000	0.000
a 9736	-400, 0	0.000	0.993	0.007
a 9843	-800, -500	0.000	0.999	0.001
a 9471	-250, -500	0.993	0.000	0.007
a 8801		0.003	0.000	0.997
a 9197		0.000	0.000	1.000

In all, a total of 32 locations were assigned to population (2), with probabilities exceeding 0.75, while 28 were assigned to population (1) with probabilities exceeding 0.90. Thus a total 60 out of 1234 locations are shown as anomalous, which is 4.9% of the whole set.

The anomalous locations on factor (1) or the Cu factor are shown on the accompanying map (Figure 6.6) by means of asterisks, and those anomalous on factor (2) by means of crosses. From a total of 60 anomalous locations identified by the classification study, as many as 30 are grouped into five clusters which are identified on the map. Cluster No. 1 is the largest of the five, comprising 10 high probability locations grouped in an area which straddles the monzonite contact, and therefore must be considered as a prime target for testing the occurrence of a porphyry-type deposit, both on the geochemical and geological standpoints.

6.3.4.3. Alligator Rivers U-Au project

Factor Output:

Two factors were disclosed by the Factor Analysis program run on radiometric readings collected from 427 locations in the Alligator Rivers

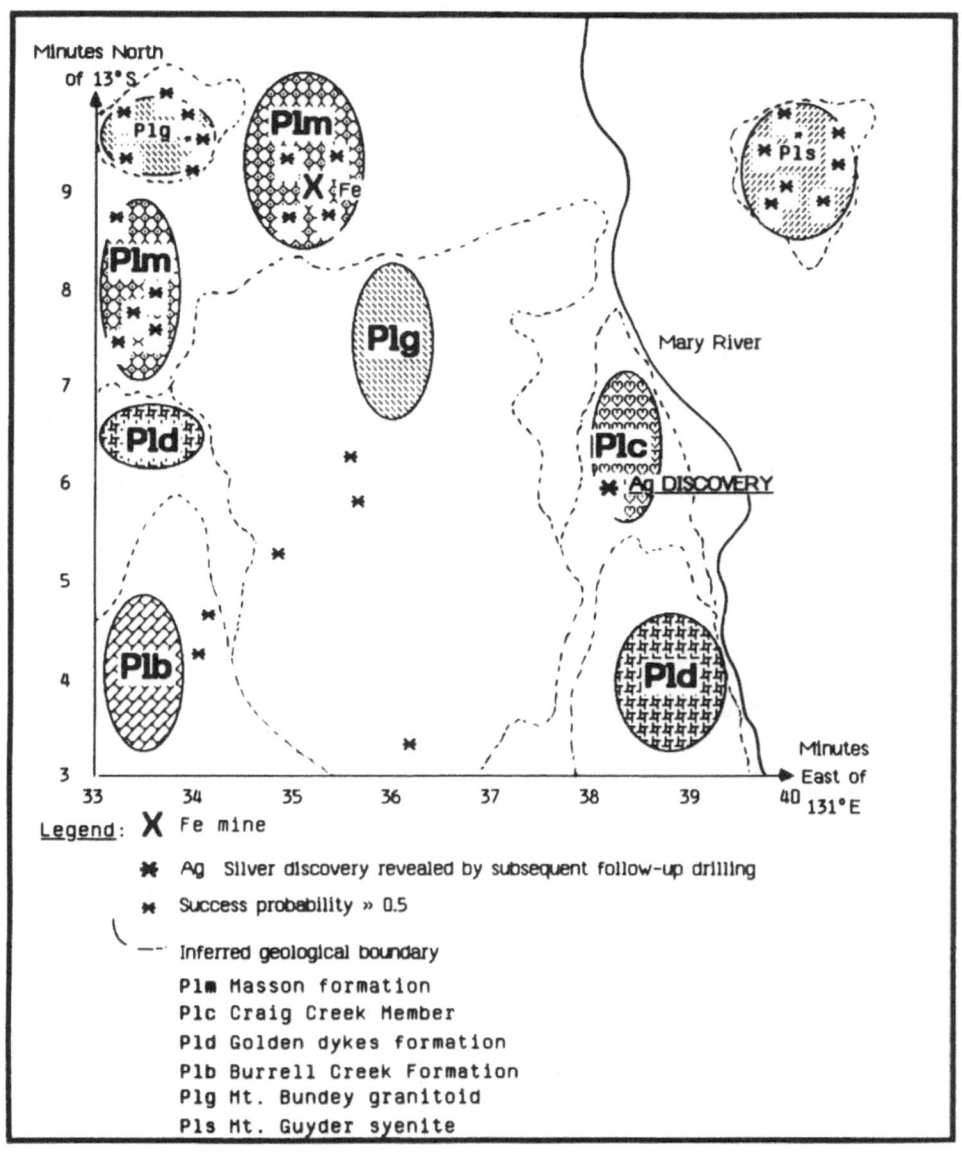

Figure 6.7. POPMIX & Classification Analysis for the
U-Au search at Alligator Rivers, Northern
Territory, Australia.

area of the Northern Territory of Australia, which was discussed in Chapter 5. The field data were radio-activity counts on Uranium, Thorium and Potassium. For geophysical reasons, a third factor of great importance defined as the ratio U/Th was added to the first two. Factor (1) is essentially a linear combination of all the geo-variates, accounting for 75% of the variation. Factor (2) is a shape factor which is useful in emphasizing the contrast between the U & U/Th ratios with positive coefficients on the one hand, and Th, K (Potassium) & total count with negative coefficients on the other hand.

POPMIX Output:

The computer output is as follows:

Geo-variates: (1): U/Th Factor, (2): Factor 1, (3): Factor 2,
Means   :   2.04   3.39   25.71

Population (1) proportion, p, on variate (1): p = 0.017
Population (2) proportion, p, on variate (2): p = 0.005
Population (1) proportion, p, on variate (3): p = 0.004

$Mean_1$, $Mean_2$, and common st. dev. on var. (1): m_1 = 13.3, m_2 = 1.9, s = 1.1
$Mean_1$, $Mean_2$, and common st. dev. on var. (2): m_1 = 288, m_2 = 33 , s = 15
$Mean_1$, $Mean_2$, and common st. dev. on var. (3): m_1 = 315, m_2 = 27 , s = 13

Note that since both variates (1) and (3) relate to U directly, they are used to define the control for population (1); while variate (2) serves as control for population (2).

Based on the results of the POPMIX analysis, a total of 11 locations with the highest factor (1) & U/Th scores were selected as control for population (1); and the top 10 on factor (2) became the control set for population (2). The remainder comprising 406 locations were used as control for the background population.

Classification Output:

The program is designed to select only those locations belonging to the anomalous populations (1) and (2), for the printing of coordinates, such that the sum of the probabilities of belonging to a population is greater then 0.5.

The Alligator Rivers output reads:

U-Au Project (Alligator Rivers), Probabilities.

Size of populations: (1): 11,   (2): 10,   (3): 406.

Mean on Variate (1):	141	412	144	(Total count)
Mean on Variate (2):	27	80	26	(K)
Mean on Variate (3):	23	41	14	(U)
Mean on Variate (4):	3	24	10	(Th)
Mean on Variate (5):	8.8	1.8	1.8	(U/Th)

Standard deviations on var.: (1): 60, (2): 13, (3): 6.5, (4): 5.3, (5): 1.1

Correlation matrix:

	(1)	(2)	(3)	(4)	(5)
(1)	1	0.883	0.759	0.772	-0.193
(2)		1	0.615	0.529	-0.103
(3)			1	0.558	0.162
(4)				1	-0.540
(5)					1

The variates are all positively correlated, except the geophysical Factor ratio which is negatively correlated with all except U.

The inverse of the variance-covariance matrix was printed, and was well-conditioned.

Comparison of population mean differences and significance levels:

Significance level on variate	Probability
(1) (total count): F = 26.7	0.00000, pops 1, 2
(2) (K, potass.) : F = 22.2	0.00000, pops 1, 2
(3) (U, uranium) : F = 10.5	0.00000, pops 1, 2
(4) (Th, thorium): F = 21.2	0.00000, pops 1, 2
(5) (U/Th ratio) : F = 52.8	0.00000, pops 1, 2
(1) (total count): F = 48.5	0.00000, pops 2, 3
(2) (K, potass.) : F = 42.6	0.00000, pops 2, 3
(3) (U, uranium) : F = 41.1	0.00000, pops 2, 3
(4) (Th, thorium): F = 16.7	0.00000, pops 2, 3
(5) (U/Th ratio) : F = 0.0	0.99955, pops 2, 3
(1) (total count): F = 0.0	0.99912, pops 1, 3
(2) (K, potass.) : F = 0.0	0.99986, pops 1, 3
(3) (U, uranium) : F = 4.3	0.00224, pops 1, 3
(4) (Th, thorium): F = 5.3	0.00060, pops 1, 3
(5) (U/Th ratio) : F =109.8	0.00000, pops 1, 3

The set of F-tests on the differences between the populations taken pairwise show highly significant differences on all variates except for variate 5, for populations (2) and (3); and variates 1, 2 for populations (1) and (3).
Clear classifications can therefore be expected.

Among the results for 427 locations only a random sample of anomalies presented here.

Loc. Ident.	Coordinates		Probabilities: Pop. 1	Pop. 2	Pop. 3
1	-41,	55	1.000	0.000	0.000
8	-32,	86	1.000	0.000	0.000
10	-18,	79	1.000	0.000	0.000
13	-43,	55	1.000	0.000	0.000
15	-45,	50	0.998	0.000	0.002
16	-49,	52	1.000	0.000	0.000
17	-52,	51	1.000	0.000	0.000
19	-49,	66	0.998	0.000	0.002
27	-63,	82	0.863	0.000	0.137
45	-89,	27	0.000	1.000	0.000
361	-85,	6	0.000	1.000	0.000
412	-91,	31	0.000	1.000	0.000
420	-91,	8	0.000	0.754	0.246
386	-88,	85	0.000	0.983	0.017
374	-88,	6	0.000	0.956	0.054

In all, a total of 18 locations were assigned to population (1), with probabilities exceeding 0.75, with two more exceeding 0.5, while 23 were assigned to population (2), with probabilities exceeding 0.70, making a total of 43 or 10% of the whole set of 427. All the known mines were assigned probabilities of 1.000 by the classification scheme, so the chances of economic success for the other anomalous locations selected by the process appear to be excellent. However, the final selection in the field should be left to expert field geologists; the presence of old mine workings and slag heaps should be taken into account. The anomalous locations are plotted on the map shown in Figure 6.7. One cluster of three locations with probabilities equal to 1.0 is noted in the southwest corner of the map area. Three clusters of two high probability locations are apparent, as follows: (a) due East of the first cluster, (b) in the central portion of the map area, and (c) in the northeast corner of the map area. They should be considered as prime targets for detailed exploration and drilling.

6.3.4.4. Mount Bundey U-Fe-Ag project

The Mount Bundey, Northern Territory, project was introduced in Volume 1, by De Geoffroy & Wignall; a more detailed account of the statistical processing of the field data follows. The variates consisted of (1) aeromagnetometer survey readings, (2) airborne radiometric total counts, (3) topography, (4) ground radiometric total counts, and (5) geological data.

Factor Output:

From the factor analysis program, which was discussed in Chapter 5, three factors were found. Factor (1) was essentially a linear combination of all the geo-variates and accounted for 63% of the variation; factor (2) is a shape factor with a contrast between the radiometric geo-variates with positive coefficients and aeromagnetomic & syenite with negative coefficients accounting for a further 27% of the variation; and factor (3) is another contrast factor between topographic and granitoid variates with positive loadings and geophysical variates with negative loadings accounting for 10% of the variation.

Factor (1) reflects the total geophysical environment and probably relates to deposits of the Mount Bundey Fe type which contain too many impurities to be commercial; it may also relate to quarries of industrial minerals. As a result, factor (1) will be taken to select the control group for population (1). Factor (2) appears to relate to U, so it is used to define the control for population (2).

POPMIX Output:

Sample size = 247 locations, sampled for each variate (no zeros or missing data)

Geovariates: (1): Factor 1,   (2): Factor 2, (3): Factor 3
Means   : (1):   29,   (2): 105,   (3):   -12.

Population$_1$ proportion, p, on variate (1): p = 0.013

Population$_2$ proportion, p, on variate (2): p = 0.043

Population$_1$ proportion, p, on variate (3): p = 0.005

Mean$_1$, Mean$_2$, and common st. dev. on var. (1): m_1 = 97, m_2 = 28.2, s = 17

Mean$_1$, Mean$_2$, and common st. dev. on var. (2): m_1 = 332, m_2 = 100 , s = 57

Mean$_1$, Mean$_2$, and common st. dev. on var. (3): m_1 = 91, m_2 = -12 , s = 11

As a result of the POPMIX resolution of the 247 locations, a total of 19 locations with the highest Factor(1) scores were selected as control for population (1); and the top 7 on Factor (2) became the control set for population (2). The remainder were control for the background population.

Classification Output:

The program is designed to select for co-ordinate printing only those locations belonging to the anomalous populations, such that the sum of the probabilities of belonging is greater than 0.5.

The Alligator Rivers output reads as follows:

U-Au Project (Alligator Rivers), Probabilities.

Size of populations: (1): 19,   (2): 7,   (3): 222.

Mean on Variate (1):	81	166	37	(Aeromag.)
Mean on Variate (2):	141	166	129	(Air: rad.)
Mean on Variate (3):	37	49	40	(Topograph.)
Mean on Variate (4):	310	378	118	(Grd: rad.)
Mean on Variate (5):	0.4	0.7	0.2	(Geo. design)

Standard deviations on variates: (1): 35,(2): 25,(3): 2,(4): 78,(5): 0.6

Correlation matrix:

	(1)	(2)	(3)	(4)	(5)
(1)	1	-0.392	0.184	0.575	0.438
(2)		1	-0.204	0.135	0.118
(3)			1	0.115	-0.069
(4)				1	-0.429
(5)					1

The inverse of the variance-covariance matrix was well-conditioned.

Comparison of population mean differences and significance levels:

Significance level	Probability
Variate (1): F = 3.4	0.00087, pops 1, 2
Variate (2): F = 0.6	0.78255, pops 1, 2
Variate (3): F = 0.2	0.99394, pops 1, 2
Variate (4): F = 0.4	0.91712, pops 1, 2
Variate (5): F = 0.2	0.99609, pops 1, 2

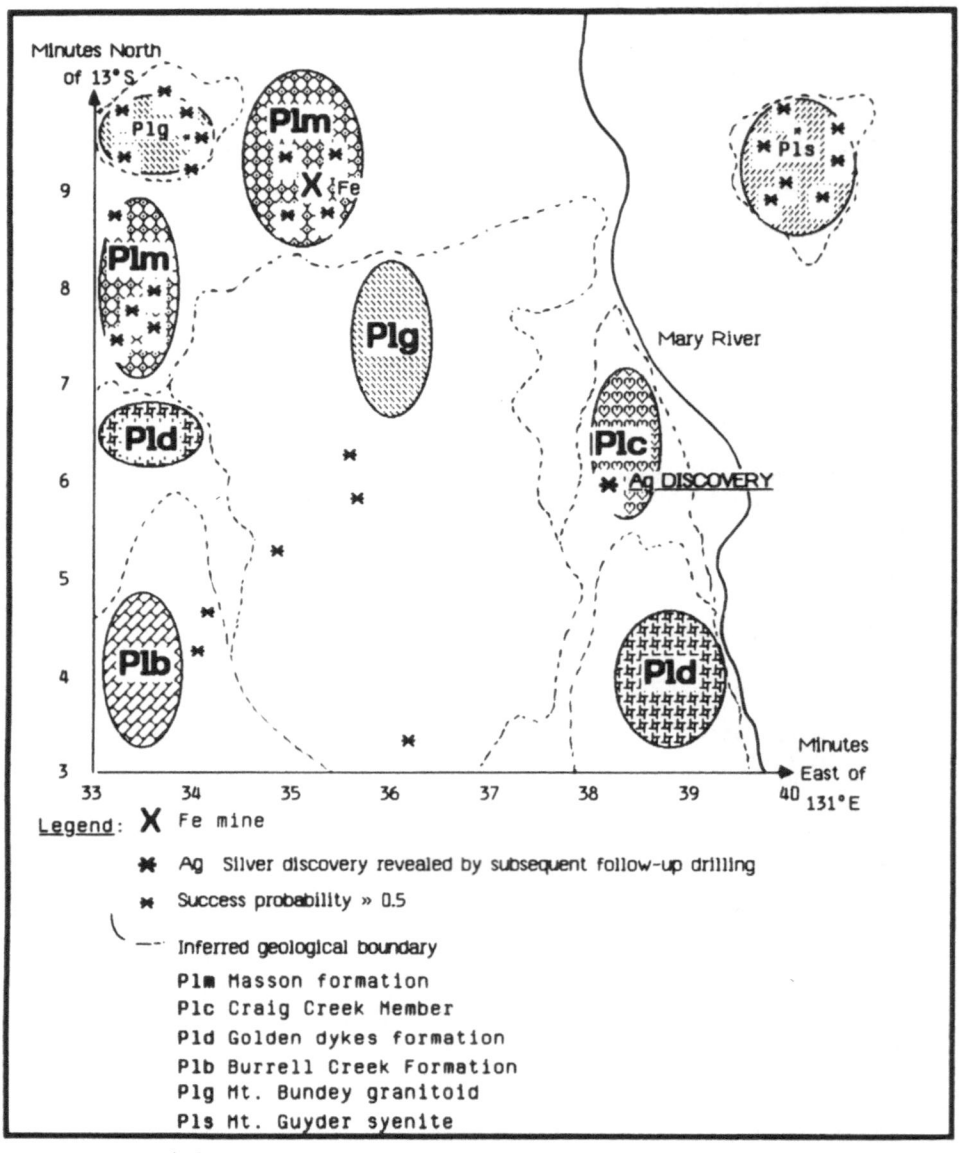

Figure 6.8. POPMIX & Classification Analysis for the contact
metasomatic U-Fe-Au search at Mt. Bundey, Northern
Territory, Australia.

```
Variate (1): F =  1.2          0.32586, pops 2, 3
Variate (2): F =  0.2          0.99493, pops 2, 3
Variate (3): F =  0.0          0.99998, pops 2, 3
Variate (4): F =  4.6          0.00006, pops 2, 3
Variate (5): F =  1.0          0.43508, pops 2, 3

Variate (1): F = 26.3          0.00000, pops 1, 3
Variate (2): F =  4.6          0.00006, pops 1, 3
Variate (3): F =  0.4          0.94437, pops 1, 3
Variate (4): F = 21.7          0.00000, pops 1, 3
Variate (5): F =  5.6          0.00001, pops 1, 3
```

The set of F-tests on the differences between the populations taken pairwise show highly significant differences only on the geophysical variates.

Classification of Locations into populations

From the results obtained at all 257 locations, only a random sample of anomalies are presented here. Co-ordinates are given only for locations assigned to populations (1) and (2) , so that they can easily be read and selected.

Loc. Ident.	Coordinates	Probabilities: Pop. 1	Pop. 2	Pop. 3
M 1		0.000	0.001	1.000
M 3	-52, - 9	0.000	0.660	0.340
M15		0.000	0.000	1.000
B47	-54, -21	0.998***	0.000	0.002
B50	-54, -20	0.989***	0.001	0.010

N.B. *** subsequent drilling lead to the discovery of an Ag mine

In all, a total of 17 locations were assigned to population (1) with probabilities exceeding 0.75, and two more locations had probabilities exceeding 0.5. At two of these locations, namely B47 & B50, which were assigned probabilities very close to unity, subsequent drilling lead to the discovery of a high-grade silver deposit. Eight locations were assigned to anomalous locations are plotted on the map shown on Figure 6.8. population (2), with probabilities exceeding the 0.70 threshold.

The 25 anomalous locations are plotted on the map shown in Figure 6.8. As many as 12 of them are clustered in the northwest corner of the map area, where an apophyse of the Mt Bundey granitoid is surrounded by metamorphic Masson formations which yielded a sub-economic contact-metasomatic iron deposit. A second cluster of 8 anomalous locations occurs in the northeast corner of the map area, east of the Mary River, coinciding with the Mt Guyder syenite intrusive from which industrial minerals are being quarried. The silver discovery lies in the eastern portion of the metamorphic aureola associated with the main body of Mt. Bundey granitoid, west of the Mary River.

6.3.4.5. Gippsland Basin on-shore petroleum project

Factor Output:

Based on the Factor Analysis program which was discussed in Chapter 5, we found that only one factor only accounted for 100% of the variation.

POPMIX Output:

Gippsland Basin on-shore petroleum project:

Sample size = 1110 locations, but only 778 on variate 4, and 925 on
        variate 5

Geovariates: (1): Factor 1, (2): variate 4, (3): variate 5
Means   : (1):   5.86 , (2):   0.123 , (3):   7.13

$Population_1$ proportion, p, on variate (1): p = 0.005
$Population_2$ proportion, p, on variate (2): p = 0.003
$Population_1$ proportion, p, on variate (3): p = 0.004

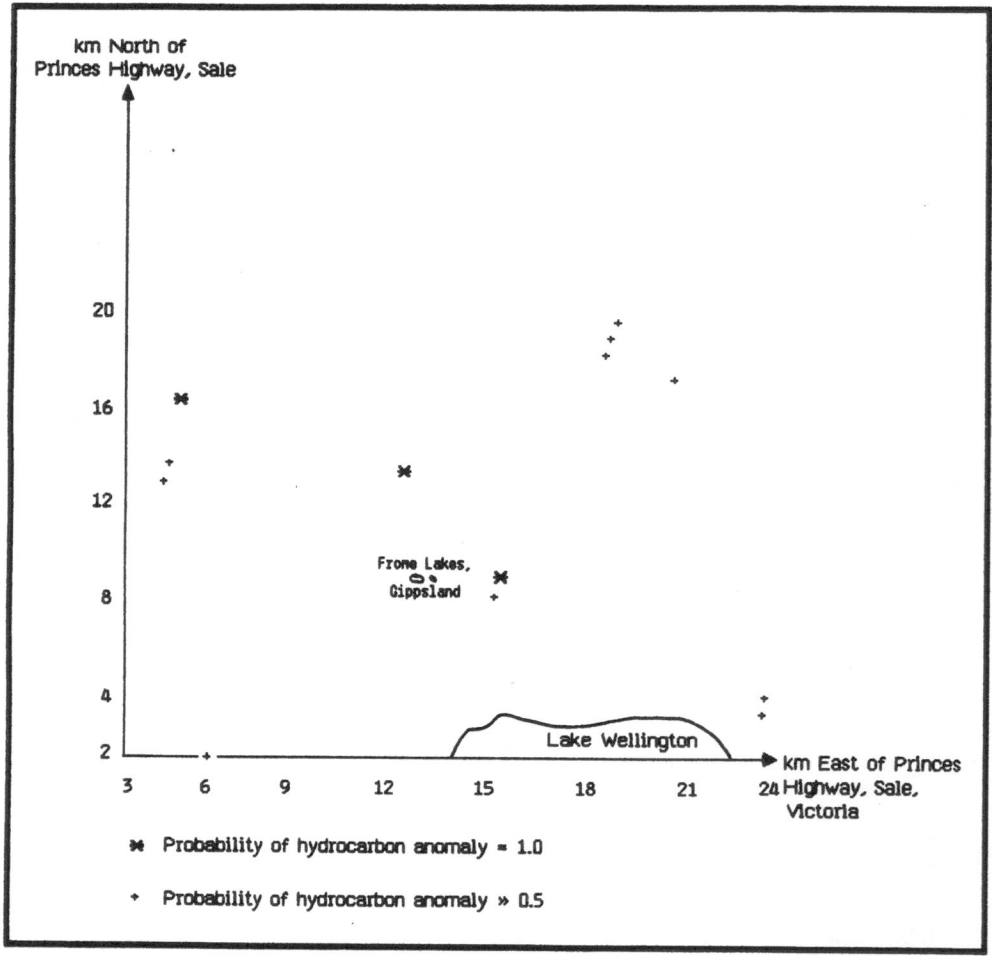

Figure 6.9. POPMIX & Classification Analysis for the hydrocarbon
        search in the on-shore Gippsland Basin (Western section),
        Victoria, Australia.

278

Mean$_1$, Mean$_2$, and common s. d. on var (1): m$_1$ = 199, m$_2$ = 5, s = 8.1

Mean$_1$, Mean$_2$, and common s. d. on var (2): m$_1$ = 7.1, m$_2$ = 0.1 , s = 0.2

Mean$_1$, Mean$_2$, and common s. d. on var (3): m$_1$ = 277, m$_2$ = 6 , s = 10

The reader should note that the factor and all variates relate to high gas counts, which infers that all the proportions relate to Population (1).

As a result of the POPMIX analysis, we can infer the occurrence of only 1110 x 0.005 $\approx$ 6 anomalies. Six locations were selected on the basis of their factor scores and were set aside to make up a control set for population (1), and the program was re-run on the remaining 1104 locations. In the second round of the procedure, in order to select a control set for population (2),the following results were obtained using the factor variate, and the same two gas variates:

Sample size = 1104 locations, but only 772 on variate 4, and 919 on variate 5

Geovariates: (1): Factor 1, (2): variate 4, (3): variate 5

Means : (1): 4.9 (2): 0.1 , (3): 5.8

Population$_1$ proportion, p, on variate (1): p = 0.015

Population$_2$ proportion, p, on variate (2): p = 0.001

Population$_1$ proportion, p, on variate (3): p = 0.011

Mean$_1$, Mean$_2$, and common st. dev. on var. (1): m$_1$ = 51, m$_2$ = 4, s = 5.2

Mean$_1$, Mean$_2$, and common st. dev. on var. (2): m$_1$ = 0.2, m$_2$ = -0.1, s = 0.1

Mean$_1$, Mean$_2$, and common st. dev. on var. (3): m$_1$ = 67, m$_2$ = 5 , s = 6

In the second round, the inferred number of anomalous locations for population (2) was 22, but in fact we chose 28 locations with factor scores exceeding 36 to make up the control set for population (2). The remaining 1076 locations were selected as control for the background population.

Classification Output:

The program is designed to select only those locations belonging to the anomalous populations, such that the sum of the probabilities of belonging is greater than 0.5. The Gippsland on-shore petroleum search output reads as follows:

Gippsland Geochemical project, Probabilities.
Size of populations: (1): 6, (2): 28, (3): 1076.

Mean on Variate (1):	181	46	4	(Total count)
Mean on Variate (2):	50	10	1	(ethane)
Mean on Variate (3):	14	4	0.3	(methane)
Mean on Variate (4):	4.2	0.9	0.1	(butane)

Standard deviates on var.: (1): 6.6, (2): 2.4, (3): 0.8, (4): 0.2

Correlation matrix:

	(1)	(2)	(3)	(4)
(1)	1	0.589	0.654	0.796
(2)		1	0.733	0.869
(3)			1	0.762
(4)				1

The inverse of the variance-covariance matrix was well-conditioned.
Comparison of population mean differences and significance levels:

Significance level on variate	Probability
(1): F =527.7	0.00000, pops 1, 2
(2): F =357.2	0.00000, pops 1, 2
(3): F =200.5	0.00000, pops 1, 2
(4): F =232.2	0.00000, pops 1, 2

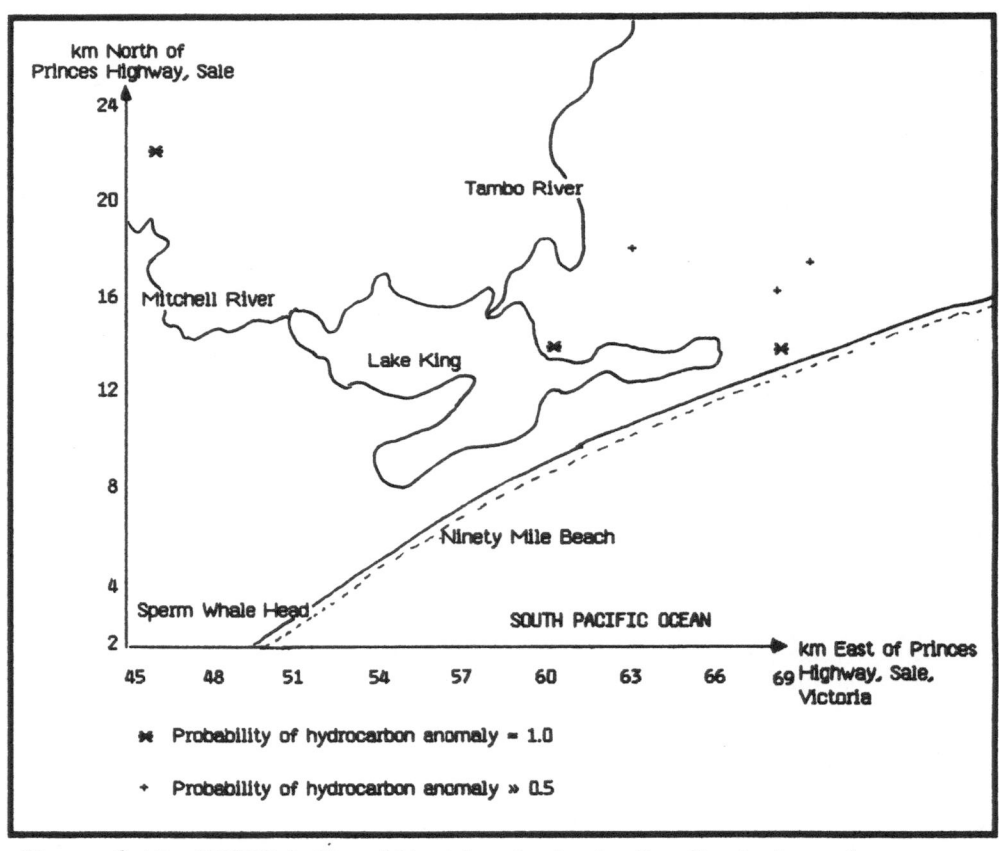

Figure 6.10. POPMIX & Classification Analysis for the hydrocarbon
search in the on-shore Gippsland Basin (Eastern section),
Victoria, Australia.

Significance level on variate	Probability
(1): F =278.5	0.00000, pops 2, 3
(2): F = 91.6	0.00000, pops 2, 3
(3): F =187.1	0.00000, pops 2, 3
(4): F = 86.7	0.00000, pops 2, 3
(1): F = 1091	0.00000, pops 1, 3
(2): F =637.0	0.00000, pops 1, 3
(3): F =482.3	0.00000, pops 1, 3
(4): F =446.3	0.00000, pops 1, 3

The set of F-tests on the differences between the populations taken pairwise show highly significant differences on all variates.

Classification of Locations into populations

The classification output was printed for all 1110 locations, but only a random sample is presented here. Co-ordinates are given only for locations assigned to anomalous populations (1) and (2) so that they can easily be read and selected.

Identification	Coordinates	Probabilities: Pop. 1	Pop. 2	Pop. 3
B 2		0.000	0.000	1.000
B 4	0, 4	0.000	0.994	0.006
B10		0.000	0.000	1.000
B20		0.000	0.000	1.000
B30		0.000	0.000	1.000
B33		0.000	0.000	1.000

As a result of the classification program, a total of 6 locations were assigned to population (1), with probabilities exceeding 0.99, while 23 were assigned to population (2), with probabilities exceeding 0.70. The total of anomalies is 29 out of 1110, which is 2.6% of the whole set. Three of the six locations with the highest probabilities appear to be aligned in a northwesterly direction, north of Lake Wellington, in the western section of the on-shore Gippsland Basin (see Figure 6.9). The other three are situated on the north and east side of Lake King, in the eastern section of the basin (see figure 6.10). None of them has been drill-tested to date as far as is known.

CHAPTER 6: Exercises

E6 Exercises

Using the Factor loadings from exercise E5.1., and
the POPMIX program, 7.1. in Appendix 7, find the following:

E6.1. (i) the significant Factor scores for each location,
    (ii) the proportion p of locations in $population_1$

       on each significant Factor,
  (iii) the expected means, m_1 & m_2, and the common

       standard deviation for populations 1 & 2
       on each significant Factor,
 (iiii) Select these proportions on $Factors_{1,2}$, from

       the Factor scores found in (i)

E6.2. Repeat exercise E6.1. using the results of the
     Factor loadings found in exercise E5.2.
E6.3. Using the sets of anomalous locations found
     in E6.1.(iiii) as control sets for the target
     $populations_{1,2}$, and the remainder as control for the

     background, based upon the CLASSIFICATION program 7.2.
     in Appendix 7, and variates 3, 4, 5 of data set 8.2.1.,
     derive the target population probabilities for each
     location, and select those locations with probabilities
     exceeding 0.5 of belonging to either $population_1$ or

     $population_2$.

E6.4. Using the sets of anomalous locations found
     in E6.2.(iiii) as control sets for the target
     $populations_{1,2}$, and the remainder as control for the

     background, based upon the CLASSIFICATION program 7.2.
     in Appendix 7, and variates 3, 4, 5, 6, 7, 8, 9, 10, 11
     of data set 8.2.2. derive the target population
     probabilities for each location, and select those
     locations with probabilities exceeding 0.5 of belonging
   to either $population_1$ or $population_2$.

BIBLIOGRAPHY FOR CHAPTER 6

Section 1: Classification based on control data

1.   ANDERSON, T.W., 1958, An introduction to multivariate
               statistical analysis; John Wiley, New York.

2.   BEAUCHAMP, J.J., BEGOVICH, C.L., KANE, V.E. and WOLF, D.A.,
               1980, Application of Discriminant Analysis and
               generalized distance measures to uranium exploration;
               Jour. Math. Geol., V.12, pp.539-558.

3.   BECKMAN, R.J. and JOHNSON, M.E., 1981, A ranking procedure for
               "partial" Discriminant Analysis; Jour. Amer. Stat.
               Assoc., V.76, pp.671-675.

4.   BOTBOL, J.M., 1971, An application of Characteristic Analysis
               to mineral exploration, in Decision-making in the
               Mineral Industry; Can. Inst. Min. and Metal., Spec.
               V.12, pp.92-99.

5.   BOTBOL, J.M., 1971, CHARAN, a FORTRAN IV Programme for
               Characteristic Analysis.

6.   BOTBOL, J.M., SINDING LARSEN, R., McCAMMON, R.B. and GOTT,
               G.B., 1978, A regionalized multivariate approach to
               target selection in geochemical exploration; Econ.
               Geol., V.73, pp.534-546.

7.   BROFFIT, J.D., RANDLES, R.H. and HOGG, R.V., 1976,
               Distribution-free "partial" Discriminant Analysis;
               Jour. Amer. Stat. Assoc., V.71, pp.934-939.

8.   BULL, A.J. and MAZUCHELLI, R.H., 1975, Applications of
               Discriminant Analysis to the geochemical evaluation
               of gossans, in "Geochemical Exploration 1974",
               Elliot, J.L. and Fletcher, W.K., Editors, Elsevier
               Publ., pp.219-226.

9.   CASSETTI, E., 1964, Classification and regional analysis by
               discriminant iterations; Tech. Rept. 12, ONR Task
               No.389-135, pp.95.

10.   CHUNG, C.F., 1977, An application of Discriminant Analysis for
               evaluation of mineral potential; Proc. 14th Int.
               Symp. Computers in Mineral Industries, pp.299-311.

11.   DAVIS, D.R., KISIEL, C.C and DUCKSTEIN, L., 1973, Bayesian
               methods for decision-making in mineral exploration;
               Proc. A.P.C.O.M. Symposium No.11, Tucson, pp.B55-B67.

12.   DE GEOFFROY, J. and WIGNALL, T.K., 1970, Statistical decision
               in regional exploration, application of regression
               and Bayesian classification analysis in the southwest
               Wisconsin zinc area; Econ. Geol., V.65, pp.769-777.

13.   DE GEOFFROY, J. and WIGNALL, T.K., 1970, Application of
               statistical decision techniques to the selection of
               prospecting areas and drilling targets in regional
               exploration; Can. Inst. M.& M. Bull., August,
               pp.893-899.

14. DE GEOFFROY, J. and WIGNALL, T.K., 1971, A probabilistic appraisal of mineral resources in a portion of the Grenville Province of the Canadian Shield; Econ. Geol., V.66, pp.466-479.

15. DE GEOFFROY, J.G. and WIGNALL, T.K., 1972, A statistical study of the geological characteristics of the porphyry-copper-molybdenum deposits of North and South America; application to the rating of porphyry prospects; Econ. Geol., v.67, pp.656-668.

16. DE ST JORRE, M.G.F. and WHITMAN, W.W., 1972, A probabilistic method of ranking underground exploration proposals; Econ. Geol., V.67, pp.789-793.

17. DIVI, S.R., THORPE, R.I., and FRANKLIN, J.M., 1979, Application of Discriminant Analysis to evaluate compositional controls of stratiform massive sulphide deposits in Canada; Jour. Math. Geol., V.11, No.4, pp.391-406.

18. DOVETON, J.H., 1976, Linear Discriminant Analysis can be used as an oil-prospect exploration guideline; Oil and Gas Jour., V.74, No.18, pp.324-328.

19.. DOWDS, J.P., 1966, Petroleum exploration by Bayesian analysis; Proc., V.2, 6th Ann.Internat. Symp. Computer Ops. Res., Penn. State Univ., pp.FF1-FF26.

20. DUBOV, R.I., 1972, A statistical approach to the classification of geochemical anomalies; Proc. London Symp. Inst. M.& M. pp.275-285.

21. FISHER, R.A., The use of multiple measurements in taxonomic problems; Annual of Eugenics, V.7, pp.179-188.

22. FUKUNAGA, K., 1972, Introduction to statistical Pattern Recognition; Academic Press, New York.

23. GOVETT, G.J.C., 1972, Interpretation of a rock geochemical exploration survey in Cyprus - statistical and graphical techniques; Jour. Geochem. Explor., V.1, No.1, pp.77-102.

24. GRIFFITHS, J.C., 1966, Application of discriminant functions as a classification tool in the Earth Sciences; Kansas Gol. Survey, Computer contr., V.7, pp.48-51.

25. HARRIS, D.P., 1965, Multivariate statistical analysis - a decision tool for mineral exploration; Short Course Symposium on Computers and Computer Applications in Mining and Exploration, V.1, Univ. of Arizona, pp.C1-C35.

26. HARRIS, D.P., 1969, Alaska's base and precious metals resources: a probabilistic regional appraisal, 7th Intl. Symp. on Operations Research and Computer Applications in the Mineral Industries, Colorado School Mines Quarterly, V.64, No.3, pp.295-327.

27. HOWARTH, R.J., 1971, Empirical discriminant classification of regional stream-sediment geochemistry in Devon and East Cornwall; Trans. Inst. Min. Metall., V.80, pp.B142-B149.

28. HOWARTH, R.J., 1973, The Pattern Recognition problem in applied geochemistry, in "Geochemical Exploration, 1972", M.J. Jones, Editor; Trans.Inst.M.& M., London, pp.B259-B273.

29. HUTCHINSON, R.I., BOWES, D.R. and SKINNER, D.L., 1976, Discriminant Analysis of trace elements in strata of the Witwatersrand system; Jour. Math. Geol., V.8, No.4, pp.413-428.

30. JOYCE, A.S. and CLEMA, J.M., 1974, An application of statistics to the chemical recognition of nickel gossans in the Yilgarn Block, Western Australia; Proc. Austral. Inst. M.& M., December, pp.21-23.

31. JZIBA, Z.V., 1964, A contribution to the statistical theory of classification, Stanford Univ. Publ. in Geol. Sc., Spec. Vol. No.9, Part 2, pp.729-756.

32. KLOVAN, J.E. and BILLINGS, G.K., 1967, Classification of geological samples by discriminant-function-analysis; Bull. Can. Pet. Geol., V.15, pp.313-330.

33. MARCOTTE, D. and DAVID, M., 1981, Target definition of Kuroko-type deposits in Abitibi by Discriminant Analysis of geochemical data; Can. Inst. M.& M., Bull. April, pp.102-107.

34. PATTERSON, D.A., PIRKLE, F.L., JOHNSON, M.E. and BEMENT, T.R., 1981, Discriminant Analysis applied to radiometric data and its application to Uranium favourability in South Texas; Jour. Math. Geol., V.13, No.6.

35. PRELAT, A.E., 1977, Discriminant analysis as a method of predicting mineral occurrence potential in Central Norway; Jour. Math. Geol., V.9, pp.345-367.

36. SABINS, F.F. Jr., 1973, Remote Sensing: principles and interpretation; W.H. Freeman, San Francisco, Calif., U.S.A.

37. SIMARD, R.G., 1980, Logistic model applied to geophysical data for sulphide nickel exploration; Can. Inst. M.& M. Bulletin, December, pp.96-100.

38. SWITZER, P., 1981, Extensions of linear Discriminant Analysis for statistical classification of remotely sensed satellite imagery; Jour. of Math. Geol., V.12, pp.367-376.

39. WHITEHEAD, R.E.S. and GOVETT, G.J.S, 1974, Exploration rock geochemistry - detection of trace element halos at Heath Steel Mines (N.B., Canada) by Discriminant Analysis; Jour. Geochem. Explor., V.3, No.4, pp.371-386.

40.   WIGNALL, T.K., 1968, Oil and gas exploration; statistical
      decision criteria, S.P.E. paper 223; A.I.M.E.,
      Houston, Texas, U.S.A.

41.   WIGNALL, T.K., 1969, Generalized Bayesian classification
      functions: K classes, - Application to a regional
      geochemical survey in southeastern Pennsylvania;
      Econ. Geol., V.64, pp.570-576.

BIBLIOGRAPHY FOR CHAPTER 6

Section 2: Classification without Control data

1. AGENO, M. and FRONTALI, C., 1963, Analysis of frequency distribution curves into overlapping Gaussians; Nature, V.198, No.4887, pp 1294-1295.

2. AHRENS, L.H., 1963, Element Distributions in Igneous Rocks (VI), Negative Skewness of SiO_2 and K; Geochem. & Cosmochim. Acta, V.27, pp 929-937.

3. BHATTACHARYA, C.G., 1967, A simple method of resolution of a distribution into Gaussian components; Biometrics, V.23, No.1, pp 115-135.

4. BRIDGES, N.J. and McCAMMON, R.B., 1980, DISCRIM: a computer program using interactive approach to dissect a mixture of Normal or Lognormal distributions; Comput. & Geosc., V.6, pp 361-396.

5. CASSIE, R.K., 1954, Some uses of probability paper in the analysis of size frequency distributions; Aust. Jour. Mar. Freshwater Res., V.5, No.3, pp 513-522.

6. CLARK, M.W., 1976, Some Methods for Statistical analysis of Multimodal Distributions and their Application to Grain-Size Data; J. Math. Geol., V.8, No.3, pp 267-281.

7. DAY, N.E., 1969, Estimating the components of a mixture of normal distributions; Biometrika, V.56, No.3, pp 463-474.

8. EISENBERGER, I., 1964, Genesis of Bi-modal Distributions; Technometrics, V.6, pp 357-363.

9. FRYER, J.G. and ROBERTSON, C.A., 1972, A comparison of some methods for estimating mixed Normal distributions; Biometrika, V.59, pp 639-648.

10. HARRIS, D., 1968, A method of separating two superimposed normal distributions using arithmetic probability paper; J. Animal Ecol., V.37, pp 315-319.

11. KLOVAN, J.E., 1968, Selection of target areas by factor analysis; Western Miner., V.41, No.2, pp 44-54.

12. KOCH, G.S. and LINK, R.F., 1970, Statistical Analysis of Geological Data, (Chpt 6); J. Wiley, N.Y., U.S.A.

13. LEPELTIER, C., 1969, A simplified statistical treatment of geochemical data by graphical representation; Econ. Geol., V.64, pp 538-550.

14. McCAMMON, R.B., BRIDGES, N.J., McCARTHY, J.H. Jr. and GOTT, G.B., 1979, Estimate of mixed geochemical populations in rocks at Ely, Nevada, in "Geochemical Exploration 1978", Watterson, P.K. and Theobald, P.K., Editors, pp 385-390, Elsevier Public.

15. PARSLOW, G.R., 1974, Determination of background and threshold in exploration Geochemistry, J. Geochem. Expl., V.3, pp 319-336.

16.   PARSLOW, G.R., 1979, Interpretation of some geochemical
           distributions in Key and Seahorse Lakes, Saskatchewan;   Can.
           Inst M.& M. Bull., V.72, No.804, pp 112-117.

17.   RAO, C.R., 1952, Advanced statistical methods in biometric research;
           John Wiley and Sons, N.Y.

18.   SINCLAIR, A.J., 1974, Selection of threshold values in geochemical
           data using probability graphs;   Jour. Geochem. Expl., V.3,
           No.2, pp 129-149.

19.   UNITED STATES NAVAL PERSONNEL RESEARCH ACTIVITY, 1967, NORMIX,
           computational methods for estimating the parameters of
           multivariate normal mixtures of distributions,   Research
           Memorandum, SRM 68-2, San Diego, Calif., U.S.A.

CHAPTER SEVEN

SUMMARY - CONCLUSION: THE EXPERT SYSTEM

7.1. INTRODUCTION

The difficulties experienced by the mineral industry in general and mineral exploration in particular since the early 1980's should induce exploration managers to consider new approaches to the planning and field implementation of ore search programs. In the previous volume, "Designing optimal strategies for mineral exploration", published by Plenum in 1985, the authors suggested a new approach based upon the theory of optimization of sequential systems. The basic theory and the methodology were described with a particular emphasis on detection and target selection as well as the application to the ore search in North America.

We had three main aims for this volume. The first one was to formulate the mathematical and statistical models, which underpin the optimization of the mineral exploration sequence. The second was to computerize the models in order to facilitate their application to the optimized search for any type of ore body in any region of the World.. Finally, the third one was to illustrate the application of the computerized models by taking as examples the optimized search for and development of six of the main types of base and precious metal deposits in five continents of the World, based upon control samples totalling 900 commercial and sub-commercial ore deposits.

Two imperatives guided the organization of the book. The first one, of a didactic nature, led to the introduction of the models in a progression of increasing complexity from Chapter 2 through to Chapter 6: from deterministic and heuristic, through univariate stochastic, to multivariate stochastic models. The second one was the necessity to follow wherever possible the logics of the mineral exploration business, from exploration planning, through field implementation, to economic evaluation and development planning.

7.2. Summary of methodology and results

Our first task was the construction of a comprehensive and reliable data base, as described in section 1.3 of Chapter 1 and detailed in Appendices 1 and 8. A total of three economic and five geometric variates as well as occurrence parameters were measured from each of 900 control deposits belonging to six main genetic types and occurring in fourteen geological regions of the World. Once the data had been collated and stored in the computer memory, the processing began by constructing statistical distributions and comparing them to the main theoretical models using the chi-squared test statistic, (Appendix 2). The kinds of models fitted included:

(a) lognormal, normal and circular normal for the economic and geometric variates, and

(b) generalized negative binomial, Poisson and exponential models for occurrence parameters. The reader will find the computer programs in Appendix 2, along with the Normal, Student-t, Chi-squared and F-test listings.

Benefits derived from model-fitting which are of immediate value to the exploration planners include firstly the construction of a comprehensive set of probability tables pertaining to the economic and occurrence parameters of the six types of deposits covered by the study. A second dividend is the construction of confidence intervals for the expected values of all economic and geometric variates, which enabled us to establish an objective economic ranking of the six selected types of deposits on a world-wide basis, followed by ranking on a continent by continent basis.

Because of the very large computational requirements of the statistical methodology used for the optimization of field exploration programs in the search for specific types of deposits, it was necessary to devise efficient and objective procedures to reduce the workload to be dealt with in the second stage of the procedure. We accomplished this through the use of a computerized procedure of sequential testing of statistical commonality of populations based on economic and geometric parameters (Chapter 3 and Appendix 3). The procedure successfully reduced the computer workload by 50% by grouping regions or ore deposit types into a smaller number of statistically homogeneous pools. Population statistics (expected values and confidence interval estimates) were re-calculated for each pool for subsequent use in the calculation of detection probabilities, and optimization of design of ore search grids.

Much emphasis was again laid on these two topics in Chapter 4 of the present book. Geometric probability theory and the Dynamic Programming model were combined to provide a comprehensive series of tables of probabilities of detection of the six types of ore deposits in the regions pooled in Chapter 3. The following step was to design optimal search grids for the detection of these deposits by means of airborne and ground geophysical surveys and drilling programs by the maximization of detection probabilities under cost constraints. An important benefit derived from the detection optimization was the ranking of ore deposit type-regions based upon the detectability and upon the exploration pay-off criteria in addition to that based solely on economic parameters which was previously given in Chapter 2.

The large amounts of field data collected by optimized field surveys in the search for specific types of ore deposits were processed by various statistical procedures in order to delineate exploration targets, and to select those with the greatest merit for further investigations. In Chapter 5, we used a series of multivariate techniques, based upon an optimizing model, the multivariate General Linear model, to reach the first goal. These procedures included factor, factor trend, and residual factor trend analyses, which are computerized in Appendix 5. In order to reduce objectively the number of exploration targets which can be tested within the prevailing budget and time constraints, an objective technique was applied to screen the targets with the best economic potential. A multivariate Bayesian classification procedure, computerized in Appendix 7, was used whether we had to deal with well-prospected regions where control deposits were available, or little-known regions without controls. In the latter case, however, computerized in Appendix 7, (POPMIX) is applied before carrying out the Bayesian classification.

Finally, in order to evaluate the economic potential of the targets which were successfully tested after their optimal screening, we relied again on the General Linear Model techniques of Chapter 5. The last step of the sequence is the optimization of various aspects of development planning, whether financial (General Linear model), or organisational in the time domain (Scheduling model), or in the space domain (Networking models), (Chapter 5 and Appendix 6).

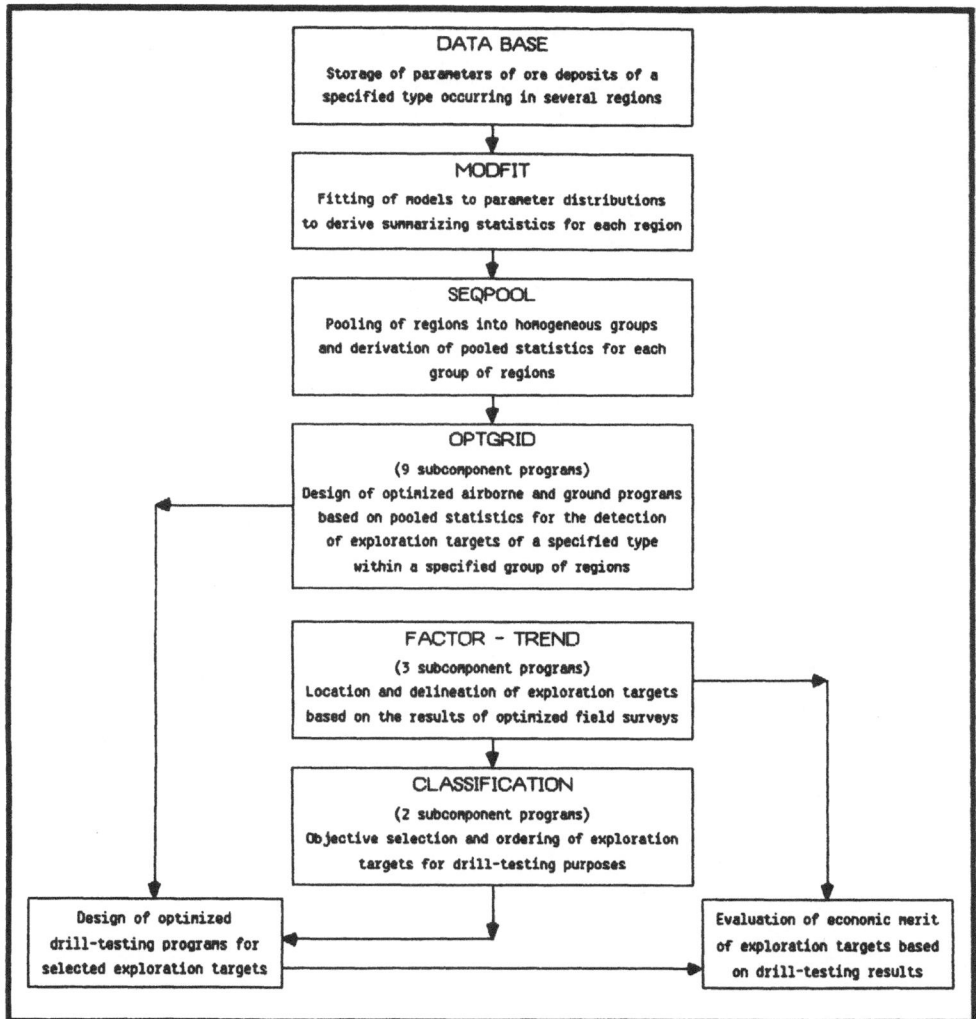

Figure 7.1. Summary diagram of the mineral exploration **Expert** system comprising 17 computer programs for the optimized detection, rating and testing of exploration targets.

7.3. Conclusion: the **Expert** Optimizing System

If we are to optimize the search for specific types of deposit within selected regions in an efficient and speedy manner, we must take full advantage of the large computational capacity and flexibility of modern computers, in order to process the large quantity of data available with the advent of Satellite sensing surveys collated as control, and that later acquired through field surveys. This is the rationale for the construction of an integrated sequential structure, an **Expert** System, which is composed of the various computer programs used for the optimization of individual stages of the ore search, as illustrated in Figure 7.1.

The MODFIT component of the system is composed of two programs, one for fitting the lognormal model to economic and geometric deposit data, and the second one to fit the generalized negative binomial models to occurrence data (Appendix 2). Appendix 3 covers the listings of the SEQPOOL component of the System which is used for comparisons and for the objective reduction of the statistical processing workload. The OPTGRID suite of nine programs (Appendix 4) covers the optimization of ore detection, and that of sampling programs required for the economic evaluation of targets successfully tested. The FACTOR-TREND suite of three programs (Appendix 5) ensures the objective delineation of exploration targets, and is called on again later for development planning purposes, when undertaking the economic evaluation of exploration targets successfully tested in the previous stage. Finally, the CLASSIFICATION program (Appendix 7) handles the optimal selection of exploration targets for testing purposes in all cases, whether control locations are available or not, in conjunction with the POPMIX program in the latter case.

In conclusion, the **Expert** sequence of computer programs provides the following results:

(1) an efficient and speedy method of optimizing the detection, delineation, evaluation and selection of exploration targets over large territories,
(2) confidence interval estimates for the expected values of economic and geometric parameters of the prizes sought, and
(3) probabilities of success at each stage of the sequence.

In addition to the listings of individual programs appearing in the **Expert** System sequence, in Appendices 2 through 7, the whole package is available on disk from the authors.

APPENDIX 1

LIST OF NAMES OF MINERAL DEPOSITS INCLUDED IN THE DATA BASE

List of Porphyry-Copper-Molybdenum Deposits of the North American Cordillera included in the data base

British Columbia (Canada)

Red Mountain	Huckleberry	Gem
Boss Mountan	Stikine	Island Copper
Sileurian Chief	Adanac	Valley Copper
Alwin	Bethlehem	Casino
Granisle	Highmont	Serb Creek
Cariboo Bell	Gibralter	Salal
Polyanna	Brenda	Galaxy
Newman Island	Hudson Bay Mtn.	Sheba
Ingerbelle	Schaft Creek	Trojan
Similkameen	Lornex	Lytton
B.C.Molybdenum	Endako	

United States

Castle Dome	Safford	Butte
Copper City	Mission	San Manuel
Mineral Park	Miami-Inspir.	Morenci
Esperanza	Bisbee	Twin Butte
Yerrington	Ely	Urad-Henderson
Glacier Peak	Ray	Santa Rita
Silver Bell	Ajo	Climax
Tyrone	Questa	Bingham
Bagdad		

List of Porphyry-Copper-Molybdenum Deposits of the South American Cordilleran Belt Region included in the data base

Argentina

Campana Mahuida
Cerro Rico
La Alumbrera
Paramillos
Mendoza

Peru

Toquepala
Michiquillay
Morococha
Cerro Verde

Mexico

Cananea

Ecuador

Chaucha

Chile

Chuquicamata
Mocha
El Salvador
Potrerillos
Los Pelambres
Cerro Blanco
El Teniente
El Abra
Mantos Blancos

List of Porphyry-Copper-Molybdenum deposits of the South Pacific Island Arc Region included in the data base

Papua New Guinea and
Outlying Islands   Philippine

Nong River   Boneng
OK Tedi (*)   Sta. Nino
Futik   Black Mountain
Olgal   Philex
Ertsberg (**)   Batong
Panguna   San Antonio
Yanderra   Mayabo
Frieda   Butilao
Arie   Tawi-tawi
                                Dizon
                                Tayzan
Fiji   Camarines
                                Quezon
Namosi   Marcopper
                                Capayang
                                Atlas Cebu
Borneo   Atlas Carmen
                                Sipalay
Mamut   Hinoboan
                                C.D.C.P.

(*)   Skarn + intrusive

(**) Skarn only

List of Porphyry-Copper-Molybdenum deposits of the Tasman Geosyncline Region of Australia included in the data base

Queensland   New South Wales

Kidston   Copper Hill
Moonmera   Parkes
Coalstoun   A.G.M.
Anduramba   Cargo
Yeppoon   Yeoval
Mt Gordon   Peak Hill

List of Porphyry-Copper-Molybdenum deposits of four regions of the USSR included in the data base

Ural

Tur'inskoe
Volkovskoe
Voznessenskoe

Caucasus

Tyrny-auz
Kedabeg
Kafan
Ankavan
Dastakert
Dzhindara
Agarak

Kazakhstan

Olginskoe - Akbiik
South Bessoki
Ken'kuduk
Kaskyrkazgan
Borly
Karabas
Kounrad
Karatas IV
Sokurkoi
Bashchekul
Kalmakkyr
Sary-cheku

Amur

Zhirekenskoe
Bugdaya

List of Contact metasomatic (Cu-Fe) deposits of the North American Cordillera Belt included in the data base

B.C. Islands (Canada)

Quatsino
F.L. mine
Iron Hill
Prescott
Paxton
Yellow Kid
Lake
Ford
Ridge
Jessie

B.C. Mainland (Canada)

Craigmont
Merry Widow
King fisher

New Mexico (U.S.A.)

Hanover

S.W. Alaska (U.S.A.)

Mamie
Poor Man
Mt Andrew

Utah (U.S.A.)

Iron Spring
McCahill

Nevada (U.S.A.)

Dayton

California (U.S.A.)

Eagle Mountain
Dale
Iron Mountain

List of Contact metasomatic (Pb-Zn-Cu) deposits of the North American Cordillera Belt included in the data base

Utah (U.S.A.)

Horn Silver
Cactus
Red Warrior
Tintic Standard
North Lily
Eureka Lily
Burgin
Gemini
Mammoth Chief
Plutus
Iron Blossom
Godiva

Nevada (U.S.A.)

Simon
Blue Stone
Mason Valley
Casting Copper
Ludwig

Arizona (U.S.A.)

Johnson
Copper World
Hardshell
Mowry

Colorado (U.S.A.)

Iron Hill
Henriett-Maid
Cord
Minnie-Helen
Sellers
Imes
Giltedge
Comstock
Ibex
Rickard-Stone
Golden Eagle
Mahala
Tucson
Carbonate Hill

Idaho (U.S.A.)

Empire

B.C. & Yukon (Canada)

Jersey
Emerald
Nickel Plate
French
Whitehorse Copper

List of Contact metasomatic (Cu-Mo-Au) deposits of North American Cordillera Belt included in the data base

Arizona (U.S.A.)

Christmas
Twin Buttes
Mission
Bisbee
Lake Shore
Silver Bell
Morenci

Utah (U.S.A.)

Bingham

Nevada (U.S.A.)

Ely

New Mexico (U.S.A.)

Santa Rita
Continental

Mexico

Cananea

British Columbia (Canada)

Red Mountain
Similkameen

Yukon (Canada)

Logtung

List of Contact metasomatic (W-Mo) deposits of the North American Cordillera Belt included in the data base

Utah (U.S.A.)

Old Hicory
Garnet
Strategic Metals
Cupric
Copper Ranch
Copper King
Daily Metal

Idaho (U.S.A.)

Yellow Pine

California (U.S.A.)

Darwin Hill
St Charles
Durham
Little Sister
Lookout
Round Valley
Pine Creek

Nevada (U.S.A.)

St Anthony
Rose Creek
Friedman
Sutton
Humbolt
Oreana
Victory
T.N.T.
Riley
Kirby
Valley View
Tip Top
Granite Creek
Pacific
Alpine
Tempiute

British Columbia and Yukon (Canada)

Mactung
Cantung
Emerald
Dodger
Feeney
Swakum Mountain

List of the contact-metasomatic deposits (W-Mo) of the East-Asian Region
included in the data base

Japan
(**Honshu Island**)

Tsumo
Fujigatani

South Korea

Chilbo
Seoseog
Bongbog
Banggye
Sangdong
Songhag
Geumseong
Dongnam

<NEW PAGE>
List of Ni-Cu ultramafic deposits of the Canadian Shield region included in
the data base

Northern Territories
and Manitoba

Rankin Inlet
Bird River
Lynn Lake
Bowden Lake
Thompson
Pipe

Ontario

Alexo
Sothman
Ajax
Langmuir
Texmont
Trebor
Temagami Lake
Gordon Lake
Werner Lake
Kenbridge
Shebandowan

Quebec

Marbridge
Blondeau Twp.
Lake Renzy
Lorraine

New Quebec

Cross Lake
Katiniq
Raglan
Expo

List of Ni-Cu ultramafic deposits of the Scandinavian Shield
region included in the data base

Bamble
Bruvann
Flat
Hitura
Hosanger
Kotalahti
Kylmakoski
Kuhmo-Suomussalmi
Lainijaur

Lappvattnet
Laukunkanges
Makola
Mjodvattnet
Ringerike
Risliden
Vakkerlien
Vammala
Velika

298

List of Ni-Cu mafic deposits of the West Australian region included in the data base

Agnew
Black Swan
Forrestania
Six Miles
Weebo Bore
Carnilya Hill
Dordie Rocks
Kambalda (Juan & Durkin Shoots)
Kambalda (Lunnon Shoot)
Kambalda (Foster Shoot)
Mt Edward
Nepean

Redross
Scotia
Spargoville (1, 2 & 3)
Wannaway
Widgie
Mt Windarra
South Windarra
Mt Scholl
Sally Mallay
Sherlock Bay
Copper Hills
Mt Hope

Carr Boyd

List of volcanogenic massive sulphide deposits of the North American Shield region included in the data base

Ontario

Ecstall
Kamkotia
Jameland
Jamieson
Munro
Coldstream
Geco
Willroy
Willecho

Big Nama
Mattabi
N.B.U.
Lyon Lake
Creek

South Bay

Quebec

Noranda (Horne)
Quemont
McDonald
Amulet "A"
Iso
Joliet
Mobrun
Corbett
Norbec
Millenbach
Aldermac
East Waite
West Waite
New Insco
Amulet "C"
Delbridge
Vauze
Amulet "F"
Waite - Default
East Sullivan
Manitou - Barvue
Louvem
Abitibi Copper
Dunraine
Quebec Manitou
Barvue

Vendome
New Formaque
Belfort
Barvallee
Normetal
Poirier
Joutel
Northern Exploration
Brouillan
Lessard
Lemoine
Scott
Jay Copper
Lynx Yellowknife
Dumagami
Pershcourt
Louvicourt
Lake Berrigan
Mattagami
Orchan
New Hosco
Norita
Phelps Dodge
Bell Allard
Garon Lake
Radiore

Manitoba

Flin-flon
Pine Bay
Coronation
Centennial
Schist Lake
West Schist
Cuprus
White Lake
Birch Lake
North Star
Mandy

Chisel
Stall
Osborne
Freeport Reed
Anderson
H.B. Reed
Wim
Dickstone
Little Stall
Rail
Ghost Lake
Lost Lake
Copperman
Ruttan
Fox Lake

Northwest Territories

Hackett
Izok
High Lake
Taki Lake

List of volcanogenic sulfide deposits of the North American Cordillera Belt included in the data base

British Columbia	California
Tulsequah	Shasta-King
Anyox	Afterthought
Britannia	Iron Mountain
Chuchua	Bully Hill + Rising Star
Sam Goosly	Mammoth
Goldtream	Balaklala
Granduc	Copper Hill
Kutcho	Keystone
Lynx	
H.W.	
Myra	Arizona (Precambrian Window)
Price	
	Iron King
	United Verde
	Copper Queen
	Old Dick

List of exhalative sulfide deposits of the North American Cordilleran region included in the data base

British Columbia	Yukon Territory
Sullivan	Tom
	Faro
	Grum
	Swim
	Dy
	Red Dog
	Howard's Pass (of possible euxenic origin)

List of volcanogenic massive sulphide deposits of the Appalachian region included in the data base

New Brunswick

Brunswick "12"
Brunswick "6"
Heath Steele
Wedge
Armstrong Brook
Caribou
Devil's Elbow
Canoe Landing
Coulee Headway
Austin Brook
Nine Mile Brook
C.M. & S. (Nepisiguit)
Stratmat
Half-mile Lake
Captain Yellowknife
Key Anacon
Clearwater
Riley Brook
Middle River
Papineau
New Calumet
Portage Lake

U.S. Appalachian

Blue Hill
Cherokee
Burra

Quebec (Eastern Townships)

Eustis
Clinton
Cupra
Suffield
Moulton Hill
Huntingdon
Weedon
Ives

Newfoundland

Buchans
Rambler Brook
Ming
Gulbridge
Whales Back

Nova Scotia

Stirling

List of volcanogenic sulfide deposits of the Scandinavian Shield included in the data base

Finland

(Karelia)

Outokumpu
Vuonos
Hammaslahti
Linkonlahti

(Oulu District)

Vihanti
Pahtavuoma
Pyhasalmi

(Orijarvi District)

Orijarvi
Ayala
Haveri
Metsamonttu
Saramaki

Sweden

(Vasterbotten Region)
(Skellefte Mining Area)

Boliden
Kristineberg
Rackjaur
Rudtjebacken
Brannmyran
Adak
Renstrom
Langsele
Menstrask

(Norbotten Region)
(Kiruna Mining District)

Viscaria
Aitik
Radnejaur
Laver

(Sulitjelma Mining District) (*)

Sulitjelma North
Sulitjelma South
Bjorkasen

N.B.   (*)   Precambrian window in the Caledonide Orogenic Belt.

List of volcanogenic massive sulfide deposits of the Scandinavian Caledonide region included in the data base

Vaddas	Joma	Stekkenjok
Bjorkasen	Bassmo	Lergruvbakken
Bleikvassli	Killingdal	Tverfjell
Mofjell	Lokken	Gjetryggen
Gjersvik	Olavs	Rodkleiv
Skorovass	Kongens	Lillebo
	Levi	

List of volcanogenic sulfide deposits of the Iberian Peninsula included in the data base

Portugal

(**Lousal Mining Area**)

Lousal
Massa Antonio
Massa Jose & Fernando

(**Adjustrel Mining Area**)

Feitais
Carrasco and Moinho
Estacao
Gaviao

Spain

(**Rio Tinto Mining Area**)

San Dionisio
South Lode
Planes Massa
Cerro Colorado
Alfredo
San Antonio

(**Tharsis Mining Area**)

Nueva Almagrera
Cantateras
San Guillermo
Tharsis North
Lagunazo

(**Other Mining Districts**)

La Zarza
Higuereta
Aznalcollar No.3
San Telmo
Sotiel
Santa Rosa
Romanera

List of volcanogenic sulfide deposits of the East Mediterranean region
included in the data base

Yugoslavia

(Bor Mining Area)

Cuka Dulkan
Tilva Mika
Tilva Ross
Brosvonik

Cyprus

(Mitsero Mining Area)

Kokkinopezoula
Agrokipia A & B
Kokkinoyia

(Tanmassos Mining Area)

Mathiati North
Mathiati South
Sha
Kapedhes

(Miscellaneous Mining Areas)

Limni
Mavroumi
Skouriotissa
Apliki
Mesoulos

Greece

(Chalkidi Region)

Maddenlakos
Stratoni

(Peloponesos Region)

Hermioni

Turkey

(Eastern Black-Sea Coast)

Lahanos-Espiye
Kizilkaya
Kure Bakibaba
Kutlular
Madenkoy
Tunca
Murgul (Damar)
Murgul (Camakkaya)

List of volcanogenic massive sulphide deposits of the West Australian Shield
region included in the data base

Mt Angelo	Copper Hill	Golden Grove
Koongie Park	Whundo-Whundo	Wendy Bore
Whim Creek	Mt Mulcahy	Freddie Well
Mons Cupri	Teutonic Bore	Horseshoe Lights
Big Stubby	Nangeroo	

List of volcanogenic massive sulphide deposits of the Tasman Geosyncline region of Australia included in the data base

QUEENSLAND

(Volcanogenic Deposits in Volcanic Piles)

Surveyor	Mt Chalmers	O.K.
Balcooma	Thalanga	Dianne
Mt Cannindan	Mt Morgan	Red Hill
		Mt Molloy

NEW SOUTH WALES

(Volcanogenic Deposits in Volcanic Piles)

Cadia	Girilambone	Cangai
Brown's Creek	Galwadgere	Woodlawn
Kaiser	Parker's Hill	Captain's Flat
Mt Bulga	Halls Peak	Gulf Creek

(Exhalative Deposits in Sedimentary Piles)

Yerranderie	Chesney	New Occidental
Elura	Gladstone	New Cobar
Nymagee	C.S.A. Cobar	Great Cobar

Tasmania

(Volcanogenic Deposits in Volcanic Piles)

Rosebery	West Lyell & Prince Lyell	Crown Lyell
Hercules	Royal Tharsis	Comstock
Farrell	North Lyell	Blow
	Cape Horn	

List of volcanogenic massive sulphide mining areas of the Kuroko Belt of Japan included in the data base

Ainai	Kamikita	Shakanai
Fukazawa	Kosaka	Shimokita
Furutobe	Kunitomi	Tashiro
Hanawana	Matsuki	Tsuchihata
Iwami		Yoshino

List of Mississippi Valley - type Pb-Zn deposits of North American Arctic region included in the data base

Greenland

Black Angel

Canadian Arctic

Polaris
Nanisivik
Goz Creek

Northwest Territories
(Pine Point Area)

Pyramid No.1
Pyramid No.2
Coronet
Newconex
X-25
X-25
R-190
M-40
N-815
A-70
K-57

List of Mississippi Valley type Pb-Zn deposits of the Missouri-Tri State region included in the data base

(S.E. Missouri Lead District)

Mine Lamotte	Indian Creek	Viburnum No.29
Bonneterre	Higdon	Buick Mine
Hayden Creek	Annapolis	Brushy Creek
Leadwood	Viburnum No.27	Ozark Lead
Flat River	Viburnum No.28	

(Tri-State Zinc-Lead District)

Picher Field	Waco Field	Aurora Field
Webb City Field	Neck City Field	Granby field
Joplin field	Wentworth Field	

List of Mississippi Valley - type Pb-Zn deposits of the Upper Mississippi Valley district included in the data base

Bautsch	Hoskins	Liberty-Big Dick
Graham-Snyder	Martin	North Unity
James	Longhorn-Monroe	Old Ida-Blende
Kennedy-Little & Big Dad	Old Winskell	Slack-Squirrel
Black Jack		Penna-Benton
	-Lucky-Six	
Blockhouse Range	Raisbeck-Trego	South Unity
Coker No.2	Birkett	Vinegar Hill
Frontier-Treganza	Blackstone	Wipet
Gensler-Winskell	Booty	Acme
Linden Range	Byrnes	Blewett
Badger	Champion	Coker No.3
Bull Moose-Middie	Clark Range	Empress
Copeland	Cleveland	Helena-Roachdale
Crawford	Coughlin	Hospital Run
Crawhall-Thompson	Dall	Jefferson
Enterprise-Empire	Dark Horse-Optimo	Lucky Hit
Federal	Graham-Ginte	Meloy
Fox	Gray	Rodham
Highland-Kennedy	Indian Mound	Trego
	Kittoe	Treloar-Harmony

N.B. Listing in order of decreasing tonnage

List of Vein-Gold deposits of the North American Shield region included in the data base

Northwest Territories	Ontario	Quebec
Giant Yellowknife	Central Patricia	Omega
Cons. Discovery	Cochenour Willans	Lamaque
Camlaren	Gold Eagle	East Malartic
	Madsen Red Lake	Central Cadillac
	McKenzie Red Lake	O'Brien
	Hasaga	Belleterre
	Little Long Lac	MacWatters
	Jellicoe	Powell Rouyn
	Tombill	Wasa
	North Empire	Beattie
	Hard Rock	Francoeur
	Renabie	Arntfield
	Jerome	Camflo
	Leitch	Canadian Malartic
	Pamour	Granada
	Broulan	Agnico
	Hallnor	
	Dome	
	Preston East Dome	
	Paymaster	
	Buffalo Ankerite	
	Aunor	
	Delnite	
	Coniaurum	
	Moneta Porcupine	
	Hollinger	
	McIntyre	
	Kerr Addison	
	Sylvanite	
	Wright Hargreaves	
	Bidgood	

List of Vein-Gold deposits of the West Australian Shield region included in the data base

Golden Mile	Westralia	Black Range
Sons of Gwalia	Edna Bay	Gladsome
Mararoa-Crown	Fraser's	Bonnievale
Wiluna	Emu	Ida H.
Hill 50	Hannan's North	Palmer's Find
Great Fingall	Brubanks	Light of Asia
Princess Royal	Princess May	Barbara - Surprise
Mt Charlotte	Bayley's	Golden Ridge
Paddy's Flat	Cosmopolitan	Ora Banda
Big Bell	Youanni	Great Eastern
Copper Head	Timoni	Bellevue
Lance Field	Triton	Tindal's
Crusoe Line	White Feather	St George
Oroya Black Ridge	Moonlight Wiluna	Comet

APPENDIX 2

Section 1: MODFIT computer programs

2.1. Listings for Lognormal computer program (BASIC)

```
2 DIM F1(50),Q(300),X(300),T(50),Y(50),U(50),E(50)
4 DIM Z(300),X1(300,5)
10 PRINT"L,B,B/L GEOMETRIC TARGET DATA"
12 PRINT "K5 IS THE NUMBER OF VARIATES"
14 INPUT K5
16 PRINT" ENTER SAMPLE SIZE: NUMBER OF OBSERVATION"
18 INPUT N
20 FOR J5= 1 TO K5
26 PRINT" ENTER NUMBER OF HISTOGRAM INTERVALS REQD, NH"
28 INPUT N3
30 PRINT" DESIRED INTERVAL WIDTH, 0.3 IS USUAL IN LOGNORMAL"
32 INPUT W3
34 PRINT" ENTER LOWEST LIMIT, 2.6 ON LENGTH IS USUAL FOR LOGNORM"
36 INPUT T(1)
44 N3 = N3+1
46 FOR I=1 TO N
48 IF J5>2 THEN 54
50 READ X1(I,1),X1(I,2)
52 X1(I,3)= X1(I,2)/X1(I,1)
54 X(I)=X1(I,J5)
56 Q(I)=LOG(X(I))/LOG(10)
58 X(I)=Q(I)
60 NEXT I
62 Q1=0
64 Q2=0
66 FOR J=1 TO N
68 Q1=Q1+Q(J)
70 Q2=Q2+Q(J)*Q(J)
72 NEXT J
74 A8=Q1/N
76 A7=((N*Q2-Q1*Q1)/N/(N-1)))^0.5
78 A6=A7*A7
80 S6=(A6/N)^0.5
82 R3=2*S6
84 PRINT"95% CONF. LIMITS =", A8-R3,A8,A8+R3,"GEOM MEAN=",10^A8
86 FOR K=1 TO N3 STEP 1
88 F(K)=0.
90 NEXT K
92 FOR K=2 TO N3 STEP 1
94 T(K)=T(K-1)+W3
96 NEXT K
98 F8=0.
100 FOR I=1 TO N STEP 1
102 K=1
104 IF X(I)<= T(K) THEN 110
106 K=K+1
108 GO TO 104
110 F(K-1)=F(K-1)+1.
112 F8=F8+1.
114 NEXT I
116 PRINT"    INTERVAL RANGE(LOG)         FREQUENCY    CUMULATIVE"
118 PRINT"    FROM            TO          VALUE        FREQUENCY"
120 L=N3 -1
122 S9=0.
124 S6=0.
126 S3=0.
```

```
128 C4=0.
130 FOR K=1 TO L
132 C4=C4+F(K)
134 PRINT T(K),T(K+1),F(K),C4
136 Y(K)=0.5*(T(K)+T(K+1))
138 S9=F(K)*Y(K)+S9
140 S6=F(K)*Y(K)^2+S6
142 S3=F(K)*Y(K)^3+S3
144 NEXT K
146 S9=S9/F8
148 S6=S6/F8
150 S3=S3/F8
152 V5=S6-S9^2
154 IF V5<0 THEN 176
156 S1=V5^0.5
158 S7= S3-3*S6*S9+2*S9^3
160 S8=S7/(S1^3)
162 PRINT "          MEAN    S-SQUARED     S      SKEWNESS"
164 PRINT S9,V5,S1,S8
166 W1=(V5/N)^0.5
168 PRINT" STANDARD ERROR OF THE MEAN =",W1
180 Q1=0
182 R1=0
184 PRINT" THE LOGNORMAL DISTRIBUTION FOR VALUES OF X FROM X1"
186 PRINT" TO X2 STEP W3 IS GIVEN BY:"
198 X1=T(1)
200 X2= X1+(N3-1)*W3
202 X3=W3
264 M=A8
268 S2=A6
270 S1=S2^0.5
272 PRINT"     X          FREQ(X)"
274 L8=0
276 R1=0
278 FOR X=X1 TO X2 STEP X3
280 L8=L8+1
282 X=(X-M)/S1
284 H=(2*3.14159296)^0.5
286 H=1/H
288 H=H*EXP(-0.5*X*X)
290 B1=0.31938153
292 B2=-0.356563782
294 B3=1.781477937
296 B4=-1.821255978
298 B5=1.330274429
300 R=0.2316419
302 T=1/(1+R*X)
304 Q= B1*T+B2*(T^2)+B3*(T^3)+B4*(T^4)+B5*(T^5)
306 Q=Q*H
308 IF Q>1 THEN 412
310 GO TO 314
312 Q=1
314 Q3=1-Q-R1
316 Z(L8)=Q3*N
318 X=X*S1+M
320 R1=1-Q
322 NEXT X
324 Z(2)=Z(1)+Z(2)
326 Z(L8)=Z(L8)+Q*N
328 FOR I=2 TO L8
330 PRINT"  ",X1+(I-2)*X3," TO ",X1+(I-1)*X3, Z(I)
332 PRINT"  ",10^(X1+(I-2)*X3)," TO ",10^(X1+(I-1)*X3), Z(I)
```

2.2. Listings for Negative Binomial computer program (BASIC)

```
2 DIM F1(50),Q(300),X(300),T(50),Y(50),U(50),E(50)
4 DIM Z(300),X1(300,5)
6 DIM A(300),B(300),C(300),F(300)
12 K5 = 1
16 PRINT" ENTER NUMBER OF CLASSES WITH POSITIVE FREQUENCIES"
18 INPUT K
20 FOR J5= 1 TO K5
46 FOR I=1 TO K
48 PRINT "ENTER FREQUENCIES, 1 AT A TIME"
50 INPUT F(I)
54 X(I)= I - 1
56 Q(I)=X(I)
58 X(I)=Q(I)
60 NEXT I
62 Q1=0
64 Q2=0
65 N =0
66 FOR J=1 TO K
68 Q1=Q1+Q(J)*F(J)
70 Q2=Q2+Q(J)*Q(J)*F(J)
71 N = N + F(J)
72 NEXT J
74 A8=Q1/N
76 A7=((N*Q2-Q1*Q1)/N/(N-1)))^0.5
78 A6=A7*A7
80 S6=(A6/N)^0.5
82 R3=2*S6
84 PRINT" MEAN =",A8
85 PRINT" VARIANCE =",A6
86 PRINT"K PARAMETER: MAXIMUM LIKELIHOOD ESTIMATES"
88 M = A8
90 S2= A6
92 S1= S2^0.5
94 K2= M*M/(S2 - M)
100 PRINT K2
122 S9=0.
124 S6=0.
126 S3=0.
128 C4=0.
130 FOR L=1 TO 5
132 Y = K2*LOG(1 + M/K2) - LOG(N/F(1))
134 H = -Y/(LOG(1 + M/K2) + K2/(K2 + M) - 1)
136 K2 = K2 + H
138 NEXT L
140 Q5 = 1 + M/K2
142 FOR J = 1 TO 5
146 B(1) = A(1)/K2
150 C(1) = -A(1)/K2/K2
155 FOR I = 2 TO K
250 A(I) = A(I - 1) - F(I)
265 B(I) = B(I - 1) + A(I)/(K2 + I - 1)
270 C(I) = C(I - 1) - A(I)/(K2 + I - 1)/(K2 + I - 1)
275 NEXT I
278 Y = B(K) - N*LOG(K2 + M) + N*LOG(K2)
280 K2=(Y +LOG(N) - LOG(F(1)))/LOG(1 + M/K2)
282 H = -Y/(N/K2 - N/(K2 + M) + C(K))
285 PRINT K2
286 K2 = K2 + H
287 PRINT K2
```

```
334 NEXT I
336 K=INT(N3/2)
338 L=1
340 PRINT"  EXPECTED LOGNORMAL FREQUENCY  OBSERVED FREQUENCY"
342 C=0
344 FOR I=2 TO K
346 IF Z(I)<2.8 THEN 350
348 GO TO 358
350 Z(I+1)=Z(I+1)+Z(I)
352 L=L+1
354 F(I)=F(I)+F(I-1)
356 GO TO 362
358 C=C+(Z(I)-F(I-1))^2/Z(I)
360 PRINT"     ",Z(I),F(I-1)
362 NEXT I
364 N2=N3
365 N3=N3-1
366 FOR I=N3 TO K STEP -1
370 IF Z(I+1)<1.8 THEN 374
372 GO TO 382
374 Z(I)=Z(I)+Z(I+1)
376 N2=N2-1
378 F(I-1)=F(I-1)+F(I)
380 GO TO 386
382 PRINT"  ";Z(I+1),F(I)
384 C=C+(Z(I+1)-F(I))^2/Z(I+1)
386 NEXT I
388 M1=EXP(M+S2/2)
390 M2=EXP(M-S2/2)
392 M3=EXP(M)
394 V2=(EXP(S2+2*M))*(EXP(S2) -1)
396 PRINT"   MEAN    MODE    MEDIAN    VARIANCE"
398 PRINT M1,M2,M3,V2
400 PRINT" CHI-SQUARED CF. LOGNORMAL =";C;"WITH D.F.=";N2-L-2
402 FOR I=K TO N2
404 PRINT"  ";Z(I+1),F(I)
406 NEXT I
408 NEXT J5
410 PRINT "CALCULATIONS COMPLETED"
412DATA800,250
414DATA850,125
416...
418...
568DATA720,410
570END
```

```
288 NEXT J
296 P = M/K2
298 Z(1) = N/((1 + M/K2)^K2)
300 R = P/(1 + P)
302 FOR I = 2 TO X2
304 Z(I) = Z(I - 1)*(I - 2 + K2)*R/(I - 1)
306 NEXT I
340 PRINT"EXPECTED NEGATIVE BINOMIAL FREQUENCY  OBSERVED FREQUENCY"
342 C = 0
344 FOR I=1 TO K
350 PRINT "X =", I - 1, Z(I), F(I)
355 NEXT I
344 FOR I=1 TO K - 1
384 C=C+(Z(I)-F(I))^2/Z(I)
385 NEXT I
400 PRINT" CHI-SQUARED CF. NEGATIVE BINOMIAL =";C;"WITH D.F.=";K - 3
408 NEXT J5
410 PRINT "CALCULATIONS COMPLETED"
450 END
```

2.3. Listings for normal computer program (BASIC)

```
 5 PRINT "z, the standard normal variate value =)"
 8 INPUT Z
10 Y = Z
20 K = 1
30 FOR I = 1 TO 50: K = K*I
35 Y = Y +((-0.5)^I)*(Z^(2*I+1)/K/(2*I+1)
40 NEXT I
45 P1 = 4*ATN(1.0)
50 P  = (P1*2.0)^0.5
46 Y  = Y/P + 0.5
48 PRINT "Probability {Z ≤ z} =; Y
50 END
```

2.4. Listings for Student-t computer program (BASIC)

```
  5 LPRINT "T-Test"
  8 INPUT N
 10 PRINT "Degrees of freedom =";N
 20 IF N = 1 THEN 30
 25 GO TO 36
 30 G = 2*((ATN(1.0))^0.5): G1 = 1: print "Gamma{(n+1)/2} =";G1
 35 GO TO 80
 36 IF N = 2 THEN 38
 37 GO TO 40
 38 G = 1
 39 GO TO 80
 40 IF (N/2 - INT(N/2)) > 0 THEN 60
 44 G = 1
 46 FOR J1 = 1 TO N/2 - 1: G = G*J1: NEXT J1
 50 GO TO 80
 60 G = 2*((ATN(1.0))^0.5)
 65 FOR J1 = 1 TO N/2 - 0.5: G = G*(J1-0.5): NEXT J1
 80 PRINT "Gamma(N/2) ="; G
 82 IF ((N+1)/2 - INT((N+1)/2)) > 0 THEN 90
 84 IF N = 1 THEN 95
 86 G1   = 1
 86 FOR J1 = 1 TO (N+1)/2 - 1: G1 = G1*J1: NEXT J1
 88 GO TO 95
 90 G1 = 2*((ATN(1.0))^0.5)
 92 FOR J1 = 1 TO (N+1)/2 - 0.5: G1 = G1*(J1-0.5): NEXT J1
 95 PRINT "Gamma((N+1)/2) ="; G1
100 PRINT "t-value =";
110 INPUT T
115 LPRINT "t ="
120 IF N = 1 OR N = 3 THEN 800
125 P7 = N^0.5: P8 = ATN(T/P7): P9 = (1 - (COS(P8))^2)^0.5
130 Y = P9: U = 1
140 FOR I = 1 TO (N-2)/2: U = U*(N-2*I)/2/I
150 Y = Y + U/(2*I+1)*(P9^(2*I+1)))
160 NEXT I
170 P3 = 4*ATN(1.0)
180 P2 = (N*P3)^-0.5
190 P5 = P2*G1/G)
200 P  = 0.5 - Y*P5*P7
300 GO TO 900
800 P = .5 -1/4/ATN(1.0)*(ATN(Z))
810 IF N =3 THEN 850
815 GO TO 900
850 Z= ATN(Z/(3^0.5)
860 P = 0.5 - 1/4/ATN(1.0)*(Z + 0.5*(SIN(2*Z)))
900 PRINT "Probability {T ≥ t} =; P
950 END
```

2.5. Listings for χ^2 computer program (BASIC)

```
  5 PRINT "Chi-squared distribution (significance levels)"
  8 INPUT N
 10 PRINT "Degrees of freedom =";N
 20 IF N = 1 THEN 30
 25 GO TO 36
 30 G = 2*((ATN(1.0))^0.5): G1 = 1: print "Gamma{(n+1)/2} =";G1
 35 GO TO 80
 36 IF N = 2 THEN 38
 37 GO TO 40
 38 G = 1
 39 GO TO 80
 40 IF (N/2 - INT(N/2)) > 0 THEN 60
 44 G = 1
 46 FOR J1 = 1 TO N/2 - 1: G = G*J1: NEXT J1
 50 GO TO 80
 60 G = 2*((ATN(1.0))^0.5)
 65 FOR J1 = 1 TO N/2 - 0.5: G = G*(J1-0.5): NEXT J1
 80 PRINT "Gamma(N/2) =";  G
 82 IF ((N+1)/2 - INT((N+1)/2)) > 0 THEN 90
 84 IF N = 1 THEN 95
 86 G1   = 1
 86 FOR J1 = 1 TO (N+1)/2 - 1: G1 = G1*J1: NEXT J1
 88 GO TO 95
 90 G1 = 2*((ATN(1.0))^0.5)
 92 FOR J1 = 1 TO (N+1)/2 - 0.5: G1 = G1*(J1-0.5): NEXT J1
 95 PRINT "Gamma((N+1)/2) =";  G1
100 PRINT "Calculated chi-squared value =";
110 INPUT Z
115 PRINT "χ2 = ";Z
120 IF N = 2 THEN 800: IF N = 1 OR N = 3 THEN 820
130 Y = -2*(Z^(N/2-1)*(EXP(-Z/2)): U = 2
140 FOR I = 1 TO (N/2 - 2): U = 2*U*(N/2-I)/Z
150 Y = Y - U*(EXP(-Z/2))*(Z^(N/2-I)))
160 NEXT I
170 P = Y/G/(2^(N/2)) + 1 - EXP(-Z/2)
190 GO TO 900
800 P   = 1 - EXP(-Z/2)
810 GO TO 900
820 Y= 2*(Z^0.5) : U = 1 : FOR I = 1 TO 50: U = -0.5*U/I:
830 Y = Y + U/(I+.5)*(Z^(I + .5)): NEXT I
800 IF N =1 THEN 860
820 Y = Y - 2*(EXP(-Z/2)*(Z^.5)
860 P = Y/G/(2^(N/2))
900 PRINT "Probability {χ2≥ Z} =; 1 - P
950 END
```

2.6. Listings for F-Test computer program (BASIC)

```
10 PRINT "F- distribution (significance levels)"
18 INPUT M, N
20 PRINT "Degrees of freedom =";M, N
21 IF N = 1 THEN 23
22 GO TO 26
23 G = 2*((ATN(1.0))^0.5)
25 GO TO 60
30 IF (N/2 - INT(N/2)) > 0 THEN 50
31 IF V = 2 THEN 33
32 GO TO 35   FOR J1 = 1 TO N/2 - 1: G = G*J1: NEXT J1
33 G = 1
34 GO TO 60
35 G = 1
40 FOR J1 = 1 TO N/2 - 1: G = G*J1: NEXT J1
45 GO TO 60
50 G = 2*((ATN(1.0))^0.5)
52 FOR J1 = 1 TO N/2 - 0.5: G = G*(J1-0.5): NEXT J1
60 PRINT "Gamma(N/2) ="; G
61 IF (M/2 - INT(M/2)) > 0 THEN 66
62 G2 = 1/G
64 FOR J3 = 1 TO M/2 - 1: G2 = G2*J1: NEXT J3
65 GO TO 72
66 IF M = 1 THEN 69
67 G2 = 2*((ATN(1.0))^0.5)
68 GO TO 71
69 G2 = 2*((ATN(1.0))^0.5)
70 GO TO 72
71 FOR J3 = 1 TO M/2 - 0.5: G2 = G2*(J1-0.5): NEXT J3
72 G2 = G2*G
74 IF M = 2 THEN 76
75 GO TO 80
76 G2 = 1
80 PRINT "Gamma(G2) ="; G2
82 IF ((N+M)/2 - INT((N+M)/2)) > 0 THEN 90
84 G1   = 1/G/G2
86 FOR J1 = 1 TO (N+M)/2 - 1: G1 = G1*J1: NEXT J1
88 GO TO 95
90 G1 = 2*((ATN(1.0))^0.5)/G/G2
92 FOR J1 = 1 TO (N+M)/2 - 0.5: G1 = G1*(J1-0.5): NEXT J1
95 PRINT "Gamma COEFF. ="; G1
132 PRINT "Calculated F-test value =";
140 INPUT Z
150 LPRINT "F-value =";Z
152 IF N = 1 THEN 700
155 IF M = 1 OR M = 2 THEN 700
165 Y= -2*(Z^(M/2 - 1))*N/M/((1+M*Z/N)^((M+N)/2-1))/(M+N-2)
170 U = 1 : W = Y: K2 = 1
171 IF (M/2 - INT(M/2)) > 0 THEN 700
172 GO TO 177
173 FOR I = 1 TO M/2-1: U = U*N*(M/2-I)*(1+M*Z/N)/((M+N)/2-I-1)/M/Z
174 Y= Y + U*W: NEXT I
700 P3 = M/N
715 P6 = P3^(M/2)
720 P5 = G1
725 IF N = 1 THEN 760
730 IF (M/2 - INT(M/2)) > 0 THEN 760
750 P = Y*P5*P6 -((1+M*Z/N)^(-N/2)) + 1
755 GO TO 900
760 X = ATN((M*Z/N)^0.5): W = SIN(X): PRINT W
```

```
765 F = (W^M)/M: U = 1
767 IF N=1 AND M=1 THEN 780
771 IF N=1 AND M>1 THEN 790
774 FOR I = 1 TO 40: U = -0.5*U*(N-2*I)/I:
775 F = F + (W^(2*I+M))*U/M+2*I): NEXT I
778 GO TO 850
780 P = ATN((Z)^0.5)/2/(ATN(1.0))
785 GO TO 900
787 W = COS(X): F = W
790 FOR I = 1 TO 40: U = -0.5*U*(M-2*I)/I
795 F = F + (W^(2*I+N))*U/N+2*I): NEXT I
799 P = -F*2*P5 + 1
800 GO TO 900
850 P = F*P5*2
900 PRINT "Probability {F ≥ Z} =; 1 - P
950 END
```

APPENDIX 3

<u>SEQPOOL computer program listings (SPSS package)</u>

```
1$$S,T(:)
2$:IDENT:MAT-SA,1 WIGNALL
3$:SELECT:SPSS/SPSS
4RUN NAME::IBE,EMED,SCNI,SCALE,SCSH
5PAGESIZE:NOEJECT
6VARIABLE LIST:GRP,X1,X2,X3,X4,X5,X6
7COMPUTE:TON=X1
8COMPUTE:LTN=LN(TON)/LN(10)
9IF::(GRP EQ 5) X5=X4
10IF::(GRP EQ 1) X5=X5/28.35
11COMPUTE:GRD=(LN(X2*14+8*(X3+X4)+10X5+400X6))/LN(10)
12COMPUTE:VAL=GRD+LTN
13MANOVA::TON,GRD,VAL BY GRP(1,5)
14READ INPUT DATA
22  1  0.5  1.6  0   0   0   0
23  1  2.5  1.8  1.44  0   0   0
           .    .    .
           .    .    .
           .    .    .
142  5  17.2  1.3  1.7  0   0   0
143END INPUT DATA
144ONEWAY::TON BY GRP(1,5):/RANGES=DUNCAN/
145ONEWAY::LTN BY GRP(1,5):/RANGES=DUNCAN/
146ONEWAY::GRD BY GRP(1,5):/RANGES=DUNCAN/
147ONEWAY::VAL BY GRP(1,5):/RANGES=DUNCAN/
148FINISH
149END
```

APPENDIX 4

Listings for OPTGRID suite of computer programs in BASIC

4.1. Detection probabilities: airborne exploration: Parallel grid

```
1 DIM U(20),V(20),P(40)
2 DIM P1(20,10),P2(20,10),A(1)
3 PRINT "INPUT TARGET SHAPE RATIO, R = B/L"
4 INPUT K1
7  IMAGE 4X,DDDDD,12X,D.DDD,12X,D.DDD,12X,D.DDD
10 PRINT  "GEOLOGIC REGION: W.A. SHIELD"
12 PRINT  "SURVEY DESIGN: PARALLEL GRID, SPACING S METERS"
13 PRINT  "ELLIPTICAL TARGETS, LENGTH = L METERS, BREADTH = B"
14 PRINT  "ORIENTATION OF FLIGHT LINES: 90 DEGREES"
17 PRINT  "+ OR - 10 DEGREES WITH EXPECTED STRIKE LINE"
18 PRINT  "INPUT CONFID.INTERV.FOR EXPECTED TARGET LENGTH IN M"
19 INPUT  X,M,Y
22 PRINT "SHAPE RATIO, AND C.I. FOR LENGTH =",K1,X,M,Y
23 PRINT "GRID SPACING    L.C.L.=        GEOM. MEAN=        U.C.L.= "
24 PRINT " IN METERS    ",X,"        ",M,"            ",Y
25 FOR S=100 TO  2000 STEP 50
27 U(1)=X/S
28 U(2)=M/S
29 U(3)=Y/S
30 FOR J=1 TO 3
31 IF U(J)*K1>1 THEN 41
32 IF U(J)<1 THEN 34
33 GO TO 36
34 P(J)= K1*U(J)+(1-K1)*U(J)/PI*6/PI*SIN(PI/9)*K1+2*U(J)/PI*(1-K1)
35 GO TO 39
36 F=ACS(1/U(J))
37 G=(U(J)*U(J)-1)^0.5
38 P(J)=(U(J)-G+F)*2/PI*(1-K1) + K1
39 IF P(J)>1 THEN 41
40 GO TO 42
41 P(J)=1
42 NEXT J
43 PRINT USING 7;S,P(1),P(2),P(3)
44 NEXT S
45 END
```

4.2. Detection probabilities: airborne exploration: Square grid

```
1 DIM U(20),V(20),P(40)
2 DIM P1(20,10),P2(20,10),A(1)
3 PRINT "INPUT TARGET SHAPE RATIO, R = B/L"
4 INPUT K1
7   IMAGE 4X,DDDDD,12X,D.DDD,12X,D.DDD,12X,D.DDD
10 PRINT  "GEOLOGIC REGION: W.A. SHIELD"
12 PRINT  "SURVEY DESIGN: SQUARE GRID, SPACING S X S METERS"
13 PRINT  "ELLIPTICAL TARGETS, LENGTH = L METERS, BREADTH = B"
18 PRINT  "INPUT CONFID.INTERV.FOR EXPECTED TARGET LENGTH IN M"
19 INPUT  X,M,Y
22 PRINT "SHAPE RATIO, AND C.I. FOR LENGTH =",K1,X,M,Y
23 PRINT "GRID SPACING    L.C.L.=       GEOM. MEAN=        U.C.L.= "
24 PRINT " IN METERS    ",X,"         ",M,"            ",Y
25 FOR S=100 TO  2000 STEP 50
27 U(1)=X/S
28 U(2)=M/S
29 U(3)=Y/S
30 FOR J=1 TO 3
31 IF U(J)*K1>1 THEN 41
32 IF U(J)<1 THEN 34
33 GO TO 36
34 P(J)= K1*U(J)*(2-U(J)) + (4 - U(J))*U(J)/PI*(1-K1)
40 GO TO 42
41 P(J)=1
42 NEXT J
43 PRINT USING 7;S,P(1),P(2),P(3)
44 NEXT S
45 END
```

4.3. Airborne exploration: determination of grid-size for specified probability levels: Parallel grid

```
1 DIM U(20),V(20),P(40)
2 DIM P1(20,10),P2(20,10),A(1)
3 PRINT "INPUT TARGET SHAPE RATIO, R = B/L"
4 INPUT K1
7  IMAGE 10X,D.DD,20X,DDDDD,18X,"$",DDDD.DD
10 PRINT  "GEOLOGIC REGION: W.A. SHIELD"
12 PRINT  "SURVEY DESIGN: PARALLEL GRID, SPACING S METERS"
13 PRINT  "ELLIPTICAL TARGETS, LENGTH = L METERS, BREADTH = B"
14 PRINT  "ORIENTATION OF FLIGHT LINES: 90 DEGREES"
17 PRINT  "+ OR - 10 DEGREES WITH EXPECTED STRIKE LINE"
18 PRINT  "INPUT EXPECTED TARGET LENGTH IN METERS"
19 INPUT  M
22 PRINT "SHAPE RATIO, AND EXPECTED LENGTH =",K1,M
23 PRINT "SPECIFIED PROBAB.  REQUIRED GRID SPACING  CORRESPONDING"
24 PRINT " DET. LEVEL             IN METERS           COST IN $/KM"
25 FOR E=0.1 TO 1.0 STEP 0.05
27 FOR S=100 TO 10000 STEP 25
28 U(2)=M/S
29 U(3)=Y/S
30 J = 2
32 IF U(J)<1 THEN 34
33 GO TO 36
34 P(J)= K1*U(J)+(1-K1)*U(J)/PI*6/PI*SIN(PI/9)*K1+2*U(J)/PI*(1-K1)
35 GO TO 39
36 F=ACS(1/U(J))
37 G=(U(J)*U(J)-1)^0.5
38 P(J)=(U(J)-G+F)*2/PI*(1-K1) + K1
39 GO TO 42
41 P(J)=1
42 IF ABS(P(J)-E)<0.01 THEN 44
43 GO TO 46
44 Q=(1609/S+2)*70/1.609
45 GO TO 47
46 NEXT S
47 PRINT USING 7;E,S,Q
48 NEXT E
49 END
```

4.4. Airborne detection: Dynamic Programming Parallel grid

```
1 DIM U(20),V(20),P(40)R(20),S(20),T(20),Q(20)
2 DIM P1(20,10),P2(20,10),A(1)
3 PRINT "INPUT TARGET SHAPE RATIO, R = B/L"
4 INPUT K1
5  IMAGE 4X,"COST PER KM SQ. IN $",DDDD,"$",DDDD,"$",DDDD
6  IMAGE 4X,"OPTIMAL DETECT. PROB.:",D.DD,",",D.DD,",",D.DD
7  IMAGE 4X,"OPTIMAL GRID SPACING:",DDDD,"M,",DDDD,"M,",DDDD,"M"
8  IMAGE 4X,"95% C.INT. FOR LENGTH:",DDDD,"M,",DDDD,"M,",DDDD,"M"
10 PRINT   "GEOLOGIC REGION: W.A. SHIELD"
12 PRINT   "SURVEY DESIGN: PARALLEL GRID, SPACING S METERS"
13 PRINT   "ELLIPTICAL TARGETS, LENGTH = L METERS, BREADTH = B"
14 PRINT   "ORIENTATION OF FLIGHT LINES: 90 DEGREES"
17 PRINT   "+ OR - 10 DEGREES WITH EXPECTED STRIKE LINE"
18 PRINT   "INPUT CONF. INT. TARGET LENGTH IN METERS"
19 INPUT   U(1),U(2),U(3)
22 PRINT USING 8;U(1),U(2),U(3)
23 PRINT "INPUT 95TH% COST FOR LCL,G.MEAN,UCL AND COST/LINE KM"
24 INPUT Q(1),Q(2),Q(3),C
25 K=K1
26 FOR I=1 TO 3
27 FOR X3=50 TO U(I)+100 STEP 25
28 P5=(1-K)*U(I)/PI*2/X3-(U(I)*U(I)/X3/X31)^.5+ACS(X3/U(I))+K
29 P5=P5/0.95
30 A1=C/Q(I)*(1609/X3+2)
31 R1=P5 - A1
32 PRINT "EFFICIENCY FUNCTION =",R1
33 PRINT "PROBABILITY, COST, GRID SPACING=",0.95*P5,Q(I)*A1,X3
34 NEXT X3
35 NEXT I
36 PRINT "N.B. OPTIMUM OCCURS AT MAX. VALUE OF EFFIC. FUNCTION"
48 END
```

4.5. Airborne detection: Dynamic Programming: Square grid

```
1 DIM U(20),V(20),P(40)R(20),S(20),T(20),Q(20)
2 DIM P1(20,10),P2(20,10),A(1)
3 PRINT "INPUT TARGET SHAPE RATIO, R = B/L"
4 INPUT K1
5  IMAGE 4X,"COST PER KM SQ. IN $",DDDD,"$",DDDD,"$",DDDD
6  IMAGE 4X,"OPTIMAL DETECT. PROB.:",D.DD,",",D.DD,",",D.DD
7  IMAGE 4X,"OPTIMAL GRID SPACING:",DDDD,"M,",DDDD,"M,",DDDD,"M"
8  IMAGE 4X,"95% C.INT. FOR LENGTH:",DDDD,"M,",DDDD,"M,",DDDD,"M"
10 PRINT  "GEOLOGIC REGION: W.A. SHIELD"
12 PRINT  "SURVEY DESIGN: SQUARE GRID, SPACING S X S METERS"
13 PRINT  "ELLIPTICAL TARGETS, LENGTH = L METERS, BREADTH = B"
18 PRINT  "INPUT CONF. INT. TARGET LENGTH IN METERS"
19 INPUT  U(1),U(2),U(3)
22 PRINT USING 8;U(1),U(2),U(3)
23 PRINT "INPUT 95TH% COST FOR LCL,G.MEAN,UCL AND COST/LINE KM"
24 INPUT Q(1),Q(2),Q(3),C
25 K=K1
26 FOR I=1 TO 3
28 X1=(2*(1-K)*U(I)/0.95/PI - C/Q(I)*5280)^2
29 V(I)=0.95*0.95*PI*PI*X1/4/(1-K)/(1-K) + U(I)*U(I)
30 X2= U(I)/(V(I)^0.5)
31 IF X2>1 THEN 33
32 GO TO 38
33 P(I)=(5280/(V(I)^0.5)+2)*50/1.609
34 F=ACS(1/X2)
35 G= (X2*X2-1)^0.5
36 S(I)=(X2-G+F)*2/PI*(1-K) + K
37 GO TO 40
38 S(I)= (1-K)*2*X2/PI + K*X2
39 P(I)= (5280/(V(I)^0.5)+2)*50/1.609
34 PRINT "OPTIMAL GRID SPACING =",V(I)^0.5
32 PRINT "COST PER LINE-KM =",P(I)
33 PRINT "PROBABILITY OF DETECTION =",S(I)
35 NEXT I
48 END
```

4.6. Listings for OPTGRID computer program in BASIC

```
2 DIM E3(2),E4(2),E5(2)
3 DIM P3(8),L(30),X(30),K(8)
4 DIM Q3(8)
5:      #####        ##.###        ##.###        ##.###
6:      #####   0.### 0.### 0.### 0.### 0.### 0.### 0.### 0.###
7:      #####   #.### #.### #.### #.### #.### #.### #.### #.###
8:    ANGLE =  ##    ##    ##    ##    ##    ##    ##    ##
10 PRINT    "PROBABILITY OF DETECTION OF ORIENTATED TARGETS"
12 PRINT    "SURVEY DESIGN: SQUARE GRID, SPACINGS SXS METERS"
13 PRINT    "ELLIPTICAL TARGETS, LENGTH = L METERS, BREADTH = B"
14 PRINT    "R IS THE RATIO L/B WITH 0 < R <= 1"
17 PRINT    "A IS THE RATIO SEMI-MAJOR AXIS/GRID SIZE = L/(2S)"
19 PRINT    "INPUT ORIENTATION ANGLE - 5 DEGREES, FROM TABLE 4.10"
20 INPUT X9
21 PRINT    "INPUT EXPECTED LENGTH OF TARGET IN METERS"
22 INPUT X8
23 PRINT    "EXPECTED SHAPE RATIO R = B/L OF TARGET"
24 INPUT K(1)
25 PRINT    "N.B. IF A < 0.5, PROBABILITIES = RANDOMLY"
26 PRINT    "ORIENTATED TARGET PROBABILITIES: SEE TABLE"
28 PRINT "SHAPE RATIO = ", K(1)
29 PRINT "A = L/2S  PROBABILITY OF DETECTION (>= 1 INTERSECTION)"
32 PRINT USING 8,X9+5, X9+6, X9+7, X9+8, X9+9, X9+10, X9+11, X9+12
34 FOR I9 = 1 TO 20
35 L(I9) = X8/100/I9
36 A = L(I9)
37 E4(2) = 2
38 E5(1) = 0.05
39 E5(2) = 0.1
41 FOR J7 = 1 TO 1
42 FOR N3 = X9 TO X9 + 7
43 N4 = 10*N3
44 L6 = N3 - X9 + 1
45 B = A*K(J7)
46 C2 = 0
47 X3 = 0
48 Y3 = 0
49 D2 = 0
50 H  = 3.14159296/1800
51 K  = K(J7)
52 N5 = N4 + 100
53 FOR I5 = N4 TO N5
54 C3 = (COS(I5*H))^2
55 D3 = (SIN(I5*H))^2
56 C2 = (1- (1-K*K)*C3)^0.5
57 D2 = (1- (1-K*K)*D3)^0.5
58 C  = C2
60 D  = D2
61 IF A <= 0.5 THEN 63
62 GO TO 68
63 P1 = A*B*3.14159296
64 P2 = 0
65 P0 = 1 - P1
66 P3(L6) = P1
```

```
67  GO TO 171
68  IF B > C/2 THEN 70
69  GO TO 72
70  A1 = 2*(B*B*ATN(((4*B*B-C*C)^.5)/C)-0.5*C*((B*B-0.25*C*C)^0.5))
71  GO TO 73
72  A1 = 0
73  IF B > D/2 THEN 75
74  GO TO 77
75  A2 = 2*(B*B*ATN(((4*B*B-D*D)^.5)/D)-0.5*D*((B*B-0.25*D*D)^0.5))
76  GO TO 78
77  A2 = 0
78  IF A <= (2^0.5)/2 THEN 80
79  GO TO 85
80  P2 = (A1 + A2)/K
81  P1 = 3.14159296*B*B/K - 2*P2
82  PO = 1 - P1 - P2
83  P3(L6) = 1 - PO
84  GO TO 171
85  Y9 = D*D*C*C - K*K
86  Y9 = (ABS(Y9))^0.5
87  IF Y9 < .00000001 THEN 92
88  T = ATN(K/Y9)
89  GO TO 90
90  IF I5 = 0 THEN 92
91  GO TO 96
92  J1 = (C*C + D*D)^0.5
93  JO = 5
94  RO = J1/2
95  GO TO 111
96  I = 1 + INT(D*COS(T)/C)
97  IF I = 1 THEN 99
98  GO TO 103
99  J1 = (C*C + D*D - 2*C*D*COS(T))^0.5
100 JO = 5
101 RO = J1/2/SIN(T)
102 GO TO 106
103 J1 = (I*I*C*C + D*D - 2*I*C*D*COS(T))^0.5
104 JO = (((I-1)^2)*C*C + D*D - 2*(I-1)*C*D*COS(T))^0.5
105 RO = J1*JO/2/D/SIN(T)
106 IF B >= RO THEN 108
107 GO TO 111
108 P3(L6) = 1
109 PO = 0
110 GO TO 171
111 IF B > J1/2 THEN 113
112 GO TO 115
113 A3 = 2*(B*B*ATN(((4*B*B-J1*J1)^.5)/J1)-J1/2*((B*B-
    J1*J1/4)^0.5))
114 GO TO 116
115 A3 = 0
116 IF B > JO/2 THEN 118
117 GO TO 120
118 A4 = 2*(B*B*ATN(((4*B*B-JO*JO)^.5)/JO)-JO/2*((B*B-
    JO*JO/4)^0.5))
119 GO TO 121
120 A4 = 0
121 M = 1 + INT(2*D*COS(T)/C)
122 IF M=1 THEN 124
```

```
123 GO TO 127
124 M1 = (C*C + D*D - 2*C*D*COS(T))^0.5
125 M0 = 5
126 GO TO 129
127 M1 = (M*M*C*C + 4*D*D - 4*M*C*D*COS(T))^0.5
128 M0 = ((M-1)*(M-1)*C*C + 4*D*D - 4*(M-1)*C*D*COS(T))^0.5
129 IF J1 = M1/2 THEN 139
130 IF J0 = M1/2 THEN 139
131 IF J1 = M0/2 THEN 139
132 IF J0 = M0/2 THEN 139
133 IF B > M1/2 THEN 136
134 A5 = 0
135 GO TO 142
136 A5 = 2*(B*B*ATN(((4*B*B-M1*M1)^.5)/M1)-M1/2*((B*B-
    M1*M1/4)^0.5))
137 GO TO 142
138 A5 = 0
139 A6 = 0
140 GO TO 145
141 IF B > M0/2 THEN 144
142 A6 = 0
143 GO TO 145
144 A6 = 2*(B*B*ATN(((4*B*B-M0*M0)^.5)/M0)-M0/2*((B*B-
    M0*M0/4)^0.5))
145 IF B > J1 THEN 148
146 A7 = 0
147 GO TO 149
148 A7 = 2*(B*B*ATN(((4*B*B-J1*J1)^.5)/J1)-J1/2*((B*B-
    J1*J1/4)^0.5))
149 IF B > J0 THEN 152
150 A8 = 0
151 GO TO 153
152 A8 = 2*(B*B*ATN(((4*B*B-J0*J0)^.5)/J0)-J0/2*((B*B-
    J0*J0/4)^0.5))
153 IF B > C THEN 156
154 A9 = 0
155 GO TO 157
156 A9 = 2*(B*B*ATN(((B*B-C*C)^.5)/C)-C/2*((B*B-C*C)^0.5))
157 IF B > D THEN 160
158 B9 = 0
159 GO TO 161
160 B9 = 2*(B*B*ATN(((B*B-D*D)^.5)/D)-D/2*((B*B-D*D)^0.5))
161 P2 = (A1+A2+A3+A4+A5+A6-A7-A8-A9-B9)/K
162 P1 = 3.14159296*B*B/K -(A7+A8+A9+B9)/K - 2*P2
163 P0 = 1 - P1 - P2
164 P3(L6) = 1 -P0
165 IF P3(L6) > 1 THEN 170
166 IF P3(L6) < 0 THEN 168
167 GO TO 169
168 P3(L6) = 0
169 GO TO 171
170 P3(L6) = 1
171 Q3(I5,L6) = P3(L6)
172 NEXT I5
173 P3(L6) = Q3(N4,L6) + Q3(N5,L6)
174 FOR I = N4 + 1 TO N5 - 1 STEP 2
175 P3(L6) = P3(L6) + 4*Q3(I,L6)
176 NEXT I
```

```
177 FOR J = N4 + 2 TO N5 - 2 STEP 2
178 P3(L6) = P3(L6) + 2*Q3(J,L6)
179 NEXT J
180 P3(L6) = P3(L6)/300
181 IF P3(L6) > 1 THEN 183
182 GO TO 184
183 P3(L6) = 1
184 NEXT N3
185 NEXT J7
186 IF A > 0.99 THEN 188
187 GO TO 190
188 PRINT USING 7, 50*I9, P3(1),P3(2),P3(3),P3(4),P3(5),
    P3(6),P3(7),P3(8)
189 GO TO 191
190 PRINT USING 6, 50*I9, P3(1),P3(2),P3(3),P3(4),P3(5),
    P3(6),P3(7),P3(8)
191 NEXT I9
192 STOP
193 END
```

4.7. Ground grid orientation & detection probabilities for oriented targets

```
2 DIM E3(2),E4(2),E5(2)
3 DIM P3(8),L(30),X(30),K(8)
4 DIM Q3(8)
5:      #####           ##.###          ##.###          ##.###
6:      #####  0.### 0.### 0.### 0.### 0.### 0.### 0.### 0.###
7:      #####  #.### #.### #.### #.### #.### #.### #.### #.###
8:      ANGLE =  ##    ##    ##    ##    ##    ##    ##    ##
10 PRINT    "PROBABILITY OF DETECTION OF ORIENTATED TARGETS"
12 PRINT    "SURVEY DESIGN: SQUARE GRID, SPACINGS SXS METERS"
13 PRINT    "ELLIPTICAL TARGETS, LENGTH = L METERS, BREADTH = B"
14 PRINT    "R IS THE RATIO L/B WITH 0 < R <= 1"
17 PRINT    "A IS THE RATIO SEMI-MAJOR AXIS/GRID SIZE = L/(2S)"
19 PRINT    "INPUT ORIENTATION ANGLE - 5 DEGREES, FROM TABLE 4.12"
20 INPUT X9
21 PRINT    "INPUT EXPECTED LENGTH OF TARGET IN METERS"
22 INPUT X8
23 PRINT    "EXPECTED SHAPE RATIO R = B/L OF TARGET"
24 INPUT K(1)
25 PRINT    "N.B. IF A < 0.5, PROBABILITIES = RANDOMLY"
26 PRINT    "ORIENTATED TARGET PROBABILITIES: SEE TABLE"
28 PRINT "SHAPE RATIO = ", K(1)
29 PRINT "A = L/2S  PROBABILITY OF DETECTION (>= 1 INTERSECTION)"
32 PRINT USING 8,X9+5, X9+6, X9+7, X9+8, X9+9, X9+10, X9+11, X9+12
34 FOR I9 = 1 TO 20
35 L(I9) = X8/100/I9
36 A = L(I9)
37 E4(2) = 2
38 E5(1) = 0.05
39 E5(2) = 0.1
41 FOR J7 = 1 TO 1
42 FOR N3 = X9 TO X9 + 7
43 N4 = 10*N3
44 L6 = N3 - X9 + 1
45 B = A*K(J7)
46 C2 = 0
47 X3 = 0
48 Y3 = 0
49 D2 = 0
50 H  = 3.14159296/1800
51 K  = K(J7)
52 N5 = N4 + 100
53 FOR I5 = N4 TO N5
54 C3 = (COS(I5*H))^2
55 D3 = (SIN(I5*H))^2
56 C2 = (1- (1-K*K)*C3)^0.5
57 D2 = (1- (1-K*K)*D3)^0.5
58 C  = C2
60 D  = D2
61 IF A <= 0.5 THEN 63
62 GO TO 68
63 P1 = A*B*3.14159296
64 P2 = 0
65 P0 = 1 - P1
66 P3(L6) = P1
```

```
67 GO TO 171
68 IF B > C/2 THEN 70
69 GO TO 72
70 A1 = 2*(B*B*ATN(((4*B*B-C*C)^.5)/C)-0.5*C*((B*B-0.25*C*C)^0.5))
71 GO TO 73
72 A1 = 0
73 IF B > D/2 THEN 75
74 GO TO 77
75 A2 = 2*(B*B*ATN(((4*B*B-D*D)^.5)/D)-0.5*D*((B*B-0.25*D*D)^0.5))
76 GO TO 78
77 A2 = 0
78 IF A <= (2^0.5)/2 THEN 80
79 GO TO 85
80 P2 = (A1 + A2)/K
81 P1 = 3.14159296*B*B/K - 2*P2
82 PO = 1 - P1 - P2
83 P3(L6) = 1 - PO
84 GO TO 171
85 Y9 = D*D*C*C - K*K
86 Y9 = (ABS(Y9))^0.5
87 IF Y9 < .00000001 THEN 92
88 T = ATN(K/Y9)
89 GO TO 90
90 IF I5 = 0 THEN 92
91 GO TO 96
92 J1 = (C*C + D*D)^0.5
93 JO = 5
94 RO = J1/2
95 GO TO 111
96 I = 1 + INT(D*COS(T)/C)
97 IF I = 1 THEN 99
98 GO TO 103
99 J1 = (C*C + D*D - 2*C*D*COS(T))^0.5
100 JO = 5
101 RO = J1/2/SIN(T)
102 GO TO 106
103 J1 = (I*I*C*C + D*D - 2*I*C*D*COS(T))^0.5
104 JO = (((I-1)^2)*C*C + D*D - 2*(I-1)*C*D*COS(T))^0.5
105 RO = J1*JO/2/D/SIN(T)
106 IF B >= RO THEN 108
107 GO TO 111
108 P3(L6) = 1
109 PO = 0
110 GO TO 171
111 IF B > J1/2 THEN 113
112 GO TO 115
113 A3 = 2*(B*B*ATN(((4*B*B-J1*J1)^.5)/J1)-J1/2*((B*B-
        J1*J1/4)^0.5))
114 GO TO 116
115 A3 = 0
116 IF B > JO/2 THEN 118
117 GO TO 120
118 A4 = 2*(B*B*ATN(((4*B*B-JO*JO)^.5)/JO)-JO/2*((B*B-
        JO*JO/4)^0.5))
119 GO TO 121
120 A4 = 0
121 M = 1 + INT(2*D*COS(T)/C)
122 IF M=1 THEN 124
```

```
123 GO TO 127
124 M1 = (C*C + D*D - 2*C*D*COS(T))^0.5
125 M0 = 5
126 GO TO 129
127 M1 = (M*M*C*C + 4*D*D - 4*M*C*D*COS(T))^0.5
128 M0 = ((M-1)*(M-1)*C*C + 4*D*D - 4*(M-1)*C*D*COS(T))^0.5
129 IF J1 = M1/2 THEN 139
130 IF J0 = M1/2 THEN 139
131 IF J1 = M0/2 THEN 139
132 IF J0 = M0/2 THEN 139
133 IF B > M1/2 THEN 136
134 A5 = 0
135 GO TO 142
136 A5 = 2*(B*B*ATN(((4*B*B-M1*M1)^.5)/M1)-M1/2*((B*B-
        M1*M1/4)^0.5))
137 GO TO 142
138 A5 = 0
139 A6 = 0
140 GO TO 145
141 IF B > M0/2 THEN 144
142 A6 = 0
143 GO TO 145
144 A6 = 2*(B*B*ATN(((4*B*B-M0*M0)^.5)/M0)-M0/2*((B*B-
        M0*M0/4)^0.5))
145 IF B > J1 THEN 148
146 A7 = 0
147 GO TO 149
148 A7 = 2*(B*B*ATN(((4*B*B-J1*J1)^.5)/J1)-J1/2*((B*B-
        J1*J1/4)^0.5))
149 IF B > J0 THEN 152
150 A8 = 0
151 GO TO 153
152 A8 = 2*(B*B*ATN(((4*B*B-J0*J0)^.5)/J0)-J0/2*((B*B-
        J0*J0/4)^0.5))
153 IF B > C THEN 156
154 A9 = 0
155 GO TO 157
156 A9 = 2*(B*B*ATN(((B*B-C*C)^.5)/C)-C/2*((B*B-C*C)^0.5))
157 IF B > D THEN 160
158 B9 = 0
159 GO TO 161
160 B9 = 2*(B*B*ATN(((B*B-D*D)^.5)/D)-D/2*((B*B-D*D)^0.5))
161 P2 = (A1+A2+A3+A4+A5+A6-A7-A8-A9-B9)/K
162 P1 = 3.14159296*B*B/K -(A7+A8+A9+B9)/K - 2*P2
163 P0 = 1 - P1 - P2
164 P3(L6) = 1 -P0
165 IF P3(L6) > 1 THEN 170
166 IF P3(L6) < 0 THEN 168
167 GO TO 169
168 P3(L6) = 0
169 GO TO 171
170 P3(L6) = 1
171 Q3(I5,L6) = P3(L6)
172 NEXT I5
173 P3(L6) = Q3(N4,L6) + Q3(N5,L6)
174 FOR I = N4 + 1 TO N5 - 1 STEP 2
175 P3(L6) = P3(L6) + 4*Q3(I,L6)
```

```
176 NEXT I
177 FOR J = N4 + 2 TO N5 - 2 STEP 2
178 P3(L6) = P3(L6) + 2*Q3(J,L6)
179 NEXT J
180 P3(L6) = P3(L6)/300
181 IF P3(L6) > 1 THEN 183
182 GO TO 184
183 P3(L6) = 1
184 NEXT N3
185 NEXT J7
186 IF A > 0.99 THEN 188
187 GO TO 190
188 PRINT USING 7,50*I9,P3(1),P3(2),P3(3),P3(4),P3(5),P3(6),P3(7),
                      P3(8)
189 GO TO 191
190 PRINT USING ,50*I9,P3(1),P3(2),P3(3),P3(4),P3(5),P3(6),P3(7),
                      P3(8)
191 NEXT I9
192 STOP
193 END
```

```
2 DIM E3(2),E4(2),E5(2)
3 DIM P3(8),L(30),X(30),K(8)
4 DIM Q3(8)
5:      #####        ##.###        ##.###         ##.###
6:      #####  0.### 0.### 0.### 0.### 0.### 0.### 0.### 0.###
7:      #####  #.### ·#.### #.### #.### #.### #.### #.### #.###
8:      ANGLE = ##      ##      ##      ##      ##      ##      ##      ##
10 PRINT  "DETECTION PROBABILITIES FOR RANDOMLY ORIENTATED
              TARGETS"
12 PRINT  "SURVEY DESIGN: SQUARE GRID, SPACINGS SXS METERS"
13 PRINT  "ELLIPTICAL TARGETS, LENGTH = L METERS, BREADTH = B"
14 PRINT  "R IS THE RATIO L/B WITH 0 < R <= 1"
17 PRINT  "A IS THE RATIO SEMI-MAJOR AXIS/GRID SIZE = L/(2S)"
21 PRINT  "INPUT EXPECTED LENGTH OF TARGET IN METERS"
22 INPUT X8
23 PRINT  "EXPECTED SHAPE RATIO, R = B/L, OF TARGET"
24 INPUT K(1)
28 PRINT "SHAPE RATIO = ", K(1)
29 PRINT "          PROBABILITY OF DETECTION (>= 1 INTERSECTION)"
30 PRINT "GRID SPACING S IN METERS"
31 E3(1) = 50
32 E4(1) = 100
33 E3(2) = 150
34 E4(2) = 1000
35 E5(1) = 25
36 E5(2) = 50
37 FOR E(6) = 1 TO 2
38 FOR W = E3(E6) TO E4(E6) STEP E5(E6)
39 FOR J7 = 1 TO 1
40 A = X8/2/W
41 J6 = J7
42 B = A*K(J7)
43 N4 = 0
46 C2 = 0
47 X3 = 0
48 Y3 = 0
49 D2 = 0
50 H  = 3.14159296/400
51 K  = K(J7)
52 N5 = N4 + 100
53 FOR I5 = N4 TO N5
54 C3 = (COS(I5*H))^2
55 D3 = (SIN(I5*H))^2
56 C2 = (1- (1-K*K)*C3)^0.5
57 D2 = (1- (1-K*K)*D3)^0.5
58 C  = C2
60 D  = D2
61 IF A <= 0.5 THEN 63
62 GO TO 68
63 P1 = A*B*3.14159296
64 P2 = 0
65 P0 = 1 - P1
66 P3(L6) = P1
67 GO TO 171
68 IF B > C/2 THEN 70
```

```
69  GO TO 72
70  A1 = 2*(B*B*ATN(((4*B*B-C*C)^.5)/C)-0.5*C*((B*B-0.25*C*C)^0.5))
71  GO TO 73
72  A1 = 0
73  IF B > D/2 THEN 75
74  GO TO 77
75  A2 = 2*(B*B*ATN(((4*B*B-D*D)^.5)/D)-0.5*D*((B*B-0.25*D*D)^0.5))
76  GO TO 78
77  A2 = 0
78  IF A <= (2^0.5)/2 THEN 80
79  GO TO 85
80  P2 = (A1 + A2)/K
81  P1 = 3.14159296*B*B/K - 2*P2
82  P0 = 1 - P1 - P2
83  P3(L6) = 1 - P0
84  GO TO 171
85  Y9 = D*D*C*C - K*K
86  Y9 = (ABS(Y9))^0.5
87  IF Y9 < .00000001 THEN 92
88  T = ATN(K/Y9)
89  GO TO 90
90  IF I5 = 0 THEN 92
91  GO TO 96
92  J1 = (C*C + D*D)^0.5
93  J0 = 5
94  R0 = J1/2
95  GO TO 111
96  I = 1 + INT(D*COS(T)/C)
97  IF I = 1 THEN 99
98  GO TO 103
99  J1 = (C*C + D*D - 2*C*D*COS(T))^0.5
100 J0 = 5
101 R0 = J1/2/SIN(T)
102 GO TO 106
103 J1 = (I*I*C*C + D*D - 2*I*C*D*COS(T))^0.5
104 J0 = (((I-1)^2)*C*C + D*D - 2*(I-1)*C*D*COS(T))^0.5
105 R0 = J1*J0/2/D/SIN(T)
106 IF B >= R0 THEN 108
107 GO TO 111
108 P3(L6) = 1
109 P0 = 0
110 GO TO 171
111 IF B > J1/2 THEN 113
112 GO TO 115
113 A3 = 2*(B*B*ATN(((4*B*B-J1*J1)^.5)/J1)-J1/2*((B*B-
        J1*J1/4)^0.5))
114 GO TO 116
115 A3 = 0
116 IF B > J0/2 THEN 118
117 GO TO 120
118 A4 = 2*(B*B*ATN(((4*B*B-J0*J0)^.5)/J0)-J0/2*((B*B-
        J0*J0/4)^0.5))
119 GO TO 121
120 A4 = 0
121 M = 1 + INT(2*D*COS(T)/C)
122 IF M=1 THEN 124
123 GO TO 127
124 M1 = (C*C + D*D - 2*C*D*COS(T))^0.5
125 M0 = 5
```

```
126 GO TO 129
127 M1 = (M*M*C*C + 4*D*D - 4*M*C*D*COS(T))^0.5
128 M0 = ((M-1)*(M-1)*C*C + 4*D*D - 4*(M-1)*C*D*COS(T))^0.5
129 IF J1 = M1/2 THEN 139
130 IF J0 = M1/2 THEN 139
131 IF J1 = M0/2 THEN 139
132 IF J0 = M0/2 THEN 139
133 IF B > M1/2 THEN 136
134 A5 = 0
135 GO TO 142
136 A5 = 2*(B*B*ATN(((4*B*B-M1*M1)^.5)/M1)-M1/2*((B*B-
        M1*M1/4)^0.5))
137 GO TO 142
138 A5 = 0
139 A6 = 0
140 GO TO 145
141 IF B > M0/2 THEN 144
142 A6 = 0
143 GO TO 145
144 A6 = 2*(B*B*ATN(((4*B*B-M0*M0)^.5)/M0)-M0/2*((B*B-
        M0*M0/4)^0.5))
145 IF B > J1 THEN 148
146 A7 = 0
147 GO TO 149
148 A7 = 2*(B*B*ATN(((4*B*B-J1*J1)^.5)/J1)-J1/2*((B*B-
        J1*J1/4)^0.5))
149 IF B > J0 THEN 152
150 A8 = 0
151 GO TO 153
152 A8 = 2*(B*B*ATN(((4*B*B-J0*J0)^.5)/J0)-J0/2*((B*B-
        J0*J0/4)^0.5))
153 IF B > C THEN 156
154 A9 = 0
155 GO TO 157
156 A9 = 2*(B*B*ATN(((B*B-C*C)^.5)/C)-C/2*((B*B-C*C)^0.5))
157 IF B > D THEN 160
158 B9 = 0
159 GO TO 161
160 B9 = 2*(B*B*ATN(((B*B-D*D)^.5)/D)-D/2*((B*B-D*D)^0.5))
161 P2 = (A1+A2+A3+A4+A5+A6-A7-A8-A9-B9)/K
162 P1 = 3.14159296*B*B/K -(A7+A8+A9+B9)/K - 2*P2
163 P0 = 1 - P1 - P2
164 P3(L6) = 1 -P0
165 IF P3(L6) > 1 THEN 170
166 IF P3(L6) < 0 THEN 168
167 GO TO 169
168 P3(L6) = 0
169 GO TO 171
170 P3(L6) = 1
171 Q3(I5,L6) = P3(L6)
172 NEXT I5
173 P3(L6) = Q3(N4,L6) + Q3(N5,L6)
174 FOR I = N4 + 1 TO N5 - 1 STEP 2
175 P3(L6) = P3(L6) + 4*Q3(I,L6)
176 NEXT I
177 FOR J = N4 + 2 TO N5 - 2 STEP 2
178 P3(L6) = P3(L6) + 2*Q3(J,L6)
```

```
179 NEXT J
180 P3(L6) = P3(L6)/300
181 IF P3(L6) > 1 THEN 183
182 GO TO 184
183 P3(L6) = 1
185 NEXT J7
186 IF A > 0.99 THEN 188
187 GO TO 190
188 PRINT USING 7,W,P3(1),P3(2),P3(3),P3(4),P3(5),P3(6),P3(7),
                P3(8)
189 GO TO 191
190 PRINT USING 6,W,P3(1),P3(2),P3(3),P3(4),P3(5),P3(6),P3(7),
                P3(8)
191 NEXT W
192 NEXT E6
193 STOP
194 END
```

4.9. Ground detection: determination of grid-size

```
 2 DIM E3(2),E4(2),E5(2)
 3 DIM P3(8),L(30),X(30),K(8)
 4 DIM Q3(8)
 5:     #####        ##.###       ##.###        ##.###
 6:     #####  0.### 0.### 0.### 0.### 0.### 0.### 0.### 0.###
 7:     #####  ##.###       ##.###       ##.###
 8:  0.##  0.###  0.###  0.###  0.###  0.###  0.###  0.###  0.###
 9:  #.##  0.###  0.###  0.###  0.###  0.###  0.###  0.###  0.###
10 PRINT   "DETERMINATION OF GRID SIZE"
12 PRINT   "SURVEY DESIGN: SQUARE GRID, SPACINGS SXS METERS"
13 PRINT   "ELLIPTICAL TARGETS, LENGTH = L METERS, BREADTH = B"
14 PRINT   "R IS THE RATIO L/B WITH 0 < R <= 1"
17 PRINT   "A IS THE RATIO SEMI-MAJOR AXIS/GRID SIZE = L/(2S)"
21 PRINT   "INPUT EXPECTED TARGET LENGTH IN METERS"
22 INPUT L(2)
23 PRINT   "INPUT EXPECTED SHAPE RATIO, R = B/L, OF TARGET"
24 INPUT K(1)
25 PRINT "SHAPE RATIO = ", K(1)
26 PRINT "PROBABILITY        GRID SIZE     DRILLING COST"
27 PRINT "   LEVEL           IN METERS    IN $1000'S/ SQ. KM"
28 PRINT "                        PERCUSSION       DIAMOND"
29 FOR E9 = 0.1 TO 0.9 STEP 0.1
31 E3(1) = 25
32 E4(1) = 200
33 E3(2) = 250
34 E4(2) = 1000
35 E5(1) = 25
36 E5(2) = 50
37 FOR E(6) = 1 TO 2
38 FOR W = E3(E6) TO E4(E6) STEP E5(E6)
39 FOR J7 = 2 TO 2
40 A = X8/2/W
41 J6 = J7
42 B = A*K(1)
43 N4 = 240
46 C2 = 0
47 X3 = 0
48 Y3 = 0
49 D2 = 0
50 H  = 3.14159296/1800
51 K  = K(1)
52 N5 = N4 + 100
53 FOR I5 = N4 TO N5
54 C3 = (COS(I5*H))^2
55 D3 = (SIN(I5*H))^2
56 C2 = (1- (1-K*K)*C3)^0.5
57 D2 = (1- (1-K*K)*D3)^0.5
58 C  = C2
60 D  = D2
61 IF A <= 0.5 THEN 63
62 GO TO 68
63 P1 = A*B*3.14159296
64 P2 = 0
```

```
65 P0 = 1 - P1
66 P3(L6) = P1
67 GO TO 171
68 IF B > C/2 THEN 70
69 GO TO 72
70 A1 = 2*(B*B*ATN(((4*B*B-C*C)^.5)/C)-0.5*C*((B*B-0.25*C*C)^0.5))
71 GO TO 73
72 A1 = 0
73 IF B > D/2 THEN 75
74 GO TO 77
75 A2 = 2*(B*B*ATN(((4*B*B-D*D)^.5)/D)-0.5*D*((B*B-0.25*D*D)^0.5))
76 GO TO 78
77 A2 = 0
78 IF A <= (2^0.5)/2 THEN 80
79 GO TO 85
80 P2 = (A1 + A2)/K
81 P1 = 3.14159296*B*B/K - 2*P2
82 P0 = 1 - P1 - P2
83 P3(L6) = 1 - P0
84 GO TO 171
85 Y9 = D*D*C*C - K*K
86 Y9 = (ABS(Y9))^0.5
87 IF Y9 < .00000001 THEN 92
88 T = ATN(K/Y9)
89 GO TO 90
90 IF I5 = 0 THEN 92
91 GO TO 96
92 J1 = (C*C + D*D)^0.5
93 J0 = 5
94 R0 = J1/2
95 GO TO 111
96 I = 1 + INT(D*COS(T)/C)
97 IF I = 1 THEN 99
98 GO TO 103
99 J1 = (C*C + D*D - 2*C*D*COS(T))^0.5
100 J0 = 5
101 R0 = J1/2/SIN(T)
102 GO TO 106
103 J1 = (I*I*C*C + D*D - 2*I*C*D*COS(T))^0.5
104 J0 = (((I-1)^2)*C*C + D*D - 2*(I-1)*C*D*COS(T))^0.5
105 R0 = J1*J0/2/D/SIN(T)
106 IF B >= R0 THEN 108
107 GO TO 111
108 P3(L6) = 1
109 P0 = 0
110 GO TO 171
111 IF B > J1/2 THEN 113
112 GO TO 115
113 A3 = 2*(B*B*ATN(((4*B*B-J1*J1)^.5)/J1)-J1/2*((B*B-
        J1*J1/4)^0.5))
114 GO TO 116
115 A3 = 0
116 IF B > J0/2 THEN 118
117 GO TO 120
118 A4 = 2*(B*B*ATN(((4*B*B-J0*J0)^.5)/J0)-J0/2*((B*B-
        J0*J0/4)^0.5))
119 GO TO 121
```

```
120 A4 = 0
121 M = 1 + INT(2*D*COS(T)/C)
122 IF M=1 THEN 124
123 GO TO 127
124 M1 = (C*C + D*D - 2*C*D*COS(T))^0.5
125 M0 = 5
126 GO TO 129
127 M1 = (M*M*C*C + 4*D*D - 4*M*C*D*COS(T))^0.5
128 M0 = ((M-1)*(M-1)*C*C + 4*D*D - 4*(M-1)*C*D*COS(T))^0.5
129 IF J1 = M1/2 THEN 139
130 IF J0 = M1/2 THEN 139
131 IF J1 = M0/2 THEN 139
132 IF J0 = M0/2 THEN 139
133 IF B > M1/2 THEN 136
134 A5 = 0
135 GO TO 142
136 A5 = 2*(B*B*ATN(((4*B*B-M1*M1)^.5)/M1)-M1/2*((B*B-
        M1*M1/4)^0.5))
137 GO TO 142
138 A5 = 0
139 A6 = 0
140 GO TO 145
141 IF B > M0/2 THEN 144
142 A6 = 0
143 GO TO 145
144 A6 = 2*(B*B*ATN(((4*B*B-M0*M0)^.5)/M0)-M0/2*((B*B-
        M0*M0/4)^0.5))
145 IF B > J1 THEN 148
146 A7 = 0
147 GO TO 149
148 A7 = 2*(B*B*ATN(((4*B*B-J1*J1)^.5)/J1)-J1/2*((B*B-
        J1*J1/4)^0.5))
149 IF B > J0 THEN 152
150 A8 = 0
151 GO TO 153
152 A8 = 2*(B*B*ATN(((4*B*B-J0*J0)^.5)/J0)-J0/2*((B*B-
        J0*J0/4)^0.5))
153 IF B > C THEN 156
154 A9 = 0
155 GO TO 157
156 A9 = 2*(B*B*ATN(((B*B-C*C)^.5)/C)-C/2*((B*B-C*C)^0.5))
157 IF B > D THEN 160
158 B9 = 0
159 GO TO 161
160 B9 = 2*(B*B*ATN(((B*B-D*D)^.5)/D)-D/2*((B*B-D*D)^0.5))
161 P2 = (A1+A2+A3+A4+A5+A6-A7-A8-A9-B9)/K
162 P1 = 3.14159296*B*B/K -(A7+A8+A9+B9)/K - 2*P2
163 P0 = 1 - P1 - P2
164 P3(L6) = 1 -P0
165 IF P3(L6) > 1 THEN 170
166 IF P3(L6) < 0 THEN 168
167 GO TO 169
168 P3(L6) = 0
169 GO TO 171
170 P3(L6) = 1
171 Q3(I5,J6) = P3(L6)
```

```
172 NEXT I5
173 P3(J7) = Q3(N4,J6) + Q3(N5,J6)
174 FOR I = N4 + 1 TO N5 - 1 STEP 2
175 P3(J7) = P3(J7) + 4*Q3(I,J6)
176 NEXT I
177 FOR J = N4 + 2 TO N5 - 2 STEP 2
178 P3(J7) = P3(J7) + 2*Q3(J,J6)
179 NEXT J
180 P3(J7) = P3(J7)/300
181 IF P3(J7) > 1 THEN 183
182 GO TO 184
183 P3(J7) = 1
184 IF (P3(J7)-E9) <= 1 THEN 188
187 GO TO 194
188 V8 = (1609/W +1)*(1609/W + 1.02)
189 V7 = V8*6/1.60934/1.60934
190 V7 = 328.084/300*V7
191 V9 = 3*V7
192 PRINT USING 7,E9,W,V7,V9)
193 GO TO 195
194 NEXT W
195 NEXT E9
196 STOP
197 END
```

```
1 DIM E3(2),E4(2),E5(2)
2 DIM P3(8),L(30),X(30),K(8),Q3(600,10)
3 PRINT "INPUT 95% COST PERCENTILE FROM COST TABLE"
4 INPUT C5
5:     ####         ##.###       ##.###      ##.###
6:     ####  0.### 0.### 0.### 0.### 0.### 0.### 0.### 0.###
7:     ####  ##.###       ##.###      ##.###
8:   ####         ##.###       ##.###      ##.###
9:   #.##  0.### 0.### 0.### 0.### 0.### 0.###  0.###  0.###
10 PRINT  "OPTIMIZATION OF EFFIENCY FUNCTION"
12 PRINT  "SURVEY DESIGN: SQUARE GRID, SPACINGS SXS METERS"
13 PRINT  "ELLIPTICAL TARGETS, LENGTH = L METERS, BREADTH = B"
14 PRINT  "INPUT EXPECTED PROBABILITY LEVEL REQUIRED"
15 INPUT P7
17 PRINT  "A IS THE RATIO SEMI-MAJOR AXIS/GRID SIZE = L/(2S)"
18 PRINT  "INPUT EXPECTED TARGET LENGTH IN METERS"
19 INPUT L(2)
20 PRINT  "INPUT EXPECTED SHAPE RATIO, R = B/L, OF TARGET"
21 INPUT Z3
22 PRINT "SHAPE RATIO = ", Z3
23 PRINT "PROBABILITY     GRID SIZE     DRILLING COST  EFFICIENCY"
24 PRINT "  LEVEL         IN METERS     IN $1000'S/SQ.KM  FUNCTION"
25 E3(1) = 25
26 E4(1) = 200
27 E3(2) = 250
28 E4(2) = 1000
29 E5(1) = 25
30 E5(2) = 50
31 FOR W5 = 35 TO 55 STEP 5
32 PRINT "ANGLE = ", W5, "DEGREES"
33 W4 = W5*3.14159296/180
34 FOR E(6) = 1 TO 2
35 FOR W = E3(E6) TO E4(E6) STEP E5(E6)
36 FOR J7 = 2 TO 2
37 A = L(J7)/2/W
38 J6 = J7
39 K(1) = Z3 + 300/L(J7)/TAN(W4)
39 B = A*K(1)
40 N4 = 140
41 C7 = 3*C5/SIN(W5*3.14159296/180)
46 C2 = 0
47 X3 = 0
48 Y3 = 0
49 D2 = 0
50 H  = ATN(1.00000000/900
51 K  = K(1)
52 N5 = N4 + 100
53 FOR I5 = N4 TO N5
54 C3 = (COS(I5*H))^2
55 D3 = (SIN(I5*H))^2
56 C2 = (1- (1-K*K)*C3)^0.5
57 D2 = (1- (1-K*K)*D3)^0.5
58 C  = C2
60 D  = D2
```

```
61  IF A <= 0.5 THEN 63
62  GO TO 68
63  P1 = A*B*3.14159296
64  P2 = 0
65  P0 = 1 - P1
66  P3(J6) = P1
67  GO TO 171
68  IF B > C/2 THEN 70
69  GO TO 72
70  A1 = 2*(B*B*ATN(((4*B*B-C*C)^.5)/C)-0.5*C*((B*B-0.25*C*C)^0.5))
71  GO TO 73
72  A1 = 0
73  IF B > D/2 THEN 75
74  GO TO 77
75  A2 = 2*(B*B*ATN(((4*B*B-D*D)^.5)/D)-0.5*D*((B*B-0.25*D*D)^0.5))
76  GO TO 78
77  A2 = 0
78  IF A <= (2^0.5)/2 THEN 80
79  GO TO 85
80  P2 = (A1 + A2)/K
81  P1 = 3.14159296*B*B/K - 2*P2
82  P0 = 1 - P1 - P2
83  P3(J6) = 1 - P0
84  GO TO 171
85  Y9 = D*D*C*C - K*K
86  Y9 = (ABS(Y9))^0.5
87  IF Y9 < .00000001 THEN 92
88  T = ATN(K/Y9)
89  GO TO 90
90  IF I5 = 0 THEN 92
91  GO TO 96
92  J1 = (C*C + D*D)^0.5
93  J0 = 5
94  R0 = J1/2
95  GO TO 111
96  I = 1 + INT(D*COS(T)/C)
97  IF I = 1 THEN 99
98  GO TO 103
99  J1 = (C*C + D*D - 2*C*D*COS(T))^0.5
100 J0 = 5
101 R0 = J1/2/SIN(T)
102 GO TO 106
103 J1 = (I*I*C*C + D*D - 2*I*C*D*COS(T))^0.5
104 J0 = (((I-1)^2)*C*C + D*D - 2*(I-1)*C*D*COS(T))^0.5
105 R0 = J1*J0/2/D/SIN(T)
106 IF B >= R0 THEN 108
107 GO TO 111
108 P3(J6) = 1
109 P0 = 0
110 GO TO 171
111 IF B > J1/2 THEN 113
112 GO TO 115
113 A3 = 2*(B*B*ATN(((4*B*B-J1*J1)^.5)/J1)-J1/2*((B*B-
        J1*J1/4)^0.5))
114 GO TO 116
115 A3 = 0
116 IF B > J0/2 THEN 118
117 GO TO 120
```

```
118 A4 = 2*(B*B*ATN(((4*B*B-J0*J0)^.5)/J0)-J0/2*((B*B-
       J0*J0/4)^0.5))
119 GO TO 121
120 A4 = 0
121 M = 1 + INT(2*D*COS(T)/C)
122 IF M=1 THEN 124
123 GO TO 127
124 M1 = (C*C + D*D - 2*C*D*COS(T))^0.5
125 M0 = 5
126 GO TO 129
127 M1 = (M*M*C*C + 4*D*D - 4*M*C*D*COS(T))^0.5
128 M0 = ((M-1)*(M-1)*C*C + 4*D*D - 4*(M-1)*C*D*COS(T))^0.5
129 IF J1 = M1/2 THEN 139
130 IF J0 = M1/2 THEN 139
131 IF J1 = M0/2 THEN 139
132 IF J0 = M0/2 THEN 139
133 IF B > M1/2 THEN 136
134 A5 = 0
135 GO TO 142
136 A5 = 2*(B*B*ATN(((4*B*B-M1*M1)^.5)/M1)-M1/2*((B*B-
       M1*M1/4)^0.5))
137 GO TO 142
138 A5 = 0
139 A6 = 0
140 GO TO 145
141 IF B > M0/2 THEN 144
142 A6 = 0
143 GO TO 145
144 A6 = 2*(B*B*ATN(((4*B*B-M0*M0)^.5)/M0)-M0/2*((B*B-
       M0*M0/4)^0.5))
145 IF B > J1 THEN 148
146 A7 = 0
147 GO TO 149
148 A7 = 2*(B*B*ATN(((4*B*B-J1*J1)^.5)/J1)-J1/2*((B*B-
       J1*J1/4)^0.5))
149 IF B > J0 THEN 152
150 A8 = 0
151 GO TO 153
152 A8 = 2*(B*B*ATN(((4*B*B-J0*J0)^.5)/J0)-J0/2*((B*B-
       J0*J0/4)^0.5))
153 IF B > C THEN 156
154 A9 = 0
155 GO TO 157
156 A9 = 2*(B*B*ATN(((B*B-C*C)^.5)/C)-C/2*((B*B-C*C)^0.5))
157 IF B > D THEN 160
158 B9 = 0
159 GO TO 161
160 B9 = 2*(B*B*ATN(((B*B-D*D)^.5)/D)-D/2*((B*B-D*D)^0.5))
161 P2 = (A1+A2+A3+A4+A5+A6-A7-A8-A9-B9)/K
162 P1 = 3.14159296*B*B/K -(A7+A8+A9+B9)/K - 2*P2
163 P0 = 1 - P1 - P2
164 P3(J6) = 1 -P0
165 IF P3(J6) > 1 THEN 170
166 IF P3(J6) < 0 THEN 168
167 GO TO 169
168 P3(J6) = 0
```

```
169 GO TO 171
170 P3(J6) = 1
171 Q3(I5,J6) = P3(J6)
172 NEXT I5
173 P3(J7) = Q3(N4,J6) + Q3(N5,J6)
174 FOR I = N4 + 1 TO N5 - 1 STEP 2
175 P3(J7) = P3(J7) + 4*Q3(I,J6)
176 NEXT I
177 FOR J = N4 + 2 TO N5 - 2 STEP 2
178 P3(J7) = P3(J7) + 2*Q3(J,J6)
179 NEXT J
180 P3(J7) = P3(J7)/300
181 IF P3(J7) > 1 THEN 183
182 GO TO 184
183 P3(J7) = 1
184 IF (P3(J7)-E9) <= 1 THEN 188
187 GO TO 194
188 V8 = (1609/W +1)*(1609/W + 1.02)
189 V7 = V8*6/1.60934/1.60934/SIN(W4)/C7
190 V7 = 328.084/300*V7
191 V9 = P3(J7)/P7-V7
192 V6 = V7*C7
193:     #.##        #####    ####.##    ####.##
194 PRINT USING 193,P3(J7),W,V6,V9
195 NEXT J7
196 NEXT W
197 NEXT E6
198 NEXT W5
199 STOP
200 END
```

APPENDIX 5

FACTOR-TREND computer programs

5.1. Listings for FACTOR computer program (SPSS package)

```
1$$S,T(:)
2$:IDENT:MAT-SA,1 WIGNALL
3$:SELECT:SPSS/SPSS
4RUN NAME::ALLIGATOR RIVER PROJECT
5PAGESIZE:NOEJECT
6VARIABLE LIST:GRP,X1,X2,X3,X4,X5,X6
7COMPUTE:X7=X5/X6
8FACTOR::VARIABLES=X3,X4,X5,X6,X7
9STATISTICS:ALL
10READ INPUT DATA
22   0.5  1.6  0    0    0    0
23   2.5  1.8  1.44 0    0    0
         .    .    .
         .    .    .
         .    .    .
142  17.2 1.3  1.7  0    0    0
143END INPUT DATA
148FINISH
149END
```

```
1    DIMENSION BI(100),AA(7,7),VARB(10),FS(7,7),PF(7,7)
2    DIMENSION Y1(20),A(20),Y(30),X(1000),SRES(20),STER(20)
3    DIMENSION YBAR(20),YSTD(20),B(7,7),P(7,7),B1(10,1)
4    DIMENSION IWORK(100),JWORK(100),AX(10),RES(10),SY(10)
5    DIMENSION YY(7,7),NA(36),AV(7,7),SA(10),SAA(10),AY(7,7)
6    READ (5,1) L,M
9  1 FORMAT(V)
10 3 WRITE(6,4) L,M
13 4 FORMAT(2H  ,2I6,' AL2O3,THICKNESS,FACTOR;V.P. BAUXITE`)
14   II = 0
15   N = 0
16   M = M+1
17   DO 50 I=1,M
18   DO 50 J=1,L
19 50 AY(I,J)=0.
20   DO 5 I=1,M
21   SA(I)=0.
22   SAA(I)=0.
23   DO 5 J=1,M
24 5 AA(I,J)=0
25   DO 6 I=1,L
26   SY(I)=0.
27   DO 6 J=1,L
28 6 YY(I,J)=0.
29 7 READ (5,8) A(2),A(3),(Y(I),I=1,3)
30 8 FORMAT(V)
31   IF (Y(1)) 100,7,100
32 100 CONTINUE
34 9 N=N+1
35   A(I)=1.
36   Y(4)=-0.164*Y(1)+0.375*Y(2)+0.411*Y(3)
37   L=4
38   DO 10 I=1,M
39   DO 10 J=1,L
40 10 AY(I,J)=AY(I,J)+A(I)*Y(J)
41   DO 11 I=1,L
42 11 SY(I)=SY(I)+Y(I)
43   DO 12 I=1,M
44   SA(I)=SA(I)+A(I)
45   SAA(I)= SAA(I)+A(I)*A(I)
46   DO 12 J=1,M
47 12 AA(I,J)=AA(I,J)+A(I)*A(J)
48   DO 14 I=1,L
49   DO 14 J=1,L
50 14 YY(I,J)=YY(I,J)+Y(I)*Y(J)
51   GO TO 7
52 90 CONTINUE
53   EN=N
54   DO 15 I=1,L
55 15 YBAR(I)=SY(I)/EN
56   DO 16 I=1,L
57   DO 16 J=1,L
58 16 YY(I,J)=YY(I,J)-EN*YBAR(I)*YBAR(J)
59   MSQ=M*M
60   K=0
61   DO 17 I=1,M
62   DO 17 J=1,M
63   K=K+1
64   BI(K)=AA(I,J)
65 17 CONTINUE
```

```
66 CALL MINVI(BI,M,MSQ,DET,IWORK,JWORK)
67 WRITE (6,18)
68 18 FORMAT(2H ,' INVERSE =')
69 K=0
70 DO 19 I=1,M
71 DO 19 J=1,M
72 K=K+1
73 AA(I,J)=BI(K)
74 WRITE (6,20) AA(I,J)
75 19 CONTINUE
76 20 FORMAT(F10.4)
77 WRITE (6,21) L,M,N
78 21 FORMAT(2H ,'L = ',I5,' M = ',I5,' N = ',I6)
79 DO 22 I=1,L
80 YSTD(I)=SQRT(YY(I,I)/(EN-1.))
81 22 WRITE(6,23) I,YBAR(I),YSTD(I)
82 23 FORMAT(2H ,' MEAN & STD. DEV. ON VARIATE ',I2,'=',2F10.4)
83 DO 24 J=1,L
84 DO 24 I=1,L
85 B(I,J)=0.
86 DO 24 K=1,M
87 B(I,J)=B(I,J)+AA(I,K)*AY(K,J)
88 24 CONTINUE
89 WRITE (6,25)
90 25 FORMAT(2H ,'REGRESSION COEFF. ON VAR. Y1,Y2,... =')
91 DO 26 J=1,L
92 26 WRITE(6,30) J,(B(I,J),I=1,M)
93 30 FORMAT(2H ,I6,8F10.4/8F10.4)
94 DO 31 I=1,L
95 DO 31 J=1,L
96 31 P(I,J)=YY(I,J)/((YY(I,I)*YY(J,J))**0.5)
97 DO 32 I=1,L
98 32 WRITE(6,33) I,(P(I,J)=J=1,L)
99 33 FORMAT(2H ,'CORRELATION ON VARIATE',I2,'=',8F10.5)
100 WRITE (6,34)
101 34 FORMAT(2H ,`LOC.A#  E.  N.   TREND ON Y1,Y2,...')
102 DO 36 I=1,L
103 SRES(I)=0.
104 36 CONTINUE
105 N7=0
106 DO 70 I=1,10
107 DO 70 J=1,9
108 X6=I-1
109 X7=1-J
110 39 A(1)=1.
111 N7=N7+1
112 DO 40 K=2,4
113 40 Y1(K)=B(1,K)+B(2,K)*X6+B(3,K)*X7
114 533 FORMAT(2H , 2F10.3)
116 EM=M
117 ENN=EN-EM
118 WRITE (6,45) X6,X7,Y1(3),Y1(2),Y1(4)
119 45 FORMAT(2H ,2F4.0,3F10.1)
120 70 CONTINUE
121 WRITE (6,79)
122 79 FORMAT(2H ,`JOB COMPLETE')
123 STOP
124 END
218 SUBROUTINE MINVI(A,N,MSQ,D,L,M)
219 INTEGER L(NSQ),M(NSQ)
220 REAL A(NSQ)
221 D=1.
```

```
222 NK=-N
223 DO 80 K=1,N
224 NK=NK+N
225 L(K)=K
226 M(K)=K
227 KK=NK+K
229 DO 20 J=K,N
230 IZ=N*(J-1)
231 DO 20 I=K,N
232 IZ=IZ+1
233 10 IF (DABS(BIGA)-DABS(A(IJ))) 15,20,20
234 15 BIGA=A(IJ)
235 L(K)=I
236 M(K)=J
237 20 CONTINUE
238 J=L(K)
239 IF (J-K) 35,35,25
240 25 KI=K-N
241 DO 30 I=1,N
242 KI=KI+N
243 HOLD=-A(KI)
244 JI=KI-K+J
245 A(KI)=A(JI)
246 30 A(JI)=HOLD
247 35 I=M(K)
248 IF (I-K) 45,45,38
249 38 JP=N*(I-1)
250 DO 40 J=1,N
251 JK=NK+J
252 JI=JP + J
253 HOLD=-A(JK)
254 A(JK)=A(JI)
255 40 A(JI)=HOLD
256 IF (BIGA) 48,46,48
257 46 D=0.
258 RETURN
259 48 DO 55 I=1,N
260 IF (I-K) 50,55,50
261 50 IK=NK+I
262 A(IK)=A(IK)/(-BIGA)
263 55 CONTINUE
264 DO 65 I=1,N
265 IK = NK+I
266 HOLD=A(IK)
267 IJ=I-N
268 DO 65 J=1,N
269 IJ=IJ+N
270 IF (I-K) 60,65,60
271 60 IF (J-K) 62,65,62
272 62 KJ=IJ-I+K
273 A(IJ)=HOLD*A(KJ)+A(IJ)
274 65 CONTINUE
275 KJ=K-N
276 DO 75 J=1,N
277 KJ=KJ+N
278 IF (J-K) 70,75,70
279 70 A(KJ)=A(KJ)/BIGA
280 75 CONTINUE
281 D=D*BIGA
282 A(KK)=1./BIGA
283 80 CONTINUE
```

```
284 K=N
285 100 K=K-1
286 IF (K) 150,150,105
287 105 I=L(K)
288 IF(I-K) 120,120,108
289 108 JQ=N*(K-1)
290 JR=N*(I-1)
291 DO 110 J=1,N
292 JK = JQ+J
293 HOLD=A(JK)
294 JI=JR+J
295 A(JK)=-A(JI)
296 110 A(JI)=HOLD
297 120 J=M(K)
298 IF (J-K) 100,100,125
299 125 KI=K-N
300 DO 130 I=1,N
301 KI=KI+N
302 HOLD=A(KI)
303 JI=KI-K+J
304 A(KI)=-A(JI)
305 130 A(JI)=HOLD
306 GO TO 100
307 150 RETURN
308 END
```

5.3. Listings for residual trend factor analysis

```
1     DIMENSION BI(100),AA(7,7),VARB(10),FS(7,7),PF(7,7)
2     DIMENSION Y1(20),A(20),Y(30),X(1000),SRES(20),STER(20)
3     DIMENSION YBAR(20),YSTD(20),B(7,7),P(7,7),B1(10,1)
4     DIMENSION IWORK(100),JWORK(100),AX(10),RES(10),SY(10)
5     DIMENSION YY(7,7),NA(36),AV(7,7),SA(10),SAA(10),AY(7,7)
6     READ (5,1) L,M
9   1 FORMAT(V)
10  3 WRITE(6,4) L,M
13  4 FORMAT(2H   ,2I6,`U, Au: ALLIGATOR RIVERS, N.T., AUSTRALIA')
14    II = 0
15    N = 0
16    M = M+1
17    DO 50 I=1,M
18    DO 50 J=1,L
19 50 AY(I,J)=0.
20    DO 5 I=1,M
21    SA(I)=0.
22    SAA(I)=0.
23    DO 5 J=1,M
24  5 AA(I,J)=0
25    DO 6 I=1,L
26    SY(I)=0.
27    DO 6 J=1,L
28  6 YY(I,J)=0.
29  7 READ (5,8) A(2),A(3),(Y(I),I=1,4)
30  8 FORMAT(V)
31    IF (Y(2)) 9,90,9
32  9 N=N+1
33    A(I)=1.
34    Y(5)=Y(3)/Y(4)
35    Y(6)=0.07*Y(1)+0.7*Y(2)+0.2*Y(3)+0.2*Y(4)-0.2*Y(5)
36    Y(7)=0.2*Y(1)-0.5*Y(2)+0.9*Y(3)-0.1*Y(4)+0.2*Y(5)
37    L=7
38    DO 10 I=1,M
39    DO 10 J=1,L
40 10 AY(I,J)=AY(I,J)+A(I)*Y(J)
41    DO 11 I=1,L
42 11 SY(I)=SY(I)+Y(I)
43    DO 12 I=1,M
44    SA(I)=SA(I)+A(I)
45    SAA(I)= SAA(I)+A(I)*A(I)
46    DO 12 J=1,M
47 12 AA(I,J)=AA(I,J)+A(I)*A(J)
48    DO 14 I=1,L
49    DO 14 J=1,L
50 14 YY(I,J)=YY(I,J)+Y(I)*Y(J)
51    GO TO 7
52 90 CONTINUE
53    EN=N
54    DO 15 I=1,L
55 15 YBAR(I)=SY(I)/EN
56    DO 16 I=1,L
57    DO 16 J=1,L
58 16 YY(I,J)=YY(I,J)-EN*YBAR(I)*YBAR(J)
59    MSQ=M*M
60    K=0
61    DO 17 I=1,M
62    DO 17 J=1,M
63    K=K+1
64    BI(K)=AA(I,J)
```

```
65 17 CONTINUE
66 CALL MINVI(BI,M,MSQ,DET,IWORK,JWORK)
67 WRITE (6,18)
68 18 FORMAT(2H  ,' INVERSE =')
69 K=0
70 DO 19 I=1,M
71 DO 19 J=1,M
72 K=K+1
73 AA(I,J)=BI(K)
74 WRITE (6,20) AA(I,J)
75 19 CONTINUE
76 20 FORMAT(F10.4)
77 WRITE (6,21) L,M,N
78 21 FORMAT(2H  ,'L = ',I5,'  M = ',I5,'  N = ',I6)
79 DO 22 I=1,L
80 YSTD(I)=SQRT(YY(I,I)/(EN-1.))
81 22 WRITE(6,23) I,YBAR(I),YSTD(I)
82 23 FORMAT(2H  ,' MEAN & STD. DEV. ON VARIATE ',I2,'=',2F10.4)
83 DO 24 J=1,L
84 DO 24 I=1,L
85 B(I,J)=0.
86 DO 24 K=1,M
87 B(I,J)=B(I,J)+AA(I,K)*AY(K,J)
88 24 CONTINUE
89 WRITE (6,25)
90 25 FORMAT(2H  ,'REGRESSION COEFF. ON VAR. Y1,Y2,... =')
91 DO 26 J=1,L
92 26 WRITE(6,30) J,(B(I,J),I=1,M)
93 30 FORMAT(2H  ,I6,8F10.4/8F10.4)
94 DO 31 I=1,L
95 DO 31 J=1,L
96 31 P(I,J)=YY(I,J)/((YY(I,I)*YY(J,J))**0.5)
97 DO 32 I=1,L
98 32 WRITE(6,33) I,(P(I,J)=J=1,L)
99 33 FORMAT(2H  ,'CORRELATION ON VARIATE',I2,'=',8F10.5)
100 WRITE (6,34)
101 34 FORMAT(2H  ,`LOC.A#   E.  N.    TREND ON Y1,Y2,...')
102 DO 36 I=1,L
103 SRES(I)=0.
104 36 CONTINUE
105 N7=0
106 37 READ(5,38) A(2),A(3),(Y(I),I=1,4)
107 38 FORMAT (V)
108 IF (Y(2)) 39,95,39
109 39 CONTINUE
110 A(I)=1.
111 N7=N7+1
112 Y(5)=Y(3)/Y(4)
113 Y(6)=0.07*Y(1)+0.7*Y(2)+0.2*Y(3)+0.2*Y(4)-0.2*Y(5)
114 Y(7)=0.2*Y(1)-0.5*Y(2)+0.9*Y(3)-0.1*Y(4)+0.2*Y(5)
115 L=7
112 DO 40 K=1,L
113 Y1(K)=0.
114 DO 40 J=1,M
113 40 Y1(K)=Y1(K)+B(J,K)*A(J)
114 533 FORMAT(2H  , 2F10.3)
115 DO 41 K=1,L
116 Y(K)=Y(K)-Y1(K)
117 41 CONTINUE
118 DO 43 I=1,L
119 43 SRES(I)=SRES(I)+Y(I)*Y(I)
```

```
120 GO TO 37
121 95 CONTINUE
126 EM=M
127 ENN=EN-EM
128 DO 44 1=1,L
129 DO 44 J=1,M
130 VARB(I)=SRES(I)*AA(J,I)/ENN
131 FS(I,J)=B(J,I)*B(J,I)/VARB(I)
132 PF(I,J)=PROBF(A(1),ENN,FS(I,J))
138 WRITE (6,45) I,J,FS(I,J),A(1),ENN,PF(I,J)
139 45 FORMAT(3H Y(,I2,`COEFF.,B(   ',I21,` F=  ',F7.4,`D.F.',
140 & F2.0,F4.0,122` SIG. LEVEL', F10.5)
141 44 CONTINUE
142 DO 46 I=1,L
143 46 STER(I)=SQRT(SRES(I)/(EN-3.))
144 WRITE (6,47) (STER(I),I=1,L)
145 47 FORMAT(2H  ,` STAND. ERROR = ',10F8.1)
150 DO 48 J=1,L
151 STER(J)=1.64*STER(J)
152 48 CONTINUE
153 WRITE (6,59)
154 59 FORMAT(2H  ,` LOCATIONS ABOVE THE TREND   ')
155 N8=0
156 61 READ(5,62) A(2),A(3),(Y(I),I=1,4)
157 62 FORMAT (V)
158 IF (Y(1)) 99,99,63
159 63 CONTINUE
160 A(I)=1.
161 N8=N8+1
162 Y(5)=Y(3)/Y(4)
163 Y(6)=0.07*Y(1)+0.7*Y(2)+0.2*Y(3)+0.2*Y(4)-0.2*Y(5)
164 Y(7)=0.2*Y(1)-0.5*Y(2)+0.9*Y(3)-0.1*Y(4)+0.2*Y(5)
165 L=7
166 DO 64 I=1,L
167 Y1(K)=0.
168 DO 64 J=1,M
169 64 Y1(K)=Y1(K)+B(J,K)*A(J)
170 DO 65 K=1,L
171 Y(K)=Y(K)-Y1(K)
172 65 CONTINUE
173 DO 70 I=1,L
174 IF(Y(I)-STER(I)) 67,66,66
175 66 CONTINUE
179 WRITE(6,68) N8,A(2),A(3), (Y(K),K=1,L)
180 GO TO 67
181 70 CONTINUE
182 68 FORMAT(2H  , `RES = ',I4,10F8.3)
183 67 CONTINUE
184 GO TO 61
185 99 CONTINUE
186 100 CONTINUE
187 STOP
188 END
218 SUBROUTINE MINVI(A,N,MSQ,D,L,M)
219 INTEGER L(NSQ),M(NSQ)
220 REAL A(NSQ)
221 D=1.
222 NK=-N
223 DO 80 K=1,N
224 NK=NK+N
225 L(K)=K
```

```
226 M(K)=K
227 KK=NK+K
229 DO 20 J=K,N
230 IZ=N*(J-1)
231 DO 20 I=K,N
232 IZ=IZ+1
233 10 IF (DABS(BIGA)-DABS(A(IJ))) 15,20,20
234 15 BIGA=A(IJ)
235 L(K)=I
236 M(K)=J
237 20 CONTINUE
238 J=L(K)
239 IF (J-K) 35,35,25
240 25 KI=K-N
241 DO 30 I=1,N
242 KI=KI+N
243 HOLD=-A(KI)
244 JI=KI-K+J
245 A(KI)=A(JI)
246 30 A(JI)=HOLD
247 35 I=M(K)
248 IF (I-K) 45,45,38
249 38 JP=N*(I-1)
250 DO 40 J=1,N
251 JK=NK+J
252 JI=JP + J
253 HOLD=-A(JK)
254 A(JK)=A(JI)
255 40 A(JI)=HOLD
256 IF (BIGA) 48,46,48
257 46 D=0.
258 RETURN
259 48 DO 55 I=1,N
260 IF (I-K) 50,55,50
261 50 IK=NK+I
262 A(IK)=A(IK)/(-BIGA)
263 55 CONTINUE
264 DO 65 I=1,N
265 IK = NK+I
266 HOLD=A(IK)
267 IJ=I-N
268 DO 65 J=1,N
269 IJ=IJ+N
270 IF (I-K) 60,65,60
271 60 IF (J-K) 62,65,62
272 62 KJ=IJ-I+K
273 A(IJ)=HOLD*A(KJ)+A(IJ)
274 65 CONTINUE
275 KJ=K-N
276 DO 75 J=1,N
277 KJ=KJ+N
278 IF (J-K) 70,75,70
279 70 A(KJ)=A(KJ)/BIGA
280 75 CONTINUE
281 D=D*BIGA
282 A(KK)=1./BIGA
283 80 CONTINUE
284 K=N
285 100 K=K-1
286 IF (K) 150,150,105
287 105 I=L(K)
288 IF(I-K) 120,120,108
```

```
289 108 JQ=N*(K-1)
290 JR=N*(I-1)
291 DO 110 J=1,N
292 JK = JQ+J
293 HOLD=A(JK)
294 JI=JR+J
295 A(JK)=-A(JI)
296 110 A(JI)=HOLD
297 120 J=M(K)
298 IF (J-K) 100,100,125
299 125 KI=K-N
300 DO 130 I=1,N
301 KI=KI+N
302 HOLD=A(KI)
303 JI=KI-K+J
304 A(KI)=-A(JI)
305 130 A(JI)=HOLD
306 GO TO 100
307 150 RETURN
308 END
392 FUNCTION(DFNUM,DFDEN,FRATIO)
393 PROBF=1.
394 IF(DFNUM*DFDEN*FRATIO) 17,16,17
395 16 RETURN
396 17 CONTINUE
397 IF(FRATIO -1.) 18,19,19
398 18 GO TO 1
399 19 CONTINUE
400 A=DFNUM
401 B=DFDEN
402 F=ABS(FRATIO)
403 GO TO 2
404 1 A=DFDEN
405 B=DFNUM
406 F=1./ABS(FRATIO)
407 2 AA=2./9./A
408 BB=2./9./B
409 Z=ABS(((1.-BB)*F**(1./3.)-1.+AA)/SQRT(BB*F**(2./3.)+AA))
410 IF (B-4.) 83,84,84
411 83 Z=Z*(1.+0.08*Z**4/B**3)
412 84 CONTINUE
413 PROBF=0.5/(1.+Z*(.196+Z*(.115+Z*(.00034+Z*.019527))))**4
414 IF (FRATIO-1.) 3,4,4
415 PROBF= 1. - PROBF
416 4 RETURN
```

LINEAR PROGRAMMING: dual simplex computer program

Listings for LINEAR PROGRAM computer program (BASIC)

```
 2 DIM A(50,50), B(50), C(50), U(50)
 5 PRINT "L IS THE NUMBER OF CONSTRAINTS"
 7 PRINT "N IS THE NUMBER OF VARIATES (INCLUDING SLACK)"
 8 INPUT L,N
10 M = L + 1
16 PRINT" ENTER CONSTRAINTS AND OBJECTIVE COEFFICIENTS"
18 FOR I = 1 TO M: FOR J = 1 TO N: INPUT A(I,J): NEXT J
19 FOR K1= 1 TO N: LPRINT A(I,K1)" ";: NEXT K1: NEXT I
20 PRINT "INPUT INITIAL BASIC SOLUTION"
24 FOR I = 1 TO M: INPUT A(I,0): NEXT I
28 LPRINT "INITIAL SOLUTION ="; : FOR I = 1 TO M: LPRINT A(I,0)"  ";
30 NEXT I
32 PRINT "INPUT INITIAL BASIS NOS.": FOR I = 1 TO L: INPUT B(I):NEXT I
34 FOR I = 1 TO L: C(B(I)) = 1: NEXT I
36 LPRINT "TABLEAUX"
44 LPRINT "ITERATION"K
46 GO SUB 700
48 H = 0.0000001
50 REM FIND SMALLEST COEFF. IN OBJECTIVE ROW
52 MI = -H
54 S = 0
56 PV = 0
58 FOR J = 1 TO N
60 IF C(J) = 1 THEN 66
62 IF A(M,J)>=MI THEN 66
64 MI = A(M,J)
65 S = J
66 NEXT J
70 IF S = 0 THEN 132
72 REM SEARCH FOR LEAVING VAR
74 MI = 10000000
76 R = 0
78 FOR I = 1 TO L
80 IF A(I, S) <= H THEN 86
82 RT = A(I,0)/A(I,S): IF RT>=MI THEN 86
84 R = I: MI = A(I,0)/A(I,S)
86 NEXT I
88 REM IF I=0 THEN UNBOUNDED
90 IF R = 0 THEN 152
92 LPRINT "PIVOT IS (";R;",";S")"
94 REM DIVIDE BY PIVOT
96 PV = A(R, S)
98  FOR I = 0 TO N
100 A(R,I) = A(R,I)/PV
102 NEXT I
104 FOR I = 1 TO M
106 U(I) = A(I,S): NEXT I
108 FOR I = 1 TO M
110 IF I = R THEN 118
112 FOR J = 0 TO N
114 A(I, J) = A(I,J) - U(I)*A(R, J)
116 NEXT J
118 NEXT I
120 C(B(R)) = 0: C(S) = 1: B(R) = S
122 K = K + 1
```

```
124 LPRINT "ITERATION"K
126 GO SUB 700
128 GO TO 52
130 REM DUAL SIMPLEX FIND LEAVING VARIABLE
132 MI = -H
134 R = 0
136 FOR I = 1 TO L
138 IF A(I, 0) > MI THEN 144
140 MI = A(I, 0)
142 R = I
144 NEXT I
148 IF R = 0  THEN 188
150 MI = 1000000
152 S = 0
154 FOR J = 1 TO N
156 IF C(J) = 1 THEN 163
158 IF A(R,J) >= -H THEN 163
159 R1 = ABS(A(M,J)/A(R,J))
160 IF RT >= MI THEN 163
161 S = J
162 MI = ABS(A(M,J)/A(R,J)
163 NEXT J
164 IF S = 0 THEN 168
166 GO TO 92
168 LPRINT "NO FEASIBLE SOLUTION"
180 GO SUB 700
182 GO TO 500
188 LPRINT" FINAL SOLUTION"
195 LPRINT "BASIS    VALUE"
200 FOR I = 1 TO L
202 LPRINT" ";I;" ";B(I);
264 P1 = A(I,0): LPRINT P1;: LPRINT" "
268 NEXT I
270 LPRINT "MIN Z =";-A(M,0)
274 GO SUB 700
500 END
700  LPRINT "BASIS     VALUE";
710  FOR J = 1 TO N:LPRINT "      X"J" ";:NEXT J
720  LPRINT " "
730  FOR I = 1 TO M
740  IF I = M THEN LPRINT "-Z";: GO TO 770
750  LPRINT B(I);
770  FOR J = 0 TO N
800  P1=A(I,J): LPRINT USING "#####.##";P1;
900  NEXT J
910  LPRINT " "
920  NEXT I:LPRINT " "
950  RETURN
```

*N.B. The input for the Linear Programming example in the text is:

4, 7

and the following are entered 1 coefficient at a time:

```
-40  -120  -60  1  0  0  0  -3  -4  -3  0  1  0  0  -1
  0    0    0  0  1  0  10  15  8  0  0  0  0  1  10  15
  8    0    0  0  0.
```

APPENDIX 7

CLASSIFICATION computer programs

7.1. Listings for POPMIX computer program (FORTRAN IV)

```
1 INTEGER WN
2 DIMENSION EC(20),EA(20),EB(20),WN(20),U(20),V(20),X(20)
3 DIMENSION Y(20),Z(20),ED(20),R(20),S(20),T(20),XB(20)
4 DIMENSION XC(20),XD(20),XE(20),XK(20)
5 40 READ (5,1) N
6 IF (N) 53,53,34
7 34 CONTINUE
8 N=N+3
9 1   FORMAT (V)
10 DO 2 I = 1, N
11 WN(I) = 0
12 Y(I) = 0
13 Z(I) = 0
14 U(I) = 0
15 2 V(I) = 0
16 WRITE (6,3) N
17 3 FORMAT(2H   ,I2,'ALLIGATOR RIVERS PROJECT')
18 WRITE (6,101)
19 101 FORMAT (2H  ,` FACTOR SCORES ')
20 4 N = N-3
21 READ (5,5) (X(I), I = 1,N)
22 N = N - 3
23 5 FORMAT (V)
24 IF (X(1) - 999) 6, 99, 6
25 6 CONTINUE
26 IF (X(3) - 0.0001) 8, 8, 7
27 7 X(7) = X(5)/X(6)
28 X(8) = 1.7*X(3) - 1.4*X(4) + 0.05*X(5) - 0.14*X(6) - 0.02*X(7)
29 X(9) = 1.1*X(3) - 0.3*X(4) - 0.84*X(5) + 0.46*X(6) - 0.03*X(7)
31 DO 11 I=1,N
32 WN(I) = WN(I) + 1
33 Z(I)  = X(I) + Z(I)
34 Y(I)  = Y(I) + X(I)*X(I)
35 U(I)  = U(I) + X(I)*X(I)*X(I)
36 11 V(I) = V(I) + (X(I))^4
37 WRITE (6,59) WN(3), X(7), X(8), X(9)
38 59 FORMAT(3H    ,I5, 3F10.2)
39 GO TO 4
40 8 CONTINUE
41 WRITE (6,14) (WN(I), I = 1,N)
42 14 FORMAT(16I6)
43 DO 15 I= 7,9
44 EA(I) = Z(I)/WN(I)
45 EB(I) = Y(I)/WN(I)
46 EC(I) = U(I)/WN(I)
47 ED(I) = V(I)/WN(I)
48 R(I)  = EB(I) - EA(I)*EA(I)
49 S(I)  = EC(I) - 3.*EA(I)*EB(I)-2.*EA(I)*EA(I)*EA(I)
50 T(I)  = ED(I) - 4.*EA(I)*EC(I)+6.*EA(I)*EA(I)*EB(I)-3.*(EA(I))^4
52 XB(I) = WN(I)*R(I)/(WN(I)-1)
53 XC(I) = WN(I)*WN(I)*S(I)/(WN(I)-1)/(WN(I)-2)
54 XD(I) = WN(I)*WN(I)*((WN(I)+1)*T(I)-3.*(WN(I)-1)*R(I))/(WN(I)-1)/
55 &   (WN(I)-2)/(WN(I)-3)
56 15 XE(I) = XC(I)*XC(I)
57 WRITE (6,16) (EA(I), XB(I), XC(I), XD(I), I = 7,9)
```

```
58 16 FORMAT (2H   ,`MEAN, K2, K3, K4 STATISTICS =',/7F16.7/7F16.7/
59 & 7F16.7/7F16.7/7F16.7)
60 DO 17 I= 7,9
61 AA = -XB(I)
62 EE = EA(I)
63 BB = XB(I)
64 CC = XC(I)
65 DD = XD(I)
66 XX = XE(I)
67 18 CONTINUE
68 YB=(-AA^3-0.5*DD*AA-0.5*XX)/(3.*AA*AA+0.5*DD)
69 AA = AA + YB
70 IF (ABS(YB) - 0.00001) 19, 19, 18
71 19 DZ = -4.*AA + XX/AA/AA
72 DY = (-CC/AA - DZ^0.5)*.5
73 XEE = DY + EE
74 FF = -DY - CC/AA
75 XG = FF + EE
76 PQ = FF/(FF-DY)
77 WRITE (6,22) PQ
78 22 FORMAT(30H PROPORTION IN POPULATION 1=       ,F10.8)
79 VAR = AA + BB
80 VAR = ABS(VAR)
81 STD = VAR^0.5
82 WRITE (6,23) STD
83 23 FORMAT(30H COMMON STANDARD DEVIATION =    ,F10.4)
84 WRITE (6,24) XEE
85 24 FORMAT(30H MEAN OF POPULATION 1 =    ,F16.4)
86 WRITE (6,26) XG
87 26 FORMAT(30H MEAN OF POPULATION 2 =    ,F16.4)
88 17 CONTINUE
89 GO TO 40
90 99 CONTINUE
91 53 STOP
92 END
```

```
1  DIMENSION EE(10),H(6,6),C(10),XXX(10),R(6,6),NL(100),ML(100)
2  DIMENSION EL(6,6,4),XX(10),A(10),YY(10),WN(10),EF(6,6)
3  DIMENSION T(1,1),P(6,6),EM(6,6,4),NAME(39),YYY(10),SS(10)
4  DIMENSION IND(16),X(10),B(10),C(10),RD(10),XD(10),TV(100)
5  DIMENSION S(6,6),AV(6,6),EX(10),XV(6,6),BV(6,6),MV(6,6)
6  DIMENSION FV(6,6),BI(100),AB(10)
7  1  CONTINUE
8  READ (5,2) N
9  2 FORMAT(V)
10 IF (N) 3,100,3
11 3 CONTINUE
12 WRITE (6,4)
13 4 FORMAT (2H   ,'MT. BUNDEY AU, AG, U')
14 DO 5 I = 1,3
15 5  WN(I) = 0
16 K = 1
17 DO 6 I = 1,N
18 DO 6 L = 1,3
19 DO 7 J = 1,N
20 7 EL(I,J,L) = 0.
21 6 EF(I,L) = 0.
22 8 READ (5,51) AB(1),AB(2),(A(I), I = 1,N),L1,L2,L3,L4,L5
23 51   FORMAT (V)
24 IF (A(3) - 0.001) 10,10,9
25 9 CONTINUE
32 WN(K) = WN(K) + 1.
33 DO 11 I = 1,N
34 EF(I,K) = EF(I,K) + A(I)
35 DO 11 J=1,N
36 11 EL(I,J,K) = EL(I,J,K) +A(I)*A(J)
37 GO TO 8
38 10 IF (K-3) 12,14,12
39 12 K = K+1
40 GO TO 8
41 14 DO 15 I=1,N
42 G(I) = EF(I,1) + EF(I,2) + EF(I,2)
43 DO 15 K = 1,3
44 15 EF(I,K) = EF(I,K)/WN(K)
46 WRITE (6,17) (K,WN(K), K = 1,3)
48 17 FORMAT(25H SIZE OF POPULATION =      ,4(I6,F8.0))
49 EN = N
52 FACT = WN(1)+WN(2)+WN(3)-EN-1
53 DO 18 I = 1,N
56 WRITE (6,19) I, (EF(I,K),K=1,3)
58 19 FORMAT (17H   MEAN ON VARIATE      ,I2,1H=,6F10.5)
60 DO 20 I= 1,N
61 DO 20 J= 1,N
62 DO 20 K= 1,3
69 20 EL(I,J,K) = EL(I,J,K)- WN(K)*EF(I,K)*EF(J,K)
70 DO 21 I = 1,N
71 DO 21 J = 1,N
72 21 S(I,J) = EL(I,J,1)+EL(I,J,2)+EL(I,J,3)
73 WN(4) = WN(1)
74 DO 22 K = 1,3
75 DO 22 I = 1,N
76 EF(I,4) = EF(I,1)
77 FACTOR = WN(K) + WN(K+1)
78 XD(I) = EF(I,K) - EF(I,K+1)
79 FVA = XD(I)*XD(I)/S(I,I)
```

```
80 FVAR = FVA*WN(K)*WN(K+1)*FACT/FACTOR/EN
81 PF = PROBF(EN,FACT,FVAR)
82 22 WRITE (6,23) I,FVAR,PF
83 23 FORMAT(21H SIG. LEVEL ON VAR  ,I2,2HF=,F9.4,5HPROB=,F6.4)
84 WRITE (6,24)
85 24 FORMAT(38H  IDENT. PROB: FOR POP 1  POP 2  POP 3          )
86 DO 61 J=1,N
87 DO 61 I=1,N
88 61 P(I,J)=S(I,J)/((S(I,I)*S(J,J))**0.5)
89 DO 62 I=1,N
90 62 WRITE(6,63) (P(I,J),J=1,N)
91 63 FORMAT(2H  ,'CORRELATION MATRIX =',/(6F10.4))
92 DO 64 I=1,N
93 DO 64 J=1,N
94 DO 64 K=1,3
95 64 EM(I,J,K) = WN(K)*EF(I,K)*EF(J,K)
96 DO 65 I=1,N
97 65 EX(I)=(S(I,I)/(FACTOR-2.))**0.5
98 WRITE(6,66) (EX(I), I=1,N)
99 66 FORMAT (2H  ,STAND. DEV. = ', 10F10.4)
100 DO 67 I=1,4
101 DO 67 J=1,N
102 67 H(I,J)=EM(I,J,1)+EM(I,J,2)-G(I)*G(J)/FACTOR
103 MSQ = N*N
104 K = 0
105 DO 68 I=1,N
106 DO 68 J=1,N
107 K= K+1
108 BI(K)=S(I,J)
109 68 CONTINUE
110 CALL MINVI(BI,N,MSQ,ML,NL)
111 K = 0
112 DO 69 I=1,N
113 DO 69 J=1,N
114 K=K+1
115 S(I,J)=BI(K)
116 69 CONTINUE
117 DO 70 I=1,N
118 DO 70 J=1,N
119 70 S(I,J)=S(I,J)*(WN(1)+WN(2)+WN(3)-3.)
120 DO 71 I=1,N
121 71 WRITE (6,72) (S(I,J),J=1,N)
122 72 FORMAT( 'INVERSE VAR. MAT. =`,(5F10.6))
123 WN(4)=WN(1)
124 DO 73 K=1,3
125 DO 74 I=1,N
126 XD(I) = EF(I,K)-EF(I,K+1)
127 RD(I)=0
128 DO 74 J=1,N
129 74 RD(I) = RD(I) + S(I,J)*XD(J)
130 UD=0.
131 DO 76 I=1,N
132 76 UD=UD +XD(I)*RD(I)
133 FACTOR = WN(K)+WN(K+1)
134 UD=UD*FACTOR/EN
135 FRA=UD*WN(K)*WN(K+1)/(WN(K)+WN(K+1)-2.)/FACTOR
136 PF=PROBF(EN,FACT,FRA)
137 WRITE (6,77) FRA,N,FACT,PF
138 73 CONTINUE
139 77 FORMAT(2H  ,'F=',F9.4,' DF= ',I2,F3.0,' SIG. LEVEL =',F12.5)
140 IJ = 0
```

360

```
141 LOC= 0
142 29 READ (5,30) AB(1),AB(2),(X(I),I=1,N),L1,L2,L3,L4,L5
143 IF (X(1) - 999.) 26,25,26
144 26 CONTINUE
145 LOC = LOC + 1
146 TOT = 0
147 DO 27 K=1,3
148 DO 28 I=1,N
149 28 B(I)=X(I)-EF(I,K)
150 G1=0
151 G2=0
152 30 FORMAT(V)
153 DO 31 J=1,N
154 GE1=0
155 DO 32 I=1,N
156 GE1=GE1 +B(I)*S(I,J)
157 G1=G1+GE1*B(J)
158 SS(K)=EXP(-0.5*G1)
159 TOT=TOT+SS(K)
160 27 CONTINUE
161 IF (TOT) 34,34,33
162 34 WRITE (6,259) LOC,AB(1),AB(2)
163 259 FORMAT(2H  ,I5,2F10.2,'= ANOMALY`)
164 33 CONTINUE
165 DO 36 I=1,3
166 SS(I)=SS(I)/TOT
167 36 CONTINUE
168 IF (SS(1) - 0.5) 38,38,39
169 39 CONTINUE
170 IJ=IJ+1
173 WRITE (6,37) LOC,AB(1),AB(2),SS(1),SS(2),SS(3)
174 GO TO 38
175 38 CONTINUE
176 IF (SS(2)-0.5) 49,49,40
177 40 CONTINUE
178 IJ=IJ+1
179 WRITE (6,37) LOC,AB(1),AB(2),SS(1),SS(2),SS(3)
182 49 CONTINUE
183 GO TO 29
184 25 CONTINUE
185 37 FORMAT(2H  ,I5,2F8.2,3F17.3)
186 GO TO 1
187 100 CONTINUE
188 WRITE (6,41)
189 41 FORMAT(2H  ,`JOB COMPLETED')
190 STOP
191 END
192 FUNCTION(DFNUM,DFDEN,FRATIO)
193 PROBF=1.
194 IF(DFNUM*DFDEN*FRATIO) 17,16,17
195 16 RETURN
196 17 CONTINUE
197 IF(FRATIO -1.) 18,19,19
198 18 GO TO 1
199 19 CONTINUE
200 A=DFNUM
201 B=DFDEN
202 F=ABS(FRATIO)
203 GO TO 2
204 1 A=DFDEN
205 B=DFNUM
```

```
206 F=1./ABS(FRATIO)
207 2 AA=2./9./A
208 BB=2./9./B
209 Z=ABS(((1.-BB)*F**(1./3.)-1.+AA)/SQRT(BB*F**(2./3.)+AA))
210 IF (B-4.) 83,84,84
211 83 Z=Z*(1.+0.08*Z**4/B**3)
212 84 CONTINUE
213 PROBF=0.5/(1.+Z*(.196+Z*(.115+Z*(.00034+Z*.019527))))**4
214 IF (FRATIO-1.) 3,4,4
215 PROBF= 1. - PROBF
216 4 RETURN
217 END
218 SUBROUTINE MINVI(A,N,MSQ,D,L,M)
219 INTEGER L(NSQ),M(NSQ)
220 REAL A(NSQ)
221 D=1.
222 NK=-N
223 DO 80 K=1,N
224 NK=NK+N
225 L(K)=K
226 M(K)=K
227 KK=NK+K
229 DO 20 J=K,N
230 IZ=N*(J-1)
231 DO 20 I=K,N
232 IZ=IZ+1
233 10 IF (DABS(BIGA)-DABS(A(IJ))) 15,20,20
234 15 BIGA=A(IJ)
235 L(K)=I
236 M(K)=J
237 20 CONTINUE
238 J=L(K)
239 IF (J-K) 35,35,25
240 25 KI=K-N
241 DO 30 I=1,N
242 KI=KI+N
243 HOLD=-A(KI)
244 JI=KI-K+J
245 A(KI)=A(JI)
246 30 A(JI)=HOLD
247 35 I=M(K)
248 IF (I-K) 45,45,38
249 38 JP=N*(I-1)
250 DO 40 J=1,N
251 JK=NK+J
252 JI=JP + J
253 HOLD=-A(JK)
254 A(JK)=A(JI)
255 40 A(JI)=HOLD
256 IF (BIGA) 48,46,48
257 46 D=0.
258 RETURN
259 48 DO 55 I=1,N
260 IF (I-K) 50,55,50
261 50 IK=NK+I
262 A(IK)=A(IK)/(-BIGA)
263 55 CONTINUE
264 DO 65 I=1,N
265 IK = NK+I
266 HOLD=A(IK)
267 IJ=I-N
```

```
268 DO 65 J=1,N
269 IJ=IJ+N
270 IF (I-K) 60,65,60
271 60 IF (J-K) 62,65,62
272 62 KJ=IJ-I+K
273 A(IJ)=HOLD*A(KJ)+A(IJ)
274 65 CONTINUE
275 KJ=K-N
276 DO 75 J=1,N
277 KJ=KJ+N
278 IF (J-K) 70,75,70
279 70 A(KJ)=A(KJ)/BIGA
280 75 CONTINUE
281 D=D*BIGA
282 A(KK)=1./BIGA
283 80 CONTINUE
284 K=N
285 100 K=K-1
286 IF (K) 150,150,105
287 105 I=L(K)
288 IF(I-K) 120,120,108
289 108 JQ=N*(K-1)
290 JR=N*(I-1)
291 DO 110 J=1,N
292 JK = JQ+J
293 HOLD=A(JK)
294 JI=JR+J
295 A(JK)=-A(JI)
296 110 A(JI)=HOLD
297 120 J=M(K)
298 IF (J-K) 100,100,125
299 125 KI=K-N
300 DO 130 I=1,N
301 KI=KI+N
302 HOLD=A(KI)
303 JI=KI-K+J
304 A(KI)=-A(JI)
305 130 A(JI)=HOLD
306 GO TO 100
307 150 RETURN
308 END
```

APPENDIX 8

<u>Data lists for exercises</u>

<u>Commodity price table</u>

Commodity price table for exercises

Commodity	Price per kg or g
Cu	$ 1.50/kg
Mo	$12.00/kg
Zn	$ 1.00/kg
Pb	$ 0.85/kg
Ni	$ 7.00/kg
W	$55.00/kg
Al	$ 1.80/kg
Au	$14.00/g
Ag	$ 0.35/g

Section 1: Economic and Geometric data

8.1.1. Deposit type: Porphyry Cu-Mo

Geologic Regions: (1) B.C. Cordillera
                  (2) S. American Cordillera
                  (3) S. Pacific Island Arc
                  (4) Tasman Geosyncline

Geologic Region	Tonnage Mil. tonnes	Cu kg/tonne	Mo kg/tonne	Ag g/tonne	Au g/tonne
1	77	5	0	0	0
1	35	18	0	0	0
1	50	5	0	0	0
1	184	2	1	0	0
1	32	15	0	0	0
1	155	5	0	0	0
1	466	4	0	0	0
1	115	10	0	0	0
1	35	10	0	0	0
1	47	0	2	0	0
1	239	1	2	0	0
1	39	2	2	0	0
1	31	1	2	0	0
1	181	5	0	0	0
1	61	1	2	0	0
1	52	0	1	0	0
1	54	4	0	0	0
1	68	3	1	0	0
1	853	5	0	0	0
1	175	3	0	0	0
1	201	3	0	0	0
1	48	3	1	0	0
1	286	4	0	0	0
1	46	3	0	0	0
1	87	4	0	0	0
1	40	6	0	0	0
1	37	2	1	0	0
1	42	1	1	0	0
1	30	0	1	0	0
1	405	4	1	0	0
1	101	1	3	0	0
1	28	0	2	0	0
1	55	3	0	0	0
1	66	1	1	0	0
1	95	2	2	0	0
1	126	11	0	0	0
1	116	4	1	0	0
1	295	4	0	0	0
1	46	6	0	0	0
1	179	1	1	0	0
1	181	1	10	0	0
1	184	4	0	0	0
2	150	4	0	0	0
2	160	4	0	0	0
2	120	6	0	0	0
2	1800	13	0	0	0
2	620	9	0	0	0
2	260	17	0	0	0

2	428	8	0	0	0
2	150	14	0	0	0
2	125	17	0	0	0
2	4200	10	0	0	0
2	310	10	0	0	0
2	3100	12	0	0	0
2	60	17	0	0	0
2	350	7	0	0	0
2	265	7	0	0	0
2	210	7	0	0	0
2	1450	7	0	0	0
2	675	7	0	0	0
2	808	8	0	0	0
2	1200	6	0	0	0
2	460	8	0	0	0
2	450	11	0	0	0
2	1720	8	0	0	0
3	61	5	0	0	0
3	327	9	0	1	2
3	16	4	0	0	1
3	53	6	0	0	1
3	43	25	0	1	9
3	921	5	0	1	2
3	345	4	0	0	3
3	165	4	0	0	2
3	766	5	0	0	2
3	453	5	0	0	1
3	98	5	0	0	0
3	53	4	0	0	1
3	27	5	0	0	1
3	122	5	0	0	0
3	51	7	0	0	0
3	28	5	0	0	0
3	45	5	0	0	0
3	32	6	0	0	0
3	177	4	0	0	0
3	57	4	0	0	1
3	60	4	0	0	0
3	18	6	0	0	1
3	60	4	0	0	0
3	18	6	0	0	1
3	100	4	0	0	1
3	108	6	0	0	1
3	75	6	0	0	1
3	699	5	0	0	0
3	267	4	0	0	0
3	342	5	0	0	2
3	98	5	0	0	1
3	111	6	0	0	0
3	79	6	0	1	0
4	41	1	0	2	1
4	66	3	0	0	0
4	86	3	0	0	0
4	17	1	0	0	0
4	11	3	0	0	0
4	13	3	0	0	0
4	22	3	0	0	0
4	167	3	0	1	1
4	253	7	0	0	1
4	43	3	0	0	0
4	41	3	0	0	0

8.1.2. Deposit type: Porphyry Cu-Mo

Regions (1) : B. C. Cordillera
        (2) : South Pacific Island Arc
        (3) : Tasman Geosyncline

Region	Length in meters	Breadth in meters
1	1591	917
1	1295	831
1	832	198
1	924	276
1	655	331
1	777	279
1	1588	69
1	652	230
1	927	206
1	838	29
1	695	475
1	475	84
1	1676	317
1	168	34
1	1951	312
1	411	186
1	930	26
1	1387	925
2	1478	658
2	1097	351
2	1113	646
2	1396	780
2	1859	360
2	1591	1326
2	1597	381
2	1591	198
2	1113	878
2	975	914
2	1603	1289
2	1113	198
2	671	564
2	1024	646
2	655	387
2	805	732
2	1381	314
2	1134	1021
3	1061	768
3	1219	259
3	366	351
3	914	366
3	366	320
3	1463	381
3	564	259

N.B. Listed dimensions are those of the horizontal section
of the deposits. The foot-note applies also to all
subsequent geometric parameter listings.

8.1.3. Deposit type: Vein Gold

Geologic Regions: (1) North American Shield
(2) Western Australian Shield

Geologic Region	Tonnage tonnes (100,000's)	Au g/tonne	Gross value ($Mil.)
1	8	8	78
1	4	7	38
1	20	4	102
1	23	6	203
1	1	13	20
1	76	15	1591
1	39	8	404
1	89	7	888
1	61	5	441
1	90	20	2484
1	21	15	454
1	42	6	370
1	3	7	25
1	41	4	231
1	11	30	461
1	385	7	3546
1	198	6	1744
1	95	24	3100
1	9	7	84
1	11	8	127
1	8	5	48
1	7	7	74
1	32	7	296
1	659	8	6848
1	7	6	54
1	167	15	3478
1	135	5	975
1	4	5	23
1	23	8	244
1	63	3	278
1	13	7	129
1	35	3	169
1	336	8	3498
1	17	3	77
1	8	9	105
1	659	7	6322
1	41	12	692
1	71	9	883
1	11	6	89
1	56	8	608
1	17	3	77
1	6	12	109
1	73	8	755
1	3	6	23
1	2	10	21
1	105	4	585
1	96	9	1224
1	2	15	50
1	2	9	27
1	51	3	205
1	100	14	1910
1	67	3	296
2	2	11	76
2	6	10	217

2	103	11	4030
2	43	6	800
2	3	12	126
2	8	19	518
2	3	21	223
2	9	4	130
2	2	15	105
2	16	4	210
2	35	5	595
2	3	10	100
2	23	30	2404
2	31	15	1610
2	7	13	319
2	3	18	188
2	13	10	471
2	3	7	75
2	5	13	232
2	14	21	992
2	3	15	156
2	8	16	428
2	3	14	146
2	5	11	186
2	4	15	201
2	12	9	373
2	8	15	423
2	5	7	120
2	112	23	8763
2	2	21	146
2	7	15	359
2	3	21	218
2	32	22	2432
2	2	15	101
2	13	16	717
2	12	10	408
2	4	26	362
2	8	15	418
2	712	7	17124
2	34	7	840
2	13	17	780

8.1.4. Deposit type: Vein Gold

Geologic Regions: (1) North American Shield
                  (2) Western Australian Shield

Region	Length in meters	Breadth in meters	Dip Angle (degrees)
1	1478	11	45
1	503	21	46
1	719	18	45
1	1981	7	75
1	991	9	72
1	433	7	73
1	381	2	63
1	341	34	90
1	686	6	85
1	198	2	50
1	2195	1	62
1	261	1	82
1	168	1	68
1	686	6	75
1	198	6	66
1	571	8	58
1	556	8	61
1	209	1	45
1	108	5	72
1	411	23	70
1	381	11	71
1	503	9	73
1	386	12	78
1	1234	64	70
1	230	38	45
1	686	2	80
1	747	15	85
1	325	8	50
1	567	6	78
1	991	37	80
1	261	27	78
1	777	5	53
1	1478	34	30
1	465	23	43
1	320	2	28
1	768	2	70
1	322	4	62
1	1905	99	76
1	442	2	81
1	384	37	58
1	777	18	73
1	189	15	87
1	219	11	65
1	558	8	45
1	564	38	74
1	567	11	75
1	1042	21	83
1	341	9	88
1	1317	3	44
1	625	30	59
1	495	4	54
1	1295	13	50
1	556	19	88
1	1652	9	50
1	198	8	40

1	975	61	71
1	1225	20	90
1	503	2	51
1	344	4	71
2	472	8	80
2	206	2	41
2	570	3	61
2	463	8	45
2	283	11	77
2	314	8	61
2	1387	11	60
2	404	23	63
2	229	30	30
2	411	8	31
2	2591	5	46
2	1067	2	45
2	920	2	47
2	869	6	33
2	1838	2	82
2	1561	2	77
2	777	2	83
2	978	2	82
2	198	3	45
2	421	24	46
2	384	11	60

8.1.5. Deposit type: Ni-Cu Ultramafic

Geologic Regions: (1) North American Shield
(2) Western Australian Shield
(3) Scandinavian Shield

Geologic Region	Tonnage Mil. tonnes	Ni kg/tonne	Cu kg/tonne	Ag g/tonne	Au g/tonne
1	1	39	7	0	0
1	1	13	3	0	0
1	1	4	15	0	0
1	1	4	7	0	0
1	1	16	3	0	0
1	4	11	2	0	0
1	1	11	2	0	0
1	30	4	1	0	0
1	1	11	5	0	0
1	1	11	4	0	0
1	29	8	5	0	0
1	40	8	3	0	0
1	3	31	8	0	0
1	3	44	10	0	0
1	4	10	10	0	0
1	1	8	7	0	0
1	3	6	10	0	0
1	1	5	9	0	0
1	5	17	9	0	0
1	26	21	10	0	0
1	1	33	10	0	0
1	2	10	4	0	0
1	1	16	7	0	0
1	4	11	6	0	0
1	1	6	16	7	1
2	45	21	200	0	0
2	4	9	260	0	0
2	14	13	200	0	0
2	1	29	80	0	0
2	1	12	120	0	0
2	31	36	130	0	0
2	2	25	110	0	0
2	1	35	160	0	0
2	1	34	140	0	0
2	1	21	160	0	0
2	1	24	110	0	0
2	5	12	140	0	0
2	1	12	100	0	0
2	7	19	110	0	0
2	5	12	210	0	0
2	5	13	210	0	0
2	1	15	30	0	0
2	4	5	12	0	0
2	5	20	25	0	0
2	62	5	50	0	0
2	4	3	3	0	0
3	1	8	0	0	0
3	142	3	0	0	0
3	19	7	0	0	0
3	62	5	0	0	0
3	4	8	0	0	0
3	163	7	0	0	0
3	6	3	0	0	0

3	4	8	0	0	0
3	3	22	0	0	0
3	10	10	0	0	0
3	15	3	0	0	0
3	3	8	0	0	0
3	3	14	0	0	0
3	2	8	0	0	0
3	4	7	0	0	0
3	4	10	0	0	0
3	90	5	0	0	0
3	11	7	0	0	0

8.1.6. Deposit type: Ni-Cu Ultramafic

Geologic Regions: (1) North American Shield
                  (2) Western Australian Shield
                  (3) Scandinavian Shield

Region	Length in meters	Breadth in meters	Dip Angle (degrees)
1	1177	190	62
1	229	15	72
1	411	38	70
1	76	11	75
1	366	15	69
1	503	11	85
1	511	11	78
1	189	5	86
1	341	41	73
1	381	24	66
1	320	8	73
1	939	17	78
1	503	5	84
1	229	7	90
1	686	11	88
1	253	5	78
1	221	15	76
1	198	8	84
1	351	107	64
1	139	38	55
1	646	5	76
1	130	111	88
1	125	26	85
1	139	56	90
1	1865	23	64
1	634	99	90
2	1280	62	82
2	564	23	80
2	207	30	54
2	238	41	66
2	1067	146	73
2	1256	29	85
2	568	72	48
2	381	56	58
2	308	31	60
2	140	47	86
2	148	25	45
2	914	46	84
2	320	24	79
2	689	61	85
2	497	98	58
2	311	23	86
2	111	20	70
2	625	43	86
2	640	24	57
2	732	20	55
2	914	61	85
2	2073	24	80
2	1664	44	45
2	1981	30	61
2	762	37	51
3	189	23	75
3	113	20	70
3	94	24	74

3	509	29	76
3	192	20	78
3	114	64	80
3	472	32	75
3	139	24	70
3	238	142	72
3	110	14	68

8.1.7. Deposit type: Volcanogenic massive sulfides

Geologic Region: (1) North American Shield

Region	Grade: US$/T.	Tonnage (Mil.)	Gross Value ($Mil.)
1	55	53	2905
1	29	6	158
1	29	1	21
1	68	13	909
1	121	3	374
1	70	3	174
1	71	2	148
1	28	3	75
1	120	1	132
1	29	2	61
1	72	16	1114
1	164	7	1167
1	72	5	377
1	46	2	97
1	102	1	55
1	79	1	28
1	65	1	81
1	61	1	73
1	66	65	4257
1	21	2	32
1	61	1	86
1	50	1	71
1	88	1	66
1	98	1	50
1	40	1	14
1	77	1	23
1	54	1	12
1	207	1	25
1	48	1	5
1	88	1	9
1	47	1	4
1	67	1	47
1	95	5	492
1	81	3	250
1	69	3	189
1	48	3	121
1	56	2	101
1	31	2	47
1	44	1	47
1	65	1	49
1	94	1	63
1	50	1	16
1	113	1	29
1	60	1	15
1	58	1	14
1	35	51	1790
1	46	13	615
1	44	9	374
1	29	4	108
1	28	2	67
1	29	3	73
1	95	1	76
1	70	1	70
1	56	1	11
1	72	4	259
1	31	1	17

1	33	60	1978
1	31	16	515
1	115	5	610
1	45	4	185
1	18	4	65
1	27	3	81
1	85	3	255
1	86	3	249
1	84	3	210
1	84	2	126
1	89	1	111
1	31	1	36
1	98	1	59
1	100	1	50
1	49	1	18
1	77	1	21
1	24	1	6
1	28	1	5
1	142	1	18
1	26	1	1
1	89	28	2456
1	87	6	487
1	31	2	58
1	71	2	114
1	123	1	32
1	38	1	8
1	47	1	6
1	29	1	3
1	17	17	281
1	26	13	338
1	44	2	95
1	14	1	14
1	20	1	18
1	20	1	15
1	24	10	230
1	65	1	79
1	25	1	13
1	59	1	15
1	63	1	13
1	22	1	3
1	71	11	781
1	40	1	4
1	30	5	156
1	30	2	48
1	76	1	38
1	58	5	290
1	48	2	72
1	91	1	66
1	23	1	7
1	150	1	105
1	89	225	20025
1	25	7	163
1	41	1	25
1	68	1	34
1	37	1	3
1	43	1	26
1	12	2	29
1	40	1	12
1	30	7	198
1	32	3	106
1	51	1	26

1	73	1	15
1	52	4	218
1	38	1	19
1	22	1	4
1	31	6	180
1	37	1	48
1	13	11	141
1	65	1	52
1	47	1	38
1	26	1	8
1	34	1	7

8.1.7.(Cont.) Deposit type: Volcanogenic massive sulfides

Geologic Regions: (2) Western Australian Shield
(3) Appalachian Belt
(4) Scandinavian Caledonides
(5) Tasman Geosyncline
(6) North American Cordillera
(7) N. American Cordillera: (exhal.)
(8) Kuroko Belt of Japan
(9) Scandinavian Shield
(10) Iberian Peninsula Pyrite Belt
(11) Eastern Mediterranean

Geologic Region	Tonnage Mil. tonnes	Cu kg/tonne	Pb kg/tonne	Zn kg/tonne	Ag g/tonne	Au g/tonne
2	1	6	1	0	0	0
2	2	1	1	0	0	0
2	3	2	1	0	9	0
2	1	2	2	0	0	0
2	1	0	8	1	0	0
2	1	0	15	0	25	0
2	1	0	9	0	0	0
2	1	3	1	0	0	0
2	1	2	1	0	0	0
2	1	2	1	0	20	0
2	24	2	8	0	50	0
2	16	4	11	0	50	0
2	3	4	10	0	150	0
3	100	0	9	4	0	0
3	7	0	5	2	0	0
3	34	1	5	2	0	0
3	3	3	2	1	0	0
3	4	0	2	0	0	0
3	51	0	4	2	0	0
3	1	1	1	1	0	0
3	4	1	2	1	0	0
3	1	1	5	3	0	0
3	1	0	3	2	0	0
3	1	0	2	1	0	0
3	4	0	2	1	0	0
3	1	0	6	4	0	0
3	5	0	6	4	0	0
3	1	0	8	3	0	0
3	2	0	6	2	0	0
3	9	0	5	3	0	0
3	4	0	1	1	0	0
3	2	0	5	3	0	0
3	1	1	6	3	0	0
3	1	0	6	3	0	0
3	4	0	7	4	0	0
3	1	1	1	0	0	0
3	4	1	2	0	0	0
3	2	2	1	0	0	0
3	2	3	2	0	0	0
3	3	1	6	1	0	0
3	1	1	0	2	2	0
3	1	2	2	0	0	0
3	1	2	1	0	0	0
3	1	3	3	0	1	0
3	1	4	0	0	0	0
3	32	1	1	0	0	0

3	18	1	16	9	4	0
3	1	0	7	1	0	0
3	1	2	7	0	0	0
3	1	3	0	0	1	0
3	3	1	2	1	0	0
3	1	2	5	0	0	0
4	1	2	0	0	0	0
4	3	2	1	0	0	0
4	21	2	2	0	0	0
4	2	0	0	0	0	0
4	17	1	2	0	0	0
4	5	1	2	0	0	0
4	5	0	1	0	0	0
4	3	3	4	0	0	0
4	7	1	2	0	0	0
4	3	1	3	0	0	0
4	2	1	5	0	0	0
4	1	1	11	0	0	0
4	1	1	8	U	0	0
4	15	1	3	0	0	0
4	2	2	6	0	U	0
4	3	0	6	0	0	0
4	4	0	11	0	0	0
4	3	0	0	0	0	0
5	1	1	11	5	70	1
5	4	3	0	0	0	2
5	3	1	0	0	0	2
5	3	2	0	0	20	2
5	1	1	0	0	0	2
5	4	2	8	3	0	0
5	83	1	0	0	1	4
5	1	2	0	0	0	0
5	1	1	0	0	0	32
5	1	3	0	0	0	2
5	35	1	0	0	6	1
5	1	1	0	0	8	7
5	1	1	0	0	0	15
5	1	1	5	1	77	1
5	1	2	31	20	85	0
5	1	8	0	0	17	1
5	3	2	0	0	0	0
5	1	3	0	0	0	1
5	3	2	0	0	2	1
5	21	1	0	0	2	1
5	1	0	2	22	2850	57
5	6	2	15	6	3	0
5	4	2	1	0	0	0
5	5	1	11	6	55	2
5	22	2	4	1	7	5
5	1	3	1	0	3	0
5	16	1	19	6	187	4
5	2	1	20	5	0	0
5	1	1	17	7	0	0
5	89	1	0	0	2	0
5	4	2	0	0	2	1
5	8	1	0	0	3	0
5	1	2	0	0	5	1
5	6	1	0	0	61	2
5	4	3	0	0	8	0
5	5	6	0	0	34	0
6	2	1	6	1	3	0

6	26	2	0	0	0	0
6	1	2	0	0	0	0
6	63	1	1	0	1	0
6	2	2	0	C	0	0
6	29	0	0	0	4	0
6	4	4	3	0	1	0
6	20	1	0	0	0	0
6	17	2	2	0	1	0
6	2	1	7	0	3	0
6	6	2	8	1	3	C
6	15	2	5	0	1	0
6	1	1	1	1	2	0
6	1	1	1	0	1	0
6	1	3	10	7	3	0
6	1	4	3	0	2	0
6	26	7	1	0	3	0
6	4	9	1	0	3	0
6	2	3	1	0	0	0
6	4	4	4	0	2	0
6	5	3	4	3	5	0
7	222	6	6	55	0	0
7	9	8	9	80	0	0
7	67	6	4	23	0	0
7	17	5	3	47	0	0
7	7	5	4	52	0	0
7	5	6	4	42	0	0
7	22	7	6	85	0	0
7	31	8	2	48	0	0
7	85	17	5	2	0	0
7	125	6	3	2	0	0
8	4	1	5	1	41	1
8	5	1	5	2	0	0
8	1	3	9	1	220	12
8	5	1	3	1	5	1
8	1	1	4	0	63	1
8	3	1	6	1	0	0
8	11	2	2	0	20	3
8	8	1	11	3	3	1
8	10	1	12	3	0	0
8	11	2	3	1	0	0
8	3	1	3	1	0	0
8	7	2	3	1	0	0
8	11	2	7	2	86	3
8	25	2	8	2	220	2
9	16	1	1	0	0	0
9	9	1	1	0	0	0
9	17	1	12	21	0	0
9	15	0	1	0	0	0
9	10	2	1	0	0	0
9	31	1	5	0	0	0
9	1	2	1	0	0	0
9	2	0	1	0	3	0
9	2	0	5	0	0	0
9	11	1	0	60	15	0
9	13	1	3	33	0	0
9	2	0	2	45	1	0
9	7	1	3	12	0	0
9	6	1	7	133	0	0
9	5	0	9	154	0	0
9	4	1	3	54	0	0
9	4	1	2	0	0	0

9	1	0	8	150	0	0
9	3	1	3	0	0	0
9	30	1	1	4	0	0
9	63	1	1	5	0	0
9	4	1	2	33	0	0
9	15	2	1	0	0	0
9	6	1	2	0	0	0
10	36	450	400	7	2	2
10	3	420	400	8	1	1
10	4	440	400	15	1	1
10	32	450	420	10	2	2
10	71	480	430	6	1	1
10	43	420	400	8	1	1
10	26	400	400	16	3	3
10	81	520	430	10	3	3
10	21	520	430	8	1	1
10	43	350	250	10	1	1
10	25	400	400	13	1	1
10	13	400	400	16	5	5
10	45	400	410	7	1	1
10	60	400	410	8	1	1
10	51	430	410	9	1	1
10	42	400	400	8	1	1
10	111	460	430	8	3	3
10	36	440	400	18	1	1
10	103	480	440	8	3	3
10	46	430	420	4	5	5
10	43	400	400	7	2	2
10	13	400	400	16	6	6
10	43	350	300	23	4	4
10	5	520	400	24	2	2
10	6	500	450	18	3	3
11	6	35	25	5	2	1
11	13	35	25	4	1	0
11	6	35	25	2	5	2
11	16	35	25	1	5	2
11	11	48	40	1	0	1
11	6	52	40	1	0	2
11	27	35	30	1	0	0
11	23	48	40	5	0	0
11	7	48	40	2	0	0
11	1	44	40	1	0	0
11	6	40	40	1	0	0
11	5	35	35	2	0	0
11	3	35	35	2	0	0
11	3	35	30	1	0	0
11	5	35	35	2	0	0
11	32	35	30	1	0	0
11	30	35	30	1	0	0
11	2	43	40	2	0	0
11	9	42	40	2	0	2
11	11	40	35	1	0	2
11	3	40	38	2	0	2
11	27	40	25	2	0	1

8.1.8. Deposit type: Volcanogenic massive sulfides

Geologic Regions: (1) North American Shield
(2) Appalachian Belt
(3) North American Cordillera
(4) Tasman Geosyncline
(5) Western Australian Shield
(6) North American Cordillera: (Ex.)
(7) Scandinavian Shield
(8) Iberian Peninsula Pyrite Belt
(9) Eastern Mediterranean
(10) Kuroko Belt of Japan

Region	Length in meters	Breadth in meters	Dip Angle
1	808	194	84
1	564	12	79
1	652	111	0
1	392	29	87
1	390	20	50
1	102	20	48
1	404	99	0
1	230	93	84
1	503	32	71
1	230	11	63
1	558	20	50
1	198	8	79
1	79	17	55
1	305	8	84
1	137	11	79
1	198	8	45
1	238	5	79
1	162	5	83
1	312	35	79
1	309	20	79
1	445	190	84
1	223	5	83
1	198	20	73
1	201	107	79
1	229	177	79
1	183	183	40
1	216	123	45
1	130	24	63
1	329	20	87
1	189	102	36
1	392	69	0
1	200	78	36
1	99	78	29
1	98	94	76
1	221	29	79
1	239	20	79
1	311	14	0
1	189	94	67
1	111	23	50
1	94	15	79
1	308	66	59
1	306	5	79
1	114	5	88
1	674	98	50
1	94	15	63
1	945	9	79
1	384	8	84

1	213	6	55
1	991	20	79
1	1295	12	0
1	777	17	0
1	687	169	73
1	1478	11	79
1	351	11	84
1	594	46	68
1	533	20	68
1	309	11	55
1	200	23	55
1	351	64	45
1	613	139	79
1	108	64	68
1	500	7	79
1	200	5	87
1	309	17	71
1	290	15	40
1	168	17	79
1	99	15	66
1	786	64	71
1	168	14	68
1	500	16	59
1	677	9	0
1	47	5	73
1	96	5	73
1	413	7	68
1	78	8	0
1	366	12	63
1	78	12	73
1	259	20	45
1	282	8	63
1	137	8	63
1	128	7	59
1	169	9	45
1	219	5	55
1	244	5	84
1	914	41	71
1	1317	5	45
1	2438	5	45
1	375	6	40
1	501	25	85
1	617	38	29
1	250	32	59
1	247	31	63
1	1009	22	45
2	250	53	63
2	201	30	45
2	200	34	25
2	160	152	35
2	433	105	51
2	1524	4	54
2	341	22	70
2	280	38	45
2	934	46	73
2	331	25	79
2	465	108	59
2	495	69	73
2	459	43	72
2	497	62	30
2	472	11	68

2	189	17	59
2	186	7	45
2	341	11	50
2	192	17	45
2	169	22	44
2	495	35	87
2	200	32	45
2	160	11	50
2	200	5	70
2	646	9	77
2	503	17	48
2	198	40	40
2	383	34	63
2	771	32	81
2	477	38	76
2	472	5	35
2	1966	38	0
2	991	37	0
2	646	98	0
2	750	47	0
2	533	78	0
2	259	27	0
2	312	29	0
2	808	47	0
3	168	11	73
3	282	8	79
3	625	66	63
3	76	11	25
3	1562	50	63
3	463	53	66
3	384	20	83
3	1387	37	61
3	472	5	44
3	917	27	72
3	1390	34	68
3	2143	11	71
3	1021	17	76
3	808	11	68
3	594	6	25
3	503	69	59
3	198	9	79
4	290	20	71
4	351	66	59
4	99	38	29
4	168	69	0
4	1387	8	0
4	678	381	0
4	56	29	0
4	991	9	68
4	320	11	59
4	421	27	0
4	311	26	0
4	421	244	0
4	472	26	68
4	229	72	0
4	381	158	0
4	411	5	59
4	76	8	68
4	914	38	0
4	732	30	0
4	305	5	40

4	366	189	71
4	259	14	79
4	366	6	76
4	203	11	79
4	169	12	79
4	381	17	84
4	433	34	39
4	799	26	79
4	280	14	79
4	198	66	0
4	1433	32	45
4	183	17	68
4	189	55	73
4	186	46	76
4	369	14	73
4	229	40	68
4	411	32	73
4	463	148	81
4	186	56	63
4	283	44	63
4	555	280	79
4	457	189	45
4	168	27	71
4	108	23	79
5	1113	229	29
5	739	78	35
5	381	29	79
5	381	20	73
5	475	14	68
5	930	38	76
5	777	30	76
5	312	17	73
5	564	15	40
5	238	23	73
6	2195	107	25
6	165	14	68
6	1113	34	84
6	981	38	29
6	312	26	34
6	838	56	0
6	686	38	0
6	453	23	24
6	1684	56	29
6	1539	38	51
6	1349	34	84
6	2606	768	0
7	3048	189	10
7	1600	50	44
7	1225	20	66
7	1103	13	80
7	1082	66	36
7	567	94	22
7	416	38	61
7	634	20	68
7	206	18	72
7	320	14	66
7	341	23	87
7	311	21	84
7	564	38	71
7	1024	17	71
7	567	66	39

7	1173	38	87
7	960	29	76
7	620	19	46
7	646	64	0
7	381	26	61
7	311	9	58
7	384	20	76
7	556	64	51
7	1844	17	0
7	1853	50	87
7	411	64	29
7	1387	94	0
7	975	50	0
8	1295	38	57
8	415	17	71
8	564	20	75
8	1173	35	76
8	1024	99	24
8	1234	50	53
8	479	47	57
8	543	87	31
8	411	189	11
8	320	67	42
8	924	75	33
8	1289	49	59
8	808	40	68
8	741	46	55
8	488	213	9
8	1326	107	21
8	564	81	30
8	1295	98	23
8	1173	56	46
8	1030	27	75
8	384	37	0
9	312	1372	0
9	625	1097	0
9	335	2591	0
9	991	869	0
9	198	1067	0
9	686	2286	0
9	448	3414	0
9	524	3139	0
9	556	1417	0
9	503	4039	0
9	198	1006	0
9	354	991	0
9	689	3216	0
9	372	2591	0
9	512	1097	0
9	445	3719	0
9	570	4023	0
10	274	111	0
10	564	475	0
10	421	56	0
10	229	98	0
10	192	111	0
10	381	366	0
10	381	102	0
10	209	186	0
10	387	56	0
10	99	21	0

10	311	107	0
10	229	148	0
10	384	209	0
10	232	38	0
10	197	84	0
10	192	183	0
10	469	372	0
10	495	111	0
10	372	94	0
10	719	229	0
10	568	192	0
10	747	111	0
10	594	142	0
10	229	139	0
10	280	114	0
10	198	139	0
10	139	102	0
10	646	283	0
10	671	372	0
10	686	119	0

8.1.9. Deposit type: Contact Metasomatic (Cu-Fe-Au)

Geologic Region: (1) North American Cordillera

Geologic Region	Tonnage Mil. tonnes	Cu kg/t.	Fe %	Ag g/t.	Au g/t.
1	2	24	0	25	3
1	2	5	52	17	1
1	2	4	43	15	0
1	4	4	39	0	0
1	1	0	40	0	0
1	21	2	50	1	0
1	36	2	52	2	0
1	4	1	45	0	0
1	2	1	42	0	0
1	20	21	20	0	3
1	13	0	55	0	0
1	17	0	57	0	0
1	20	0	51	0	0
1	11	0	48	0	0
1	7	0	47	0	0
1	7	0	48	0	0
1	1	0	56	0	0
1	0	0	58	0	0
1	35	0	48	0	0
1	30	0	46	0	0
1	16	0	43	0	0
1	2	0	56	0	0
1	3	0	55	0	0
1	4	0	54	0	0
1	8	0	53	0	0
1	5	0	50	0	0

8.1.9.(Cont.) Deposit type: Contact Metasomatic (Pb-Zn-Cu-Ag)

Geologic Region: (2) North American Cordillera

Geologic Region	Tonnage Mil. tonnes	Cu kg/t.	Pb kg/t.	Zn kg/t.	Ag g/t.	Ag g/t.
2	1	0	82	93	151	0
2	2	32	12	0	0	3
2	2	25	0	13	35	0
2	1	32	0	0	0	3
2	0	42	13	15	0	0
2	1	54	0	0	96	3
2	0	0	38	0	928	1
2	1	18	0	0	7	2
2	1	13	32	104	328	1
2	2	4	104	4	522	4
2	0	0	88	25	290	0
2	0	0	110	35	232	0
2	1	0	120	13	261	0
2	1	0	153	123	290	0
2	4	8	48	5	435	0
2	3	0	53	12	232	4
2	1	0	102	14	348	0
2	12	0	21	63	16	0
2	1	0	50	12	319	0
2	3	6	45	12	290	3
2	2	5	43	8	232	0
2	1	24	0	85	0	3
2	1	32	0	0	29	2
2	1	0	65	15	189	0
2	1	0	82	12	1160	0
2	2	0	15	58	319	28
2	1	0	12	0	145	2
2	0	28	0	42	725	0
2	3	8	48	150	145	0
2	1	0	25	83	348	0
2	1	35	0	33	1508	6
2	2	0	42	53	145	1
2	3	0	14	56	290	0
2	1	0	48	153	145	1
2	1	12	46	0	151	0
2	0	28	0	43	696	0
2	6	0	43	38	145	0
2	5	0	48	46	232	0
2	4	5	0	0	23	12
2	0	62	0	0	29	7
2	5	23	0	0	0	1

8.1.9.(Cont.) Deposit type: Contact Metasomatic (Cu-Au-Mo)

Geologic Regions: (3) North American Cordillera &
South Pacific Island Arc

Geologic Region	Tonnage Mil. tonnes	Cu kg/t.	Ag g/t.	Au g/t.	Mo kg/t.	W kg/t.
3	61	11	15	1	0	0
3	410	9	12	1	0	0
3	81	8	23	1	0	0
3	305	8	0	1	0	0
3	310	8	0	0	0	0
3	98	25	35	1	0	0
3	17	9	35	0	0	0
3	17	9	35	0	0	0
3	35	11	0	1	0	0
3	72	9	0	1	0	0
3	47	11	0	1	0	0
3	23	11	0	1	0	0
3	25	13	0	1	0	0
3	33	12	0	1	0	0
3	230	1	0	0	1	1
3	27	23	0	2	0	0
3	50	22	9	1	0	0

8.1.9.(Cont.) Deposit type: Contact Metasomatic (W-Mo)

Geologic Region: (4) North American Cordillera
(5) East Asia

Geologic Region	Tonnage tonnes (100,000's)	W kg/tonne
4	2	6
4	3	8
4	4	10
4	2	10
4	4	12
4	6	15
4	5	8
4	8	6
4	9	5
4	3	4
4	7	5
4	4	5
4	5	5
4	5	3
4	4	6
4	13	8
4	10	4
4	3	11
4	3	8
4	3	9
4	6	5
4	2	6
4	8	5
4	5	5
4	4	5
4	24	7
4	26	5
4	5	5
4	13	5
4	3	3
4	2	4
4	3	5
4	6	5
4	2	4
4	2	5
4	6	6
4	71	14
4	59	17
4	32	8
4	16	6
4	6	5
5	2	13
5	2	12
5	2	21
5	5	5
5	285	5
5	3	5
5	84	9
5	10	6
5	8	8
5	12	13

8.1.10. Deposit type: Contact metasomatic

Geologic Region: (1) North American Cordillera (Cu-Fe-Au)
                 (2) N. American Cordillera (Zn-Pb-Cu-Ag)
                 (3) North American Cordillera
                     & South Pacific Island Arc (Cu-Au-Mo)
                 (4) North American Cordillera (W-Mo)
                 (5) East Asia (W-Mo)

Region	Length in m.	Breadth in m.	Dip angle	Strike angle
1	244	76	80	10
1	259	38	70	15
1	290	35	65	10
1	216	146	0	0
1	274	128	0	0
1	198	137	0	0
1	488	35	45	25
1	152	30	45	15
1	366	73	35	30
1	229	110	45	10
1	671	70	80	90
1	335	94	75	80
1	686	107	0	0
1	768	61	0	0
1	975	67	0	0
1	564	61	0	0
1	311	94	0	0
1	354	98	0	0
1	158	11	60	100
1	130	30	85	120
1	1871	123	68	110
1	1609	160	0	100
1	1536	114	80	100
1	264	41	65	150
1	381	76	68	120
1	445	64	65	110
1	422	85	0	90
1	317	53	0	150
1	219	125	0	85
2	140	26	70	105
2	128	64	45	115
2	381	62	75	0
2	198	23	70	20
2	494	11	63	45
2	259	23	80	120
2	494	23	85	145
2	253	17	75	50
2	355	64	40	0
2	244	26	65	20
2	488	24	50	0
2	198	56	45	0
2	2359	26	90	150
2	2301	20	85	130
2	341	46	55	0
2	259	69	0	0
2	1280	5	85	170
2	2179	9	80	150
2	2225	3	80	165
2	259	76	0	0
2	247	34	50	120
2	198	9	30	0

2	168	37	75	40
2	738	34	0	52
2	280	128	0	40
2	137	69	0	160
2	945	70	0	45
2	445	67	0	50
2	335	61	0	0
2	381	38	0	140
2	610	104	0	0
2	372	15	0	0
2	290	37	0	0
2	244	46	0	0
2	442	26	0	130
2	503	91	35	15
2	360	67	36	20
2	259	15	30	0
2	98	6	45	100
2	158	107	90	130
2	250	198	85	120
3	914	41	76	120
3	747	108	0	80
3	1097	381	0	90
3	770	85	86	115
3	604	62	86	115
3	878	145	75	125
3	610	366	0	110
3	488	91	0	110
3	930	125	0	0
3	1478	130	0	120
3	991	290	0	0
3	1082	189	0	145
3	837	558	0	0
3	1000	163	0	0
4	914	41	76	120
4	747	108	0	80
4	1097	381	0	90
4	770	85	86	115
4	604	62	86	115
4	878	145	75	125
4	610	366	0	110
4	488	91	0	110
4	930	125	0	0
4	1478	130	0	120
4	991	290	0	0
4	1082	189	0	145
4	837	558	0	0
4	1000	163	0	0
4	229	8	55	160
4	168	6	45	80
4	213	12	75	35
4	152	3	80	40
4	320	2	70	45
4	198	8	70	160
4	98	15	45	100
4	311	5	50	0
4	472	7	55	0
4	171	5	60	15
4	140	2	60	60
4	192	8	55	65
4	162	7	68	10
4	186	8	60	0

4	808	168	0	40
4	250	12	70	50
4	107	11	75	60
4	137	8	65	0
4	411	9	55	60
4	76	8	60	60
4	76	61	35	50
4	158	46	0	0
4	259	30	45	120
4	366	20	70	0
4	343	15	65	10
4	128	14	65	10
4	259	17	60	10
4	320	12	85	15
4	183	9	85	20
4	311	8	60	50
4	186	69	0	0
4	184	55	0	0
4	131	5	35	75
4	114	8	35	110
4	838	279	20	120
4	198	84	10	0
4	780	168	20	0
4	472	15	45	25
4	610	11	43	10
4	107	38	0	0
4	172	14	65	25
5	549	2	45	0
5	457	2	75	170
5	223	24	20	0
5	1524	5	35	100
5	168	17	80	0
5	579	30	60	70
5	198	34	75	10
5	387	9	60	150
5	494	8	50	0

8.1.11. Deposit type: Mississipi Valley-type Pb-Zn

Geologic Regions: (1) Missouri – Tri-State
                  (2) North American Arctic
                  (3) Upper Mississippi Valley

Region	Tonnage (Mil. tonnes)	Pb kg/t.	Zn kg/t.	Ag g/t.
1	3	3	32	3
1	9	2	29	6
1	5	1	33	7
1	5	2	34	4
1	8	2	29	7
1	9	3	32	6
1	69	2	28	4
1	81	2	28	3
1	19	3	28	6
1	24	2	31	3
1	39	2	33	5
1	7	3	33	7
1	37	2	29	5
1	38	2	30	4
1	206	23	·4	0
1	185	22	5	0
1	126	28	4	0
1	29	26	5	0
1	12	28	3	0
1	4	32	6	0
1	22	28	5	0
1	15	26	8	0
2	8	135	42	58
2	23	131	41	15
2	7	117	12	1804
2	3	112	15	827
2	15	78	25	0
2	3	76	27	0
2	2	72	33	0
2	2	98	34	0
2	3	103	28	0
2	1	116	65	0
2	2	55	14	0
2	3	144	74	0
3	6	3	159	0
3	4	22	764	0
3	3	23	594	0
3	7	5	265	0
3	5	1	50	0
3	8	4	210	0
3	6	4	179	0
3	8	2	132	0
3	14	2	162	0
3	8	1	60	0
3	10	4	297	0
3	16	2	275	0
3	7	5	314	0
3	6	0	25	0
3	10	2	124	0
3	13	1	60	0
3	8	0	18	0
3	10	0	33	0
3	7	7	0	0

3	5	0	0	0
3	11	0	0	0
3	8	6	0	0
3	6	3	0	0
3	7	0	0	0
3	5	10	0	0
3	6	3	0	0
3	9	5	0	0
3	16	6	80	0
3	8	5	0	0
3	8	11	10	0
3	8	8	10	0
3	8	13	10	0
3	20	11	0	0
3	10	1	0	0
3	13	5	0	0
3	12	6	0	0
3	9	1	90	0
3	9	1	10	0
3	9	0	10	0
3	9	0	10	0
3	7	32	10	0
3	10	2	10	0
3	4	1	10	0
3	5	5	10	0

8.1.12. Deposit type: Mississippi Valley-type Pb-Zn

Geological regions: (1) Upper Mississippi Valley
(2) Missouri - Tri-State
(3) North American Arctic

Region	Length in meters	Breadth in meters
1	152	18
1	183	21
1	366	91
1	152	61
1	396	91
1	1372	76
1	975	91
1	396	46
1	457	61
1	244	46
1	183	30
1	213	30
1	366	18
1	457	23
1	1829	46
1	366	18
1	366	46
1	137	14
1	457	8
1	305	30
1	244	18
1	213	15
1	366	18
1	152	15
1	76	6
1	244	15
1	183	9
1	366	15
1	128	15
1	457	8
1	91	21
1	610	18
1	183	30
1	305	30
1	305	18
1	274	15
1	1524	46
1	91	18
1	701	61
1	792	49
1	411	37
1	91	11
1	732	64
1	213	20
1	183	30
1	152	18
1	122	24
1	610	37
1	518	15
1	610	18
1	914	15
1	610	24
1	366	15
1	1219	37
1	610	229

1	671	18
1	305	9
1	610	12
1	457	40
2	693	64
2	285	99
2	1666	167
2	1237	102
2	1905	99
2	381	69
2	1835	72
2	392	69
2	2027	239
2	5633	442
2	7044	716
2	7772	256
2	3109	1234
2	8656	500
2	1631	198
2	2603	1113
2	2637	1451
2	4470	3505
2	5608	1414
2	3840	3109
2	1676	960
2	1920	1082
2	1280	808
2	2896	1600
3	604	567
3	331	293
3	3219	107
3	503	99
3	747	329
3	209	107
3	190	98
3	521	107
3	553	346
3	198	99
3	573	372
3	503	259
3	314	117
3	160	69
3	255	145
3	285	116

8.1.13.   Deposit type: Volcanogenic massive sulfides

Geologic Regions: (1) Iberian Peninsula Pyrite Belt
                  (2) Scandinavian Shield
                  (3) North American Shield
                  (4) Appalachian Belt
                  (5) Tasman Geosyncline

Region	Strike direction from True N.
1	60
1	90
1	90
1	90
1	100
1	90
1	90
1	120
1	100
1	110
1	90
1	100
1	90
1	80
1	90
1	90
1	90
1	90
1	90
1	115
1	120
2	80
2	60
2	20
2	60
2	75
2	65
2	70
2	25
2	90
2	90
2	80
2	80
2	90
2	90
2	90
2	130
2	120
2	160
2	0
2	30
2	90
2	90
2	130
2	0
2	160
2	30
2	0
2	0
3	148
3	125
3	50

3	140
3	0
3	160
3	160
3	160
3	10
3	10
3	0
3	135
3	45
3	45
3	160
3	45
3	65
3	0
3	90
3	95
3	45
3	135
3	15
3	0
3	70
3	65
3	125
3	90
3	130
3	135
3	95
3	105
3	130
3	130
3	100
3	145
3	20
3	55
3	135
3	130
3	70
3	60
3	90
3	95
3	90
3	120
3	65
3	120
3	80
3	90
3	125
3	120
3	45
3	100
3	145
3	135
3	180
3	105
3	115
3	65
3	135
3	90
3	160
3	95

3	100
3	125
3	105
3	90
3	140
3	135
3	85
3	165
3	135
3	85
3	90
3	135
3	40
3	55
3	55
3	45
3	90
3	95
3	90
3	95
3	45
3	20
3	100
3	90
3	60
3	0
3	95
3	90
4	95
4	150
4	130
4	60
4	10
4	0
4	110
4	50
4	160
4	50
4	65
4	60
4	95
4	40
4	45
4	90
4	55
4	50
4	70
4	35
4	30
4	65
4	45
4	55
4	65
4	65
4	45
4	35
4	55
4	75
5	0
5	10
5	160

5	20
5	45
5	80
5	175
5	170
5	125
5	115
5	60
5	165
5	10
5	55
5	70
5	52
5	15
5	30
5	130
5	0
5	135
5	125
5	10
5	20
5	25
5	10
5	160
5	10
5	175
5	150
5	165

Section 2: Selection of targets

8.2.1. Alligator Rivers Project

North	East	Total Count	Thorium	Uranium	Ratio
-24	56	120	22	17	4
-28	11	130	28	22	5
-32	3	70	16	12	3
-32	19	90	20	13	3
-32	32	90	16	15	4
-31	49	90	12	16	3
-32	86	40	8	12	1
-37	72	130	26	15	3
-41	55	790	146	116	7
-41	55	1460	271	221	14
-42	55	100	36	15	3
-43	55	310	56	50	5
-41	16	190	32	22	5
-45	50	200	44	38	3
-49	52	100	12	12	2
-51	53	120	24	16	2
-52	51	345	60	56	5
-49	66	40	8	12	2
-50	67	40	18	13	3
-50	64	55	8	11	2
-52	65	70	16	9	2
-58	4	150	34	19	4
-58	5	130	32	20	5
-58	16	50	10	15	4
-60	55	40	8	14	4
-63	82	80	16	11	2
-63	81	90	14	8	2
-68	82	100	18	15	4
-68	78	70	6	8	2
-68	71	120	24	17	4
-69	71	170	22	17	4
-69	83	130	22	17	4
-69	85	130	20	15	2
-70	82	150	26	24	4
-74	77	50	10	8	1
-75	76	110	20	16	2
-78	75	90	26	9	2
-78	78	150	40	18	2
-78	79	100	20	12	3
-78	80	90	16	12	2
-78	57	150	30	29	5
-71	9	160	32	19	5
-81	14	100	14	16	4
-89	27	600	170	84	24
-1	78	110	32	4	7
-1	79	120	31	7	5
-1	76	70	17	6	3
-1	79	70	20	5	3
-4	78	90	14	10	7
-6	89	150	36	9	6
-7	88	80	16	7	5
-7	90	100	22	10	3
-7	88	80	16	7	5
-8	59	90	15	8	9
-8	59	60	7	8	4
-8	63	120	21	11	11

-9	59	90	16	7	9
-10	69	110	17	13	10
-10	71	130	17	16	14
-12	66	110	20	12	11
-12	69	160	20	19	16
-12	72	110	19	16	9
-13	67	140	27	13	10
-13	70	100	16	14	10
-13	70	150	22	20	12
-13	71	140	20	15	15
-13	69	120	21	13	9
-13	70	140	23	16	13
-13	70	140	25	13	9
-18	83	240	36	25	27
-18	84	290	50	26	24
-18	84	290	51	27	22
-18	36	80	15	10	5
-18	82	280	46	26	28
-18	83	240	50	18	22
-19	35	140	23	20	7
-21	35	100	16	11	7
-21	55	130	23	19	12
-23	42	130	17	18	14
-23	55	130	20	17	9
-24	43	100	14	15	9
-24	49	130	20	15	11
-24	74	80	22	7	3
-24	83	130	20	15	10
-25	40	140	23	17	11
-25	46	130	19	16	13
-25	71	140	33	13	5
-25	84	90	13	12	7
-25	90	130	24	17	10
-27	42	100	15	11	9
-27	46	210	34	25	20
-27	49	170	30	16	16
-27	52	90	16	8	8
-27	49	120	22	19	9
-27	54	110	20	12	8
-27	34	160	33	15	11
-27	42	90	10	12	9
-27	46	150	20	20	12
-27	51	100	17	9	7
-27	54	120	22	12	10
-27	85	180	35	21	15
-28	45	150	20	21	10
-28	67	120	20	13	9
-28	73	125	30	9	6
-28	80	220	48	19	12
-28	82	150	33	10	7
-28	84	230	44	22	14
-29	33	140	20	23	8
-29	66	170	31	18	12
-29	76	100	26	4	5
-29	76	120	24	10	8
-29	78	110	26	10	5
-29	41	210	34	11	12
-29	52	100	14	13	10
-29	84	300	57	32	16
-29	84	300	54	29	20
-29	89	120	20	16	6

-31	81	150	30	13	9
-31	84	160	36	13	8
-31	88	140	24	15	9
-31	88	140	24	15	9
-32	45	140	23	16	11
-32	87	120	26	11	4
-32	87	130	31	12	8
-32	87	120	28	9	3
-32	36	160	25	14	14
-32	39	120	17	12	12
-32	83	120	27	11	4
-32	88	130	22	8	4
-32	82	90	18	9	4
-32	87	90	19	10	3
-33	89	110	28	8	4
-34	90	90	24	6	1
-35	36	140	26	17	6
-35	38	150	35	11	6
-35	39	150	25	16	9
-35	41	190	36	21	14
-35	42	190	25	20	16
-36	36	220	56	15	9
-36	39	220	42	22	16
-36	41	170	25	19	15
-37	31	180	26	21	12
-39	43	200	37	17	14
-39	77	100	10	15	9
-40	78	120	17	15	8
-40	86	130	17	17	9
-40	87	80	12	12	6
-40	89	80	13	11	7
-40	53	100	11	15	11
-40	59	70	12	9	7
-40	79	150	25	19	11
-40	80	150	25	19	11
-41	87	120	16	15	10
-41	53	120	21	12	8
-41	85	110	19	9	12
-42	89	140	23	14	14
-42	24	120	15	8	7
-42	37	120	15	8	7
-42	38	110	22	10	5
-42	42	160	25	16	8
-42	47	110	26	9	5
-42	51	100	21	7	5
-42	85	120	16	16	9
-42	86	120	20	15	8
-42	87	80	15	9	4
-42	88	90	21	10	4
-43	56	140	26	14	9
-43	84	100	20	11	3
-43	87	80	21	5	4
-43	89	60	16	3	3
-43	2	120	16	18	6
-43	37	130	20	17	7
-43	72	90	24	5	3
-43	73	120	29	9	3
-44	44	130	19	12	11
-45	12	60	9	9	3
-45	32	130	18	16	8
-46	7	80	13	13	5
-46	31	170	25	19	16

-48	38	90	12	14	4
-48	41	170	32	18	12
-48	42	150	25	17	9
-48	44	100	14	13	8
-48	37	190	28	21	18
-48	44	130	22	12	12
-49	48	150	14	13	9
-49	48	120	21	11	11
-50	5	210	35	24	16
-50	10	90	14	15	6
-50	12	190	28	23	21
-50	26	60	8	10	4
-50	57	100	15	14	7
-50	64	80	14	10	6
-51	5	170	32	15	13
-51	15	100	14	11	9
-51	68	90	15	10	2
-53	12	140	23	19	8
-53	14	130	25	15	10
-53	68	50	9	7	3
-54	6	210	35	29	9
-54	80	70	14	11	4
-54	41	150	20	20	13
-54	64	70	22	4	3
-56	41	120	19	15	9
-57	43	100	12	16	7
-57	62	70	24	4	2
-57	5	170	32	18	11
-57	16	120	18	18	6
-57	33	130	27	14	9
-58	4	150	20	11	2
-58	14	140	35	8	6
-58	24	120	30	7	5
-58	29	130	20	20	9
-58	30	130	20	19	11
-58	33	120	22	13	6
-59	28	160	38	17	8
-59	30	160	30	20	9
-59	41	180	42	15	7
-60	28	150	26	6	4
-60	33	130	20	20	9
-60	40	170	40	15	5
-61	7	160	30	20	10
-61	32	90	10	11	9
-61	34	130	16	13	13
-61	40	770	110	64	80
-61	6	110	20	15	7
-61	7	130	21	20	7
-61	32	170	17	19	15
-61	39	300	47	28	25
-62	39	230	32	29	21
-62	39	150	26	12	14
-62	84	120	22	14	5
-63	0	250	46	35	10
-65	4	150	27	16	9
-65	4	170	32	17	12
-65	9	110	22	10	9
-65	9	110	20	16	6
-65	19	170	35	15	11
-65	22	120	31	15	6
-65	30	130	32	8	9
-65	35	210	31	19	21

-65	44	130	23	13	12
-65	46	150	23	13	13
-65	3	140	25	17	7
-65	3	150	35	10	6
-65	9	130	32	11	6
-65	10	120	25	11	7
-65	45	130	24	15	10
-65	46	130	24	15	10
-66	2	200	30	17	14
-66	3	210	42	18	13
-66	11	160	33	18	9
-66	20	160	36	12	12
-66	37	150	23	16	12
-66	39	130	17	17	15
-66	47	140	23	16	12
-67	38	140	22	17	11
-67	83	110	22	13	5
-68	35	150	21	15	16
-68	37	140	24	13	12
-68	43	160	23	16	12
-68	88	90	13	12	8
-68	79	80	13	12	4
-68	79	80	15	10	4
-69	9	120	17	14	9
-69	9	160	26	16	10
-69	17	200	38	18	14
-69	35	140	21	15	14
-69	36	120	17	15	10
-69	44	140	25	14	10
-69	52	120	15	19	8
-69	83	110	16	18	6
-69	78	80	12	13	5
-69	88	90	11	12	7
-69	9	160	25	16	11
-69	10	140	25	15	8
-69	24	125	15	15	11
-69	36	120	18	14	9
-69	37	130	19	15	9
-69	54	110	19	16	5
-69	54	100	20	15	4
-70	10	160	42	12	6
-70	12	180	34	15	11
-70	16	210	37	20	14
-70	38	150	21	18	12
-70	43	200	28	21	22
-70	53	120	26	10	5
-71	3	160	36	12	11
-71	13	140	33	9	8
-71	47	140	33	8	9
-72	11	140	24	16	10
-72	13	130	28	16	8
-72	63	120	18	18	13
-73	12	160	31	18	13
-73	33	170	44	9	7
-73	38	170	27	16	18
-73	42	100	16	11	9
-73	44	110	22	10	9
-73	13	150	33	9	9
-73	38	170	33	14	14
-73	40	150	32	12	8
-74	39	220	40	14	17
-74	40	260	51	19	21

-74	20	120	15	18	7
-75	1	160	29	17	6
-75	41	90	17	7	7
-75	42	140	30	10	10
-75	59	100	18	8	9
-76	18	150	26	16	10
-76	22	110	22	9	9
-76	40	70	15	5	5
-76	41	140	26	11	11
-76	45	140	21	14	14
-76	56	110	20	9	10
-77	58	110	18	13	10
-77	59	80	19	5	5
-77	22	150	35	10	7
-77	4	120	21	15	6
-77	6	80	19	7	4
-77	7	120	24	9	8
-77	19	150	31	10	11
-77	27	100	20	10	5
-77	48	90	13	8	11
-77	50	90	16	8	8
-78	21	190	42	13	14
-78	42	130	33	8	6
-78	49	110	19	12	10
-79	21	150	33	13	6
-79	31	90	11	9	11
-79	49	120	21	11	10
-79	75	80	15	9	5
-80	19	160	36	13	10
-80	21	180	42	11	8
-81	19	180	37	12	12
-81	21	180	39	13	9
-81	21	160	36	15	7
-82	2	160	45	19	14
-82	3	160	39	12	8
-82	4	160	39	11	9
-82	5	150	29	15	8
-82	15	100	15	12	8
-82	19	160	35	14	8
-82	20	200	49	11	11
-82	87	180	21	17	20
-82	89	140	20	13	14
-82	4	230	20	20	21
-82	4	240	36	14	11
-82	21	190	36	18	10
-82	87	150	22	17	13
-82	89	130	18	11	11
-83	2	230	45	20	16
-83	4	190	34	15	13
-83	5	450	80	40	30
-83	21	140	31	12	7
-83	87	190	26	19	17
-84	4	290	57	25	18
-84	5	320	58	30	23
-84	88	150	23	18	12
-85	1	190	33	18	12
-85	6	450	83	36	32
-85	49	250	48	18	16
-86	5	250	43	22	16
-86	6	330	61	29	20
-86	57	90	19	5	6
-86	84	110	17	10	13

-87	5	300	56	25	19
-87	6	480	88	43	29
-87	58	90	19	7	5
-87	83	100	15	11	10
-87	85	200	25	20	24
-88	1	260	54	19	17
-88	5	210	38	17	16
-88	6	340	57	29	24
-88	48	110	30	17	12
-88	85	190	27	19	16
-88	88	90	11	12	8
-88	5	180	31	15	15
-88	6	310	54	29	23
-88	20	170	39	10	10
-88	34	100	16	10	9
-88	45	200	38	17	15
-88	46	220	44	18	18
-88	47	230	42	21	18
-88	83	90	15	10	6
-88	85	250	33	27	23
-88	87	130	20	12	10
-89	3	200	36	18	18
-89	5	260	48	21	21
-89	18	180	35	17	9
-89	34	120	16	13	12
-89	82	80	10	8	10
-89	86	120	15	14	12
-90	4	390	78	28	26
-90	5	300	55	25	23
-90	6	320	60	22	25
-90	7	340	65	23	22
-90	19	220	44	20	16
-90	23	140	30	12	7
-90	24	230	50	20	11
-90	30	210	28	22	19
-90	36	100	14	11	10
-90	46	150	34	10	8
-90	82	70	8	12	10
-90	87	150	14	11	11
-90	88	100	13	13	10
-91	1	280	57	19	20
-91	3	240	53	15	11
-91	6	370	68	31	11
-91	7	300	56	25	22
-91	20	220	44	16	12
-91	31	460	80	60	20
-91	34	140	32	8	10
-91	36	140	31	11	9
-91	37	140	33	7	10
-91	41	140	30	12	9
-91	53	130	33	9	5
-91	83	120	20	12	8
-91	3	250	52	17	13
-91	8	330	61	28	23
-91	13	150	28	14	8
-91	22	220	48	15	15
-91	28	100	20	11	5
-91	32	130	21	15	11
-91	35	150	32	11	10
-91	37	130	30	13	6
-91	83	200	28	20	20
-91	84	100	14	12	9

8.2.2. Mount Bundey U-Fe-Ag Project

North	East	Radiomet.	Aeromag.	Topogr.	Aerorad.		Geology			
-50	-27	90	160	40	400	1	0	0	0	0
-50	-27	130	160	20	460	1	0	0	0	0
-51	-27	200	160	35	460	1	0	0	0	0
-51	-27	60	145	10	290	0	0	1	0	0
-51	-27	95	145	10	290	0	0	1	0	0
-50	-26	120	160	35	460	1	0	0	0	0
-50	-26	140	160	30	400	1	0	0	0	0
-51	-26	55	160	10	360	1	0	0	0	0
-51	-26	160	155	20	360	1	0	0	0	0
-50	-26	30	220	25	200	-1	-1	-1	-1	-1
-51	-26	230	150	80	300	1	0	0	0	0
-50	-26	60	200	20	220	1	0	0	0	0
-51	-26	90	180	20	220	1	0	0	0	0
-51	-25	200	250	60	400	1	0	0	0	0
-51	-25	60	230	40	380	0	0	1	0	0
-51	-25	140	250	70	380	0	0	1	0	0
-51	-25	60	240	40	350	0	0	1	0	0
-52	-26	30	145	50	200	0	0	1	0	0
-52	-26	150	150	45	300	1	0	0	0	0
-52	-26	250	150	50	350	1	0	0	0	0
-52	-26	120	160	60	350	1	0	0	0	0
-52	-26	320	160	60	350	1	0	0	0	0
-53	-27	390	160	70	350	1	0	0	0	0
-53	-26	120	95	45	350	1	0	0	0	0
-53	-25	100	95	60	380	0	1	0	0	0
-53	-23	100	90	70	220	0	1	0	0	0
-53	-24	110	90	70	220	0	1	0	0	0
-53	-24	100	90	80	230	0	1	0	0	0
-53	-24	80	90	70	220	0	1	0	0	0
-53	-24	85	100	40	220	0	1	0	0	0
-53	-24	90	100	20	220	0	1	0	0	0
-54	-24	80	110	60	300	0	1	0	0	0
-54	-24	60	105	40	250	0	1	0	0	0
-54	-24	50	105	50	270	0	1	0	0	0
-54	-27	130	100	60	210	0	1	0	0	0
-55	-25	180	115	80	400	0	1	0	0	0
-54	-24	80	105	100	300	0	1	0	0	0
-54	-24	60	100	80	220	0	1	0	0	0
-54	-24	90	100	70	220	0	1	0	0	0
-54	-23	75	100	70	230	0	1	0	0	0
-55	-24	60	105	80	250	0	1	0	0	0
-54	-22	70	115	40	270	0	0	1	0	0
-54	-23	75	115	35	270	0	0	1	0	0
-54	-22	50	105	30	270	0	0	1	0	0
-54	-22	45	110	30	270	0	0	1	0	0
-55	-22	40	110	30	270	0	0	1	0	0
-54	-21	110	110	40	270	0	0	1	0	0
-54	-22	80	110	40	270	0	0	1	0	0
-54	-21	40	115	40	270	-1	-1	-1	-1	-1
-55	-21	30	115	40	270	-1	-1	-1	-1	-1
-55	-26	80	120	40	280	0	1	0	0	0
-55	-26	100	120	40	280	0	1	0	0	0
-56	-26	80	115	40	280	0	1	0	0	0
-56	-26	100	115	60	280	0	1	0	0	0
-56	-26	60	100	20	280	0	0	0	0	1
-56	-26	40	100	90	280	0	0	0	0	1
-55	-22	40	125	30	250	0	0	0	1	0
-57	-24	110	125	140	280	0	1	0	0	0
-51	-20	100	190	30	400	1	0	0	0	0

-52	-20	88	180	15	300	1	0	0	0	0
-52	-20	54	170	10	240	1	0	0	0	0
-50	-19	50	235	30	200	1	0	0	0	0
-51	-19	100	190	35	270	1	0	0	0	0
-51	-20	65	185	30	300	1	0	0	0	0
-52	-20	75	125	15	210	1	0	0	0	0
-52	-19	85	120	15	220	1	0	0	0	0
-52	-19	85	130	20	240	1	0	0	0	0
-51	-19	130	185	40	280	1	0	0	0	0
-52	-19	110	180	35	280	1	0	0	0	0
-51	-19	95	180	50	290	1	0	0	0	0
-52	-19	70	150	35	260	1	0	0	0	0
-51	-19	130	170	45	340	1	0	0	0	0
-51	-19	95	180	50	350	1	0	0	0	0
-50	-19	30	160	25	140	-1	-1	-1	-1	-1
-50	-18	15	160	30	120	-1	-1	-1	-1	-1
-51	-19	45	170	30	170	1	0	0	0	0
-52	-18	35	160	25	140	0	0	0	0	1
-51	-18	10	150	20	100	-1	-1	-1	-1	-1
-50	-17	20	170	35	90	-1	-1	-1	-1	-1
-50	-17	10	180	35	90	-1	-1	-1	-1	-1
-50	-18	15	180	30	90	-1	-1	-1	-1	-1
-50	-18	20	190	30	100	-1	-1	-1	-1	-1
-51	-18	25	180	30	100	-1	-1	-1	-1	-1
-51	-18	15	170	20	80	-1	-1	-1	-1	-1
-51	-18	30	160	25	100	-1	-1	-1	-1	-1
-51	-17	14	160	30	90	0	0	0	0	1
-51	-18	14	150	25	90	0	0	0	0	1
-51	-18	30	145	20	100	-1	-1	-1	-1	-1
-52	-18	10	140	20	80	-1	-1	-1	-1	-1
-51	-17	10	145	20	90	-1	-1	-1	-1	-1
-50	-17	20	125	30	80	-1	-1	-1	-1	-1
-50	-17	20	120	35	80	-1	-1	-1	-1	-1
-50	-16	15	120	30	70	-1	-1	-1	-1	-1
-51	-17	15	125	20	70	-1	-1	-1	-1	-1
-51	-17	20	125	20	80	-1	-1	-1	-1	-1
-51	-17	10	125	20	70	-1	-1	-1	-1	-1
-52	-17	15	125	10	60	-1	-1	-1	-1	-1
-52	-17	15	125	15	65	0	0	0	0	1
-52	-17	30	120	25	100	-1	-1	-1	-1	-1
-52	-16	30	120	25	100	-1	-1	-1	-1	-1
-52	-17	20	118	25	80	-1	-1	-1	-1	-1
-52	-17	20	118	25	80	-1	-1	-1	-1	-1
-52	-16	25	120	30	80	-1	-1	-1	-1	-1
-52	-16	30	122	35	70	-1	-1	-1	-1	-1
-51	-16	13	125	15	60	0	0	0	0	1
-51	-16	30	127	20	80	0	0	0	0	1
-51	-15	23	127	25	70	-1	-1	-1	-1	-1
-51	-15	30	127	30	70	-1	-1	-1	-1	-1
-52	-16	30	123	30	60	-1	-1	-1	-1	-1
-50	-14	22	127	35	40	-1	-1	-1	-1	-1
-51	-15	30	125	40	70	0	0	1	0	0
-52	-15	35	125	35	80	0	0	1	0	0
-52	-14	25	126	25	80	0	0	0	0	1
-51	-12	26	126	60	80	-1	-1	-1	-1	-1
-52	-12	27	126	60	70	-1	-1	-1	-1	-1
-52	-13	28	126	60	60	-1	-1	-1	-1	-1
-50	-12	14	126	70	40	-1	-1	-1	-1	-1
-50	-12	17	126	40	40	0	0	0	0	1
-51	-12	22	126	70	80	-1	-1	-1	-1	-1
-51	-12	19	126	80	80	-1	-1	-1	-1	-1

-51	-11	25	126	50	100	0	0	0	0	1
-52	-11	16	126	70	90	-1	-1	-1	-1	-1
-51	-11	28	126	70	80	-1	-1	-1	-1	-1
-52	-11	35	126	70	90	0	0	0	1	0
-52	-10	25	127	70	100	0	0	0	1	0
-51	-10	14	126	60	80	0	0	0	0	1
-51	-10	30	126	70	90	0	0	1	0	0
-51	-10	17	126	50	80	0	0	0	0	1
-51	-10	20	126	60	80	0	0	0	1	0
-51	-10	25	126	80	80	0	0	0	1	0
-51	-10	25	126	80	80	0	0	0	1	0
-51	-9	19	125	70	80	0	0	0	0	1
-51	-9	30	126	70	85	0	0	1	0	0
-51	-9	27	126	50	100	0	0	0	0	1
-52	-9	48	127	70	180	0	0	1	0	0
-52	-9	35	125	60	150	0	0	1	0	0
-51	-9	20	126	70	80	0	0	1	0	0
-51	-8	10	126	60	50	-1	-1	-1	-1	-1
-51	-8	15	126	50	50	0	0	0	0	1
-51	-8	24	126	60	60	-1	-1	-1	-1	-1
-53	-8	24	125	60	80	0	0	1	0	0
-53	-8	30	125	60	90	0	0	1	0	0
-53	-8	23	125	60	80	0	0	0	1	0
-53	-18	14	100	20	100	0	0	0	0	1
-53	-19	38	115	100	8	0	0	1	0	0
-53	-16	45	126	40	100	0	0	1	0	0
-52	-12	30	126	40	70	0	0	0	1	0
-53	-11	30	125	40	70	0	0	0	1	0
-54	-13	25	125	30	60	0	0	0	1	0
-54	-12	25	125	60	70	0	0	0	1	0
-53	-18	40	115	20	80	0	0	1	0	0
-53	-18	25	115	25	60	0	0	0	1	0
-53	-19	30	120	40	70	-1	-1	-1	-1	-1
-54	-19	30	120	40	80	0	0	0	1	0
-55	-19	35	123	20	90	0	0	1	0	0
-56	-19	30	125	20	80	0	0	1	0	0
-54	-19	45	120	40	80	0	0	1	0	0
-55	-19	40	120	40	80	0	0	1	0	0
-56	-19	25	125	30	80	0	0	0	1	0
-56	-19	30	125	10	150	0	0	0	0	1
-58	-20	30	120	20	80	0	0	1	0	0
-59	-20	38	125	20	80	0	0	0	1	0
-59	-19	40	125	10	80	0	0	0	0	1
-54	-15	30	125	50	70	0	0	0	1	0
-55	-15	30	125	40	70	0	0	0	1	0
-55	-15	30	125	30	60	0	0	0	1	0
-56	-15	28	125	20	60	0	0	0	1	0
-57	-16	32	125	20	70	0	0	0	1	0
-57	-16	45	120	15	70	0	0	0	0	1
-58	-17	20	120	20	70	0	0	0	1	0
-58	-17	35	120	20	60	0	0	0	1	0
-57	-12	30	125	40	70	0	0	0	1	0
-54	-14	30	125	30	60	0	0	0	0	1

8.2.3. Vrilya Point Bauxite Project

East (Origin 0,0: V.P.)	North	Overburden Meters	Econ. Zone Thickness Meters	Alumina Al2O3%
16	14	0	8	38
16	14	0	2	42
17	14	1	3	22
18	14	2	2	29
20	14	4	5	40
20	14	4	2	48
21	14	2	4	33
23	14	1	7	46
23	15	0	9	54
22	16	4	2	30
23	16	2	8	47
24	16	3	7	43
23	17	0	6	42
24	17	3	4	35
20	18	2	4	28
22	18	1	6	49
23	18	4	6	33
24	18	1	4	42
23	19	2	6	44
24	19	2	4	31
26	20	12	8	46
25	21	2	4	43
28	21	4	2	38
23	22	1	9	50
25	22	12	6	44
24	23	10	8	43
25	23	1	9	42
26	23	0	8	40
27	23	4	4	44
21	13	0	8	37
23	13	1	4	52
24	13	1	5	46
18	12	2	4	42
19	12	2	4	36
19	12	2	2	40
20	12	4	2	32
21	12	8	9	24
19	11	14	2	28
20	11	1	2	41
22	11	4	2	39
24	11	0	7	38
19	10	2	2	40
24	10	3	3	35
22	9	6	2	39
22	9	4	3	35
24	9	2	8	45
27	9	2	6	42
19	8	1	7	42
22	8	10	2	41
22	8	10	2	40
27	8	6	6	45
19	7	2	4	41
20	7	4	6	40
21	7	6	4	45
22	7	1	7	35
24	7	2	4	44
13	6	6	2	35

14	6	2	2	36
15	6	0	4	40
22	6	0	10	34
27	5	2	3	42
19	4	3	3	27
20	4	3	3	32
21	4	6	2	30
24	4	3	2	40
24	4	3	3	36
25	4	4	2	30
27	4	6	4	36
24	3	4	1	37
26	3	11	5	37
25	2	5	6	41
27	1	10	6	37
26	0	1	3	35
27	0	2	4	33
26	-2	3	3	39
25	-4	3	3	39
26	-4	4	2	34
25	-5	2	2	35
26	-5	4	3	45
26	-5	2	2	41
26	-6	1	2	21
4	1	0	2	52
4	1	0	4	46
4	4	6	2	40
4	4	6	4	34
3	2	1	5	48
3	1	2	4	48
4	1	4	2	45
2	1	2	8	51
1	1	2	6	44
4	0	4	2	45
4	0	4	4	33
3	0	1	4	48
2	0	6	6	48

Measurement conversion table and statistical tables

9.1. Measurement conversion table

TYPE OF MEASUREMENT	CONVERSION FACTORS	
	IMPERIAL SYSTEM TO METRIC	METRIC SYSTEM TO IMPERIAL
DISTANCE	1 mile = 1.61 kilometer 1 yard = 0.914 meter 1 foot = 0.305 meter	1 kilometer = 0.621 mile 1 meter = 1.09 yards 1 meter = 3.28 feet
AREA	1 sq. mile = 2.60 sq. km. 1 acre = 0.405 hectare 1 sq. foot = 0.093 sq. m.	1 sq. km. = 0.386 sq. ml. 1 hectare = 2.47 acres 1 sq. meter = 10.8 sq. feet
VOLUME	1 cu. yard = 0.765 cu. m. 1 cu. foot = 0.028 cu. m. 1 gallon = 4.55 litres	1 cu. meter = 1.31 cu. yds. 1 cu. meter = 35.3 cu. feet 1 litre = 0.220 gallons
MASS	1 short ton = 1.02 metric tonne 1 pound = 0.454 kilogram 1 ounce = 28.3 grams 1 oz/sh. ton = 27.8 gm/metr. tonne	1 met. tonne= 0.980 sht. ton 1 kilogram = 2.20 pounds 1 gram = 0.035 ounces 1 gm/met. t.= 0.036 oz/sht. t.

9.2. The Normal Probability Distribution table

Z incr.	0.00	0.01	0.02	0.03	0.04	0.05	0.06	0.07	0.08	0.09
-2.9	0.002	0.002	0.002	0.002	0.002	0.002	0.002	0.001	0.001	0.001
-2.8	0.003	0.002	0.002	0.002	0.002	0.002	0.002	0.002	0.002	0.002
-2.7	0.003	0.003	0.003	0.003	0.003	0.003	0.003	0.003	0.003	0.003
-2.6	0.005	0.005	0.004	0.004	0.004	0.004	0.004	0.004	0.004	0.004
-2.5	0.006	0.006	0.006	0.006	0.006	0.005	0.005	0.005	0.005	0.005
-2.4	0.008	0.008	0.008	0.008	0.007	0.007	0.007	0.007	0.007	0.006
-2.3	0.011	0.010	0.010	0.010	0.010	0.009	0.009	0.009	0.009	0.008
-2.2	0.014	0.014	0.013	0.013	0.013	0.012	0.012	0.012	0.011	0.011
-2.1	0.018	0.017	0.017	0.017	0.016	0.016	0.015	0.015	0.015	0.014
-2.0	0.023	0.022	0.022	0.021	0.021	0.020	0.020	0.019	0.019	0.018
-1.9	0.029	0.028	0.027	0.027	0.026	0.026	0.025	0.024	0.024	0.023
-1.8	0.036	0.035	0.034	0.034	0.033	0.032	0.031	0.031	0.030	0.029
-1.7	0.045	0.044	0.043	0.042	0.041	0.040	0.039	0.038	0.038	0.037
-1.6	0.055	0.054	0.053	0.052	0.051	0.049	0.048	0.047	0.046	0.046
-1.5	0.067	0.066	0.064	0.063	0.062	0.061	0.059	0.058	0.057	0.056
-1.4	0.081	0.079	0.078	0.076	0.075	0.074	0.072	0.071	0.069	0.068
-1.3	0.097	0.095	0.093	0.092	0.090	0.089	0.087	0.085	0.084	0.082
-1.2	0.115	0.113	0.111	0.109	0.107	0.106	0.104	0.102	0.100	0.099
-1.1	0.136	0.133	0.131	0.129	0.127	0.125	0.123	0.121	0.119	0.117
-1.0	0.159	0.156	0.154	0.152	0.149	0.147	0.145	0.142	0.140	0.138
-0.9	0.184	0.181	0.179	0.176	0.174	0.171	0.169	0.166	0.164	0.161
-0.8	0.212	0.209	0.206	0.203	0.200	0.198	0.195	0.192	0.189	0.187
-0.7	0.242	0.239	0.236	0.233	0.230	0.227	0.224	0.221	0.218	0.215
-0.6	0.274	0.271	0.268	0.264	0.261	0.258	0.255	0.251	0.248	0.245
-0.5	0.309	0.305	0.302	0.298	0.295	0.291	0.288	0.284	0.281	0.278
-0.4	0.345	0.341	0.337	0.334	0.330	0.326	0.323	0.319	0.316	0.312
-0.3	0.382	0.378	0.374	0.371	0.367	0.363	0.359	0.356	0.352	0.348
-0.2	0.421	0.417	0.413	0.409	0.405	0.401	0.397	0.394	0.390	0.386
-0.1	0.460	0.456	0.452	0.448	0.444	0.440	0.436	0.433	0.429	0.425
-0.0	0.500	0.496	0.492	0.488	0.484	0.480	0.476	0.472	0.468	0.464
0.0	0.500	0.504	0.508	0.512	0.516	0.520	0.524	0.528	0.532	0.536
0.1	0.540	0.544	0.548	0.552	0.556	0.560	0.564	0.567	0.571	0.575
0.2	0.579	0.583	0.587	0.591	0.595	0.599	0.603	0.606	0.610	0.614
0.3	0.618	0.622	0.626	0.629	0.633	0.637	0.641	0.644	0.648	0.652
0.4	0.655	0.659	0.663	0.666	0.670	0.674	0.677	0.681	0.684	0.688
0.5	0.691	0.695	0.698	0.702	0.705	0.709	0.712	0.716	0.719	0.722
0.6	0.726	0.729	0.732	0.736	0.739	0.742	0.745	0.749	0.752	0.755
0.7	0.758	0.761	0.764	0.767	0.770	0.773	0.776	0.779	0.782	0.785
0.8	0.788	0.791	0.794	0.797	0.800	0.802	0.805	0.808	0.811	0.813
0.9	0.816	0.819	0.821	0.824	0.826	0.829	0.831	0.834	0.836	0.839
1.0	0.841	0.844	0.846	0.848	0.851	0.853	0.855	0.858	0.860	0.862
1.1	0.864	0.867	0.869	0.871	0.873	0.875	0.877	0.879	0.881	0.883
1.2	0.885	0.887	0.889	0.891	0.893	0.894	0.896	0.898	0.900	0.901
1.3	0.903	0.905	0.907	0.908	0.910	0.911	0.913	0.915	0.916	0.918
1.4	0.919	0.921	0.922	0.924	0.925	0.926	0.928	0.929	0.931	0.932
1.5	0.933	0.934	0.936	0.937	0.938	0.939	0.941	0.942	0.943	0.944
1.6	0.945	0.946	0.947	0.948	0.949	0.951	0.952	0.953	0.954	0.954
1.7	0.955	0.956	0.957	0.958	0.959	0.960	0.961	0.962	0.962	0.963
1.8	0.964	0.965	0.966	0.966	0.967	0.968	0.969	0.969	0.970	0.971
1.9	0.971	0.972	0.973	0.973	0.974	0.974	0.975	0.976	0.976	0.977
2.0	0.977	0.978	0.978	0.979	0.979	0.980	0.980	0.981	0.981	0.982
2.1	0.982	0.983	0.983	0.983	0.984	0.984	0.985	0.985	0.985	0.986
2.2	0.986	0.986	0.987	0.987	0.987	0.988	0.988	0.988	0.989	0.989
2.3	0.989	0.990	0.990	0.990	0.990	0.991	0.991	0.991	0.991	0.992
2.4	0.992	0.992	0.992	0.992	0.993	0.993	0.993	0.993	0.993	0.994
2.5	0.994	0.994	0.994	0.994	0.994	0.995	0.995	0.995	0.995	0.995
2.6	0.995	0.995	0.996	0.996	0.996	0.996	0.996	0.996	0.996	0.996
2.7	0.997	0.997	0.997	0.997	0.997	0.997	0.997	0.997	0.997	0.997
2.8	0.997	0.998	0.998	0.998	0.998	0.998	0.998	0.998	0.998	0.998
2.9	0.998	0.998	0.998	0.998	0.998	0.998	0.998	0.999	0.999	0.999

9.3. Table of percentiles of the Normal Probability Distribution

Percentile			z
0.05 % ile	=	-3.290
0.10 % ile	=	-3.090
0.50 % ile	=	-2.576
1.00 % ile	=	-2.326
2.50 % ile	=	-1.960
5.00 % ile	=	-1.645
7.50 % ile	=	-1.440
10.00 % ile	=	-1.282
12.50 % ile	=	-1.150
15.00 % ile	=	-1.036
17.50 % ile	=	-0.935
20.00 % ile	=	-0.842
25.00 % ile	=	-0.674
30.00 % ile	=	-0.524
35.00 % ile	=	-0.385
40.00 % ile	=	-0.253
45.00 % ile	=	-0.126
50.00 % ile	=	0.000
55.00 % ile	=	0.126
60.00 % ile	=	0.253
65.00 % ile	=	0.385
70.00 % ile	=	0.524
75.00 % ile	=	0.674
80.00 % ile	=	0.842
85.00 % ile	=	1.036
87.50 % ile	=	1.150
90.00 % ile	=	1.282
92.50 % ile	=	1.440
95.00 % ile	=	1.645
97.50 % ile	=	1.960
99.00 % ile	=	2.326
99.50 % ile	=	2.576
99.90 % ile	=	3.090
99.95 % ile	=	3.290

9.4. Critical values of the Student T Distribution

degrees of freedom	$t_{0.10}$	$t_{0.05}$	$t_{0.025}$	$t_{0.010}$	$t_{0.005}$
1	3.078	6.314	12.706	31.812	63.657
2	1.886	2.920	4.303	6.965	9.925
3	1.638	2.353	3.182	4.541	5.841
4	1.533	2.132	2.776	3.747	4.604
5	1.476	2.015	2.571	3.365	4.032
6	1.440	1.943	2.447	3.143	3.707
8	1.397	1.860	2.306	2.896	3.355
9	1.383	1.833	2.262	2.821	3.250
10	1.372	1.812	2.228	2.764	3.169
11	1.363	1.796	2.201	2.718	3.106
12	1.356	1.782	2.179	2.681	3.055
13	1.350	1.771	2.160	2.650	3.012
14	1.345	1.761	2.135	2.624	2.977
15	1.341	1.753	2.131	2.602	2.947
16	1.337	1.746	2.120	2.583	2.921
17	1.333	1.740	2.110	2.567	2.898
18	1.330	1.734	2.101	2.552	2.878
19	1.328	1.729	2.093	2.539	2.861
20	1.325	1.725	2.086	2.528	2.845
21	1.323	1.721	2.080	2.518	2.831
22	1.321	1.717	2.074	2.508	2.819
23	1.319	1.714	2.069	2.500	2.807
24	1.318	1.711	2.064	2.492	2.797
25	1.316	1.708	2.060	2.485	2.787
26	1.315	1.706	2.056	2.479	2.779
27	1.314	1.703	2.050	2.473	2.771
28	1.313	1.701	2.048	2.467	2.763
29	1.311	1.699	2.045	2.462	2.756
30	1.310	1.697	2.042	2.445	2.750
40	1.303	1.684	2.021	2.423	2.704
50	1.299	1.676	2.009	2.403	2.678
60	1.296	1.671	2.000	2.390	2.660
70	1.294	1.667	1.994	2.381	2.648
75	1.292	1.665	1.991	2.378	2.643
80	1.291	1.664	1.989	2.374	2.639
90	1.289	1.662	1.986	2.369	2.632
100	1.287	1.660	1.984	2.365	2.626
110	1.285	1.658	1.982	2.361	2.621
120	1.284	1.656	1.980	2.358	2.617
130	1.284	1.654	1.978	2.355	2.614
140	1.284	1.652	1.976	2.353	2.611
150	1.284	1.650	1.974	2.351	2.609
200	1.284	1.648	1.972	2.348	2.605
250	1.284	1.647	1.970	2.345	2.600
300	1.284	1.647	1.968	2.342	2.596
350	1.284	1.647	1.967	2.340	2.593
400	1.283	1.647	1.966	2.338	2.588
450	1.283	1.647	1.965	2.335	2.587
500	1.283	1.646	1.964	2.334	2.586
600	1.283	1.645	1.963	2.333	2.584
800	1.283	1.645	1.962	2.331	2.582
1000	1.282	1.645	1.961	2.330	2.581
1500	1.282	1.645	1.961	2.329	2.580
2000	1.282	1.645	1.960	2.328	2.579
5000	1.282	1.645	1.960	2.328	2.578
10000	1.282	1.645	1.960	2.327	2.577
50000	1.282	1.645	1.960	2.327	2.577
∞	1.282	1.645	1.960	2.326	2.576

9.5. Critical values of the χ^2 Distribution

degrees of freedom	$\chi^2_{0.10}$	$\chi^2_{0.05}$	$\chi^2_{0.025}$	$\chi^2_{0.001}$	$\chi^2_{0.005}$
1	2.706	3.841	5.024	6.635	7.880
2	4.605	5.991	7.378	9.210	10.597
3	6.251	7.815	9.348	11.343	12.838
4	7.780	9.487	11.143	13.277	14.862
5	9.237	11.071	12.833	15.087	16.750
6	10.645	12.592	14.449	16.813	18.550
7	12.016	14.068	16.014	18.477	20.276
8	13.362	15.507	17.536	20.092	21.956
9	14.684	16.919	19.024	21.668	23.590
10	15.988	18.309	20.485	23.210	25.189
11	17.277	19.677	21.922	24.726	26.758
12	18.550	21.028	23.338	26.218	28.300
13	19.813	22.363	24.736	27.689	29.820
14	21.065	23.685	26.119	29.141	31.319
15	22.308	24.996	27.489	30.578	32.801
20	28.412	31.410	34.170	37.566	39.997
25	34.348	37.652	40.646	44.314	46.928
30	40.256	43.773	46.979	50.892	53.672
40	51.805	55.758	59.342	63.691	66.766
50	63.166	67.504	71.420	76.153	79.480
60	74.386	79.080	83.297	88.379	91.951
100	118.497	124.341	129.560	135.806	140.168

degrees of freedom: numerator	1	2	3	4	5	6	7
denominator							
1	161.45	199.50	215.76	224.58	230.16	233.99	236.77
2	18.51	19.00	19.16	19.25	19.30	19.33	19.34
3	10.13	9.55	9.28	9.12	9.01	8.94	8.88
4	7.71	6.94	6.59	6.39	6.26	6.16	6.09
5	6.61	5.79	5.40	5.19	5.05	4.95	4.88
6	5.99	5.14	4.76	4.53	4.39	4.28	4.21
7	5.59	4.74	4.35	4.12	3.97	3.87	3.79
8	5.32	4.46	4.07	3.84	3.69	3.58	3.50
9	5.12	4.26	3.86	3.63	3.48	3.37	3.29
10	4.96	4.10	3.71	3.48	3.33	3.22	3.14
11	4.84	3.98	3.59	3.36	3.20	3.09	3.01
12	4.75	3.89	3.49	3.26	3.11	3.00	2.91
13	4.67	3.81	3.41	3.18	3.03	2.92	2.83
14	4.60	3.74	3.34	3.11	2.96	2.85	2.76
15	4.54	3.68	3.29	3.06	2.90	2.79	2.71
16	4.49	3.63	3.24	3.01	2.85	2.74	2.66
17	4.45	3.59	3.20	2.96	2.81	2.70	2.61
18	4.41	3.55	3.16	2.93	2.77	2.66	2.58
19	4.38	3.52	3.13	2.90	2.74	2.63	2.54
20	4.35	3.49	3.10	2.87	2.71	2.60	2.51
21	4.32	3.47	3.07	2.84	2.68	2.57	2.49
22	4.30	3.44	3.05	2.82	2.66	2.55	2.46
23	4.28	3.42	3.03	2.80	2.64	2.53	2.44
24	4.26	3.40	3.01	2.78	2.62	2.51	2.42
25	4.24	3.39	2.99	2.76	2.60	2.49	2.40
26	4.23	3.37	2.98	2.74	2.59	2.47	2.39
27	4.21	3.35	2.96	2.73	2.57	2.46	2.37
28	4.20	3.34	2.95	2.71	2.56	2.45	2.36
29	4.18	3.33	2.93	2.70	2.55	2.43	2.35
30	4.17	3.32	2.92	2.69	2.53	2.42	2.33
40	4.08	3.23	2.84	2.61	2.45	2.34	2.25
50	4.03	3.18	2.79	2.56	2.40	2.29	2.21
60	4.00	3.15	2.76	2.53	2.37	2.25	2.17
70	3.98	3.13	2.74	2.50	2.35	2.23	2.15
80	3.96	3.11	2.72	2.49	2.33	2.21	2.13
90	3.95	3.10	2.71	2.47	2.32	2.20	2.11
100	3.94	3.09	2.70	2.46	2.31	2.19	2.10
120	3.92	3.07	2.68	2.45	2.29	2.17	2.09
140	3.91	3.06	2.67	2.44	2.28	2.16	2.08
150	3.90	3.06	2.66	2.43	2.27	2.15	2.07
180	3.89	3.05	2.65	2.42	2.26	2.14	2.06
200	3.89	3.04	2.65	2.42	2.26	2.13	2.06
250	3.88	3.03	2.64	2.41	2.25	2.12	2.05
300	3.87	3.03	2.63	2.40	2.24	2.11	2.04
400	3.86	3.02	2.63	2.39	2.24	2.11	2.03
500	3.86	3.01	2.62	2.39	2.24	2.11	2.02
1000	3.85	3.00	2.61	2.38	2.22	2.10	2.02
5000	3.84	3.00	2.60	2.37	2.21	2.10	2.01
∞	3.84	3.00	2.60	2.37	2.21	2.10	2.01

degrees of freedom: numerator	8	9	10	11	12	14	15	16
denominator								
1	238.88	240.54	241.88	242.98	243.91	245.00	245.95	246.90
2	19.37	19.38	19.40	19.41	19.42	19.43	19.43	19.44
3	8.85	8.81	8.79	8.76	8.74	8.72	8.70	8.68
4	6.04	6.00	5.96	5.93	5.91	5.87	5.86	5.84
5	4.82	4.77	4.74	4.71	4.68	4.64	4.62	4.60
6	4.15	4.10	4.06	4.03	4.00	3.96	3.94	3.92
7	3.73	3.68	3.64	3.60	3.57	3.53	3.51	3.49
8	3.44	3.39	3.35	3.31	3.28	3.24	3.22	3.20
9	3.23	3.18	3.14	3.10	3.07	3.03	3.01	2.99
10	3.07	3.02	2.98	2.94	2.91	2.86	2.85	2.83
11	2.95	2.90	2.85	2.82	2.79	2.74	2.72	2.70
12	2.85	2.80	2.75	2.69	2.64	2.62	2.61	2.60
13	2.77	2.71	2.67	2.63	2.60	2.55	2.53	2.51
14	2.70	2.65	2.60	2.56	2.53	2.48	2.46	2.44
15	2.64	2.59	2.54	2.51	2.48	2.42	2.40	2.38
16	2.59	2.54	2.49	2.45	2.42	2.37	2.35	2.33
17	2.55	2.49	2.45	2.41	2.38	2.33	2.31	2.29
18	2.51	2.46	2.41	2.37	2.34	2.29	2.26	2.25
19	2.48	2.42	2.38	2.34	2.31	2.26	2.23	2.21
20	2.45	2.39	2.35	2.31	2.28	2.22	2.20	2.18
21	2.42	2.37	2.32	2.28	2.25	2.20	2.17	2.16
22	2.40	2.34	2.30	2.26	2.23	2.17	2.15	2.13
23	2.37	2.32	2.29	2.24	2.20	2.15	2.13	2.11
24	2.36	2.30	2.25	2.22	2.18	2.13	2.11	2.09
25	2.34	2.28	2.24	2.20	2.16	2.11	2.09	2.07
26	2.32	2.27	2.22	2.18	2.15	2.09	2.07	2.05
27	2.31	2.25	2.20	2.16	2.13	2.08	2.06	2.04
28	2.29	2.24	2.19	2.15	2.12	2.06	2.04	2.02
29	2.27	2.23	2.18	2.14	2.10	2.05	2.03	2.01
30	2.25	2.21	2.16	2.12	2.09	2.04	2.01	1.99
40	2.18	2.13	2.08	2.04	2.00	1.95	1.93	1.90
50	2.13	2.08	2.04	1.99	1.95	1.89	1.87	1.85
60	2.10	2.05	1.99	1.95	1.92	1.86	1.84	1.82
70	2.07	2.02	1.97	1.93	1.89	1.84	1.81	1.79
80	2.06	1.99	1.95	1.91	1.88	1.82	1.78	1.77
100	2.03	1.98	1.93	1.89	1.85	1.79	1.77	1.75
120	2.02	1.96	1.91	1.87	1.83	1.78	1.75	1.73
140	2.01	1.95	1.90	1.86	1.82	1.76	1.74	1.72
150	2.00	1.94	1.89	1.85	1.82	1.76	1.73	1.71
180	1.99	1.93	1.88	1.84	1.81	1.75	1.72	1.70
200	1.98	1.93	1.88	1.84	1.80	1.74	1.71	1.69
250	1.98	1.92	1.87	1.83	1.79	1.73	1.70	1.68
300	1.97	1.91	1.86	1.82	1.78	1.72	1.70	1.68
400	1.96	1.90	1.85	1.81	1.78	1.72	1.69	1.67
500	1.96	1.90	1.85	1.81	1.77	1.71	1.68	1.66
1000	1.95	1.89	1.84	1.80	1.76	1.71	1.68	1.66
5000	1.94	1.88	1.83	1.79	1.75	1.70	1.67	1.65
∞	1.94	1.88	1.83	1.79	1.75	1.70	1.67	1.65

degrees of freedom: numerator	18	20	21	24	25	30	35
denominator							
1	247.00	248.01	248.30	249.05	249.26	250.10	250.69
2	19.44	19.45	19.45	19.45	19.45	19.46	19.46
3	8.67	8.66	8.64	8.64	8.64	8.62	8.60
4	5.82	5.80	5.79	5.77	5.77	5.75	5.73
5	4.58	4.56	4.55	4.53	4.52	4.50	4.48
6	3.90	3.87	4.86	3.84	3.83	3.81	3.79
7	3.47	3.44	3.44	3.41	3.40	3.38	3.36
8	3.17	3.15	3.14	3.12	3.11	3.08	3.06
9	2.96	2.94	2.93	2.90	2.89	2.86	2.84
10	2.80	2.77	2.76	2.74	2.74	2.70	2.67
11	2.65	2.63	2.62	2.60	2.60	2.57	2.54
12	2.57	2.54	2.53	2.51	2.50	2.47	2.44
13	2.48	2.46	2.45	2.42	2.41	2.38	2.36
14	2.41	2.39	2.38	2.35	2.36	2.31	2.29
15	2.35	2.33	2.32	2.29	2.29	2.25	2.23
16	2.30	2.28	2.27	2.24	2.26	2.19	2.18
17	2.26	2.23	2.22	2.19	2.22	2.15	2.13
18	2.22	2.20	2.18	2.15	2.15	2.11	2.09
19	2.18	2.16	2.14	2.11	2.09	2.07	2.05
20	2.15	2.12	2.10	2.08	2.06	2.04	2.03
21	2.12	2.09	2.07	2.05	2.04	2.01	1.99
22	2.10	2.07	2.05	2.03	2.02	1.98	1.96
23	2.08	2.05	2.03	2.01	2.00	1.96	1.95
24	2.05	2.03	2.00	1.99	1.98	1.94	1.93
25	2.03	2.01	1.98	1.96	1.95	1.92	1.90
26	2.02	1.99	1.96	1.95	1.94	1.90	1.88
27	2.00	1.97	1.95	1.93	1.92	1.88	1.86
28	1.99	1.96	1.94	1.91	1.90	1.87	1.85
29	1.97	1.95	1.93	1.90	1.89	1.85	1.83
30	1.96	1.93	1.90	1.89	1.88	1.84	1.82
40	1.87	1.85	1.83	1.82	1.77	1.74	1.72
50	1.81	1.79	1.78	1.77	1.72	1.69	1.67
60	1.78	1.75	1.74	1.72	1.68	1.65	1.63
80	1.75	1.71	1.70	1.68	1.66	1.62	1.59
100	1.73	1.68	1.66	1.65	1.64	1.59	1.55
120	1.71	1.66	1.64	1.63	1.62	1.57	1.53
150	1.67	1.63	1.62	1.61	1.61	1.55	1.51
200	1.66	1.62	1.60	1.59	1.59	1.53	1.49
250	1.65	1.61	1.59	1.57	1.57	1.52	1.48
500	1.62	1.58	1.57	1.55	1.55	1.50	1.44
1000	1.61	1.57	1.56	1.54	1.54	1.46	1.43
5000	1.61	1.56	1.55	1.53	1.53	1.45	1.41
∞	1.61	1.56	1.55	1.53	1.53	1.45	1.41

APPENDIX 10

Answers to selected exercises

Chapter 2

E2.1.   Expected frequencies: E(0) = 287, E(1) = 153,
                                  E(3) = 67, E(4) = 28,
                                  E(5) = 11, E(4) =   4,
                                  E(6) =   2, E(7) =   1,

             Chi-sq. = 4.5 ≤ Chi-sq (0.05, 4 deg. f.), which
             implies the negative binomial is a significantly
             good fit.

E2.2.   Expected frequencies: E(0) = 571, E(1) = 65,
                                  E(3) = 29, E(4) = 16,
                                  E(5) = 10, E(4) =   7,
                                  E(6) =   4, E(7) =   3,
                                  E(8) =   2, E(9) =   1,
                                  E(10)=   1,

             Chi-sq. = 3.0 ≤ Chi-sq (0.05, 7 deg. f.), which
             implies the negative binomial is a significantly
             good fit.

E2.3. (i)   Chi-sq. = 2.6 ≤ Chi-sq (0.05), which implies that
             the lognormal distribution is a significantly good
             fit.

             95% Confidence Interval {Log(length in meters)}:
                               L. c. l. = 2.5,
                          Geometric mean = 2.6,
                               U. c. l. = 2.7.
             95% Confidence Interval {length in meters}:
                               L. c. l. = 300,
                          Geometric mean = 390,
                               U. c. l. = 500.

E2.3. (ii)   Chi-sq. = 2.0 ≤ Chi-sq (0.05), which implies that
             the lognormal distribution is a significantly good
             fit.

             95% Confidence Interval {Log(shape ratio, B/L)}:
                               L. c. l. = -0.86
                          Geometric mean = -0.74
                               U. c. l. = -0.63
             95% Confidence Interval {shape ratio, B/L}:
                               L. c. l. = 0.14
                          Geometric mean = 0.18
                               U. c. l. = 0.23

E2.4. (i) Chi-sq. = 0.4 ≤ Chi-sq (0.05), which implies that
the lognormal distribution is a significantly good
fit.

95% Confidence Interval {Log(length in meters)}:
                    L. c. l. = 2.8,
              Geometric mean = 2.9,
                    U. c. l. = 3.0.
95% Confidence Interval {length in meters}:
                    L. c. l. = 610,
              Geometric mean = 750,
                    U. c. l. = 930.

E2.4. (ii) Chi-sq. = 0.1 ≤ Chi-sq (0.05), which implies that
the lognormal distribution is a significantly good
fit.

95% Confidence Interval {Log(shape ratio, B/L)}:
                    L. c. l. = -1.28
              Geometric mean = -1.12
                    U. c. l. = -0.96
95% Confidence Interval {shape ratio, B/L}:
                    L. c. l. = 0.05
              Geometric mean = 0.08
                    U. c. l. = 0.11

E2.5. (i) Chi-sq. = 1.8 ≤ Chi-sq (0.05), which implies that
the lognormal distribution is a significantly good
fit.

95% Confidence Interval {Log(length in meters)}:
                    L. c. l. = 2.6,
              Geometric mean = 2.7,
                    U. c. l. = 2.8.
95% Confidence Interval {length in meters}:
                    L. c. l. = 450,
              Geometric mean = 530,
                    U. c. l. = 640.

E2.5. (ii) Chi-sq. = 1.1 ≤ Chi-sq (0.05), which implies that
the lognormal distribution is a significantly good
fit.

95% Confidence Interval {Log(shape ratio, B/L)}:
                    L. c. l. = -1.92
              Geometric mean = -1.79
                    U. c. l. = -1.66
95% Confidence Interval {shape ratio, B/L}:
                    L. c. l. = 0.012
              Geometric mean = 0.016
                    U. c. l. = 0.022

E2.6. (i) Chi-sq. = 0.1 ≤ Chi-sq (0.05), which implies that
the lognormal distribution is a significantly good
fit.

95% Confidence Interval {Log(length in meters)}:
L. c. l. = 2.4,
Geometric mean = 2.5,
U. c. l. = 2.7.
95% Confidence Interval {length in meters}:
L. c. l. = 250,
Geometric mean = 330,
U. c. l. = 450.

E2.6. (ii) Chi-sq. = 0.1 ≤ Chi-sq (0.05), which implies that
the lognormal distribution is a significantly good
fit.

95% Confidence Interval {Log(shape ratio, B/L)}:
L. c. l. = -1.46
Geometric mean = -1.25
U. c. l. = -1.03
95% Confidence Interval {shape ratio, B/L}:
L. c. l. = 0.03
Geometric mean = 0.06
U. c. l. = 0.09

Chapter 3

E3.1. $F = 11.5 \geq F(3, 89, 0.05)$ implies that the
differences between the populations are significant.
Univariate tests with (1, 91 degrees of freedom):

$F = 24.8$: (Grade) :   significant differences,
$F = 3.0$: (Tonnage):   insignificant differences,
$F = 3.3$: (Gross-value): insignificant differences.

E3.2. $F = 2.0 \leq F(4, 75, 0.05)$ implies that the
differences between the populations are
insignificant.
Univariate tests are inappropriate, since the
MANOVA result is not significant.

E3.3. $F = 2.8 \geq F(6, 78, 0.05)$ implies that the
differences between the populations are significant.
Univariate tests with (2, 41 degrees of freedom):

$F = 3.3$: (Length) :   significant differences,
$F = 3.5$: (B/L ratio):   significant differences,

Duncan's Range test result:
The three regions form one homogeneous group.

E3.4. F = 5.4 ≥ F(6, 120, 0.05) implies that the
differences between the populations are significant.
Univariate tests with (2, 62 degrees of freedom):

F = 1.1: (Tonnage):   insignificant differences,
F = 8.8: (Grade):   significant differences,
F = 7.4: (Gross value): significant differences,

Duncan's Range test result:
Group 1: N. American Cord., S. Pacific Is. Arc.
Group 2: Scandinavian Shield.

E3.5. F = 6.1 ≥ F(12, 342, 0.05) implies that the
differences between the populations are significant.
Univariate tests with (4, 131 degrees of freedom):

F = 10.9: (Length):   significant differences,
F = 6.4: (Breadth):   significant differences,
F = 1.4: (B/L ratio):   insignificant differences,

Duncan's Range test result:
Group 1: N. American Cord. & S. Pacific (Cu-Au-Mo)
Group 2: N. American Cordillera (Cu-Fe-Au),
         (Zn-Pb-Cu-Ag), & East Asia (W-Mo)
Group 3: N. American Cordillera (W-Mo).

Chapter 4

E4.1. Airborne parallel grid probabilities:

Grid spacing (m)	Detection probability
200	0.904
300	0.853
400	0.800
600	0.690
1000	0.414

Chapter 5

E5.5. Minimum cost = $120,000: by operating:

Oil and gas field number 1: 3 hours per day
Oil and gas field number 2: 9 hours per day